Food Security and Environmental Quality in the Developing World

Rattan Lal
David Hansen
Norman Uphoff
Steven Slack

T0203907

Foreword by Lester R. Brown

CRC Press
Taylor & Francis Group
Boca Raton London New York

CRC Press is an imprint of the
Taylor & Francis Group, an **informa** business

CRC Press
Taylor & Francis Group
6000 Broken Sound Parkway NW, Suite 300
Boca Raton, FL 33487-2742

First issued in hardback 2019

© 2003 by Taylor & Francis Group, LLC
CRC Press is an imprint of Taylor & Francis Group, an Informa business

No claim to original U.S. Government works

ISBN-13: 978-1-56670-594-3 (hbk)

Visit the Taylor & Francis Web site at
http://www.taylorandfrancis.com

and the CRC Press Web site at
http://www.crcpress.com

Foreword

The new century has begun with some of the lowest grain prices in recent memory. From an economist's vantage point, this is a sure sign of excess production capacity. However, there may be more here than meets the economist's eye.

Natural scientists, many of whom have contributed to this volume, see something very different. They see reason to be concerned about such issues as the overplowing of land and the overpumping of aquifers. They look at sustainable production and see a worrisome fraction of world food output being produced with the unsustainable use of land and water.

They see countries abandoning rapidly eroding cropland, much of it land that should never have been plowed. Kazakhstan, the site of the Soviet Union's virgin lands project in the 1950s, has abandoned half its grainland since 1980. In north-western China, agriculture is retreating southward and eastward. In an effort to stem the encroachment of the desert on its cropland, Algeria is abandoning the production of grain on the southernmost 20% of its cropland, converting this land to orchard crops such as olive orchards and vineyards. To the south of the Sahara, Nigeria is losing 200 square miles of productive agricultural land each year.

The situation with water, the other basic resource used in food production, is no more encouraging. My Worldwatch colleague Sandra Postel, using data for China, India, the Middle East and the United States, estimates that we are overpumping aquifers by 160 billion tons of water per year. Using the rule of thumb of 1000 tons of water to produce 1 ton of grain, this suggests that 160 million tons of grain, or some 8% of the global harvest, are being produced with the unsustainable use of water. At the average world consumption level of a third of a ton of grain per person per year, this means that 480 million of the world's 6.1 billion people are being fed with grain that is produced with an unsustainable supply of water.

We've made impressive gains in raising world grainland productivity over the last half century, raising it from just over 1 ton of grain per hectare worldwide to nearly 3 tons per hectare today. We now need to think about systematically raising water productivity. Today it is water, not land, that is the principal constraint on our efforts to expand the world food supply. Just as India began to systematically raise land productivity with the new high-yielding wheats and rices 35 years ago, it must now devote similar energies to raising water productivity if it is to feed its 1 billion-plus people.

Over the last half century, the world added 3.4 billion people. During that period, we reduced the share of people in the world who were hungry, but the absolute number who were hungry increased. Now we are facing the addition of 3 billion more people over the next half-century. There is one difference, however, in that these 3 billion will all be added in developing countries, most of them already facing water shortages.

Given the dimension of the challenge the world faces on the food front, not only do we need this book for India, but many more like it if we are to keep focused on the effort to secure food supplies for all of humankind.

Lester R. Brown
President
Earth Policy Institute

Preface

The second half of the 20th century witnessed great advances in science and its application to enhance agricultural production in the world. Success in this endeavor is illustrated by the following data: the per capita cereal production in developed countries was 678 Kg/person/yr in 1980 and is expected to be 722 kg/person/yr in 2010. Per capita cereal production also increased in developing countries, but the total volume was less than one third of that in developed countries. Per capita cereal production in developing countries was 200 Kg/person/yr in 1980 and is projected to be 229 Kg/person/yr in 2010.

India is a microcosm of developing countries when considering biophysical, social, economic and political concerns. Per capita cereal production in India has increased steadily since the 1960s and achieved the level of 232 Kg/person/yr in 2000. Using 1980 as a baseline (1980 = 100 index), the relative index of agricultural production in India grew to 105 in 1982, 121 in 1984, 125 in 1986, 138 in 1988, 149 in 1990 and 160 in 1993. Comparable advances in total agricultural production were made in the 1990s. However, per capita cereal production remained either constant or increased at only a modest rate. While the increase in total food production was impressive, it was achieved at a high cost to environmental quality, reflected in severe soil degradation, widespread pollution and contamination of natural waters, deteriorating air quality in both rural and urban areas and increases in emissions of greenhouse gases into the atmosphere from the agricultural and industrial sectors.

Despite the impressive gains, about 300 million inhabitants of India are food insecure because of their low purchasing power. As the population of developing countries in general, and of India in particular, continues to grow, numerous relevant questions need to be addressed:

- Can developing countries meet the food requirements of their growing population without jeopardizing a natural resource base that is already stressed?
- Can the rate of food production achieved in the last two decades of the 20th century be sustained in the first 2 or 3 decades of the 21st century or until the population is stabilized?
- Can developing countries achieve freedom from hunger and malnutrition for all of their population (including children under 5 and nursing mothers)?
- How can food security be reconciled with environment quality in an industrialized society?

Food security and sustainability are interdependent. In fact, adoption of sustainable systems of agricultural production can minimize risks of soil and environmental degradation. Technological know-how to achieve food security and improve environmental quality exists, is scale-neutral, and can be adopted by resource-poor small landholders of developing countries. However, a need exists to validate and adopt such technology in the context of site-specific biophysical conditions, and socioeconomic, cultural and political factors.

The context reflected in the above discussion formed a background for a 1-day workshop that took place at The Ohio State University on 7 March 2001. The workshop was jointly organized by The Ohio State University and Cornell University. It was preceded by a public lecture by Dr. M.S. Swaminathan entitled Century of Hope. This volume represents the proceedings of this workshop. In addition to the papers presented, several authors were invited to write manuscripts on specific topics (e.g., biotechnology, energy use in agriculture, water harvesting, soil degradation, etc.).

The book is thematically divided into five sections. Section A, entitled Food Demand and Supply, contains eight chapters. As the title suggests, these chapters deal with the state of natural resources (e.g., soil, water, climate), fertilizer and energy needs and the importance of biotechnology. Section B, entitled Environment Quality consists of five chapters that address issues pertaining to water quality and the use of agricultural chemicals, and pesticide residues on food. Section C deals with Technological Options and contains eight chapters. It addresses issues related to water harvesting, post-harvest food losses, storage and processing of animal products, and sustainability and inequality issues. Section D, entitled Poverty and Equity, consists of five chapters and deals with issues of poverty alleviation, microfinance and gender equity. There are four chapters in Section E addressing policy issues and the role of the public sector. Emerging issues and priorities are discussed in the concluding chapter, which is found in Section F.

The organization of the symposium and publication of this volume were made possible by close cooperation between The Ohio State University and Cornell University. Funding support was received from the Ohio Agricultural Research and Development Center (OARDC) and the College of Food, Agriculture & Environmental Sciences (FAES) of The Ohio State University. The editors thank all authors for their outstanding efforts to document, organize and present pertinent information on topics of great concern related to the major theme of the workshop. Their efforts have contributed substantially to enhancing the overall understanding of issues pertaining to food security and environment quality in developing countries.

We offer a special vote of thanks to the staff of CRC Press for their timely efforts to publish this volume, thereby making the information contained herein available to the world community. We also recognize the invaluable contributions by numerous colleagues, graduate students and OSU staff. In particular, we thank Ms. Lynn Everett for her help in organizing the workshop and Ms. Patti Bockbrader for helping with the editorial process. We offer special thanks to Ms. Brenda Swank for her help in organizing the flow of the manuscripts from the authors and for her support in helping with all jobs related to preparing this volume for publication

The Editors

Editors

Rattan Lal is a professor of soil science in the School of Natural Resources at The Ohio State University. Prior to joining Ohio State in 1987, he served as a soil scientist for 18 years at the International Institute of Tropical Agriculture, Ibadan, Nigeria. In Africa, Professor Lal conducted long-term experiments on soil erosion processes as influenced by rainfall characteristics, soil properties, methods of deforestation, soil tillage and crop residue management, cropping systems including cover crops and agroforestry, and mixed or relay cropping methods. He also assessed the impact of soil erosion on crop yield and related erosion-induced changes in soil properties to crop growth and yield. Since joining The Ohio State University in 1987, he has continued research on erosion-induced changes in soil quality and developed a new project on soils and global warming. He has demonstrated that accelerated soil erosion is a major factor affecting emission of carbon from soil to the atmosphere. Soil erosion control and adoption of conservation-effective measures can lead to carbon sequestration and mitigation of the greenhouse effect.

Professor Lal is a fellow of the Soil Science Society of America, American Society of Agronomy, Third World Academy of Sciences, American Association for the Advancement of Sciences, Soil and Water Conservation Society and Indian Academy of Agricultural Sciences. He is the recipient of the International Soil Science Award, the Soil Science Applied Research Award of the Soil Science Society of America, the International Agronomy Award of the American Society of Agronomy, and the Hugh Hammond Bennett Award of the Soil and Water Conservation Society. He is the recipient of an honorary degree of Doctor of Science from Punjab Agricultural University, India. He received the Distinguished Scholar Award of the Ohio State University in 1994, Distinguished University Lecturer in 2000, and Distinguished Senior Faculty of OARDC in 2001. He is past president of the World Association of the Soil and Water Conservation and the International Soil Tillage Research Organization. He is a member of the U.S. National Committee on Soil Science of the National Academy of Sciences. He has served on the Panel on Sustainable Agriculture and the Environment in the Humid Tropics of the National Academy of Sciences. He has authored and co-authored about 1000 research publications.

David O. Hansen has worked in rural and institutional development for 35 years. His work has involved more than 10 years of overseas residence, including Peace Corps volunteer experience in Bolivia, research assignments in Costa Rica, Brazil and the Dominican Republic, long-term university Agency for International Development (A.I.D.) contract assignments in Brazil, short-term consulting A.I.D. assignments in the Dominican Republic, Bolivia, El Salvador, Peru, Brazil and Nicaragua; program development, administration and development experience in

India, China and Eastern and Southern Africa; and a 3-year Joint Career Corps assignment with A.I.D./Washington's Bureau for Science and Technology. His tenure with The Ohio State University includes extensive academic experience, including teaching of development-related courses, advising foreign graduate student thesis and dissertation research and Latin American field research. In addition, Dr. Hansen has had considerable experience with the administration of A.I.D., World Bank and other donor-sponsored university contracts, with administration of the Ohio State rural sociology graduate program, activities of the Rural Sociological Society, the International Rural Sociology Association, the Association for International Agriculture and Rural Development, and other national and international organizations impacting Third World development policies and programs.

Norman Uphoff is director of the Cornell International Institute for Food, Agriculture and Development (CIIFAD) and professor of government at Cornell University. He is also a member of the Steering Committee for Cornell University's Poverty and Inequality in Development Initiative. From 1970–1990, he served as chair of the Rural Development Committee in the Center for International Studies at Cornell and as a member of the Research Advisory Committee of USAID.

Having consulted for USAID, the World Bank, the Ford Foundation, FAO, the U.N., CARE and other organizations, most of Uphoff's research and outreach activities have centered on participatory approaches to development, particularly for agricultural innovation, irrigation improvement, and natural resource management. Geographically, his work has focused most on Ghana, Nepal, Sri Lanka, Indonesia and Madagascar, with current involvement in China and South Africa. His present writing and interests are in addressing agroecology, rice intensification and social capital.

Steven A. Slack has been at The Ohio State University since 1999 as associate vice president for agricultural administration and director of the Ohio Agricultural Research and Development Center. Dr. Slack received his B.S. and M.S. degrees from the University of Arkansas, Fayetteville and his Ph.D. from the University of California, Davis. In 1975, he joined the faculty of the Plant Pathology Department at the University of Wisconsin at Madison and in 1988 he joined the Cornell University faculty as the Henry and Mildred Uihlein Professor of Plant Pathology. He was department chair from 1995–1999. His major area of research interest has been seed potato pathology, especially the epidemiology of viral and bacterial diseases and tissue culture propagation techniques. He is a fellow and past president of the American Phytopathological Society, and is an honorary life member and past president of the Potato Association of America. In 1995, he and colleagues received a USDA Group Honor Award for Excellence for work on a nonpesticidal control strategy for the potato golden nematode. In 1996, he received the Outstanding Alumnus award from the Dale Bumpers College of Agricultural, Food and Life Sciences at the University of Arkansas.

Contributors

R.S. Antil
Department of Soil Science
CCS Haryana Agricultural University
Hisar, Haryana, India

Lopamudra Basu
Department of Animal Sciences
The Ohio State University
Columbus, Ohio

Nirali Bora
Cornell University

Courtney Carothers
Cornell University

Lester R. Brown
Earth Policy Institute
Washington, D.C.

S. K. De Datta
Office of International Research and
  Development
College of Agriculture and Life
  Sciences
Virginia Technological Institute
Blacksburg, Virgina

Rachel Doughty
Cornell University

Clive A. Edwards
Department of Entomology
The Ohio State University
Columbus, Ohio

Gary W. Frasier
USDA-ARS
Rangelands Resources Research Unit
Fort Collins, Colorado

Richard R. Harwood
Plant and Soil Science
Michigan State University
East Lansing, Michigan

David O. Hansen
International Programs
The Ohio State University
Columbus, Ohio

Poul Hansen
The Ohio State University
Columbus, Ohio

Peter Hazell
IFPRI
Washington, D.C.

Fred J. Hitzhusen
Agricultural Administration
The Ohio State University
Columbus, Ohio

Prem P. Jauhar
USDA-ARS
Northern Crop Science Laboratory
Fargo, North Dakota

Ramesh S. Kanwar
Department of Agricultural and
  Biosystems Engineering
Iowa State University
Ames, Iowa

Gurdev S. Khush
International Rice Research Institute
Manila, The Philippines

Laura R. Lacy
M.I.N.D. Institute Research Program
University of California
Davis, California

William Lacy
University Outreach and International
  Programs
University of California
Davis, California

Rattan Lal
School of Natural Resources
The Ohio State University
Columbus, Ohio

Sonja Lamberson
Cornell University

Katherine Lee
Cornell University

Richard L. Meyer
Agricultural Administration
The Ohio State University
Columbus, Ohio

Judith A. Narvhus
Department of Food Science
Agricultural University of Norway
Aas, Norway

R.P. Narwal
Department of Soil Science
CCS Haryana Agricultural University
Hisar, Haryana, India

Herbert W. Ockerman
Department of Animal Sciences
The Ohio State University
Columbus, Ohio

David Pimentel
College of Agriculture and Life
  Sciences
Cornell University
Ithaca, New York

K.V. Raman
Deptartment of Plant Breeding
Cornell University
Ithaca, New York

Alan Randall
Agricultural Administration
The Ohio State University
Columbus, Ohio

Paul Robbins
Department of Geography
The Ohio State University
Columbus, Ohio

Amit H. Roy
International Fertilizer
  Development Co.
Muscle Shoals, Alabama

G. Edward Schuh
HHH Institute of Public Affairs
University of Minnesota
Minneapolis, Minnesota

Sara J. Scherr
Agricultural and Resource Economics
  Department
University of Maryland
College Park, Maryland

Ashok Seth
Headley, Bordon
Hampshire, U.K.

Shahla Shapouri
USDA-ERS
Washington, D.C.

B.R. Singh
Department of Soil and Water Sciences
Agricultural University of Norway
Aas, Norway

Steve Slack
Ohio Agricultural Research and
  Development Center
Wooster, Ohio

M.S. Swaminathan
M.S. Swaminathan Research
  Foundation
Madras, India

Luther Tweeten
Agricultural Administration
The Ohio State University
Columbus, Ohio

Dina Umali-Deininger
World Bank
Washington, D.C.

Norman Uphoff
Cornell University
Ithaca, New York

Gurneeta Vasudeva
Tata Energy and Resources Institute
Arlington, Virginia

Contents

PART II
Environment Quality

PART III
Technological Options

PART IV
Poverty and Equity

PART V
Policy Issues

PART VI
Issues and Priorities

Part One

Food Demand and Supply

1 The Century of Hope

M.S. Swaminathan

CONTENTS

INTRODUCTION

The content of this chapter is based on a book I wrote 2 years ago, also titled *The Century of Hope*. During the same time frame, I also wrote a book about hope's becoming despair. First I will deal with despair and say why there are people who feel that this century will not be a bright one, and then discuss why I believe the reverse will happen. I will use the terms "despair" and "hope" as they relate to the food security front, i.e., sustainable food security. This chapter will be confined to sustainable food security and the prospects of eliminating hunger from this planet, as there are many other aspects of hope or despair. People like Lester Brown, centers such as the Worldwatch Institute, and books like *Who Will Feed China*, reiterate the wide concern regarding the future prospects of sustainable food security.

We can identify numerous global issues that, if ignored, will affect whether we can achieve sustainable food security. First is the issue of continued population growth. China alone has a population of 1.25 billion and India a population of 1 billion, with many other developing countries still having high growth rates. Second, there is environmental degradation as good soil and fertile arable land are removed from agricultural use. Third, there is the problem of water pollution, with groundwater being overexploited and aquifers rapidly disappearing, making water a critical constraint. Biodiversity is also vanishing, largely because of habitat destruction; as Dr. Wilson of Harvard said, "We have entered an era of mass extinctions. Then there are issues such as global climate change. These are all elements that contribute to environmental degradation. Soil, water, climate, biodiversity and forests are the ecological foundations essential for sustainable advances in agriculture. The president of Maldives says, "We talk about endangered species but not about endangered nations. The island I reside on would go down and our nation, Maldives, would cease to exist if the sea level rises by a meter or so." There seems to be distinct prospects of this occurring.

Then, of course, there are serious social needs to be addressed, both in terms of inequity and poverty. The cover page of the United Nations Development Programme (UNDP) human development report shows a champagne glass, its top representing a small percentage of people who have more and more income, and the bottom of the glass representing the large proportion that is being squeezed more and more. According to the World Bank, 1.3 billion people live on $1.00 a day or less. Poverty is increasing in the world along with overall unemployment or jobless economic growth, i.e., there is more economic growth, but the numbers of jobs are not growing commensurately. Although the U.S. is not currently experiencing this problem, many European countries are. Then, too, there is the question of proprietorship in science, exclusivity at a time when we need to be inclusive, either in terms of society or knowledge. We classify everything as "my" intellectual property right, and consider that everything developed requires a "patent." To indigenous communities, also known as tribal societies, the concept of intellectual property is quite alien; they do not understand what this means. They believe, as I do, that knowledge is something that comes down from earlier generations, and therefore, must be shared. The gene revolution is covered by proprietary science, while the Green Revolution was public research largely funded by public money and by philanthropic foundations.

BASIS OF OPTIMISM

Why then, in the midst of all these problems, do I consider this a century of hope? First, science is fortunately advancing very fast. The new frontiers of science include biotechnology, space technology and even weather forecasting. Who ever thought we could have such accurate weather forecasting? Even in India, the weatherman used to be the butt of all ridicule, but today everyone trusts the weatherman because of modern tools and technology, which have made it possible to predict short- and long-term weather conditions. Space technology has many other applications, such as information and communication technology; reaching the unreachable is possible today. It is not necessary to be exclusive; you can include the excluded in terms of information and knowledge empowerment. New kinds of virtual colleges involving U.S. and Indian institutions can be established where the latest developments in the U.S. can immediately be transferred across long distances to the poorest of the poor in the villages across the world.

The new frontiers of science include biotechnology, genomics or functions of genomics, proteomics, biochips, the Internet and nanotechnology. Many of these emerging concepts are as yet unfamiliar; new concepts are emerging every day and new technologies are going into what we call the new biovision for agriculture. What role that biovision and other new technologies are going to play, we still do not know; we are still investigating them and some controversy about them remains. In the next few years, there will be a new biovision that is backed by completely new biotechnologies — not only conventional genomics, but a whole sea of biotechnologies. For example, there is genetic enhancement for salinity tolerance in development of transgenic tobacco, brassica, vigna and rice brought about by the "gene revolution." There are designer potatoes and golden rice for better nutrition. The total projected population of India in 2001 is 1011 million, of which the rice-eating population is 366 million, or roughly 37%. Therefore, development of rice rich in micronutrients has a tremendous potential in the Indian scenario.

For these reasons, I have some confidence in the 21st century. Especially in the 1950s and '60s, the last century was considered to be a hopeless century as far as food production was concerned. In fact, as early as the 1960s, Paul and William Paddock wrote a book called *Famine 1975* in which they completely wrote off my country, India, and others as hopeless, never capable of feeding themselves. In *The Population Bomb* (1968), the much respected population experts Paul and Anne Erlich stated that, unless a nuclear bomb controls population, the population–food supply equation is hopeless. They believed that the ability to produce food for the increasing human numbers just did not exist.

But then things changed. We had new plant types: Nobel Peace Prize winner Norman Borlaug and Dr. Orville Vogel, along with others, developed new varieties. There were numerous other genetic and agronomic discoveries and major developments in the whole area of engineering. The start of the Green Revolution in 1968 initiated an era of hope on the food front. "Green" refers to the color of chlorophyll, and the name was coined to describe new plant types' ability to harvest more sunlight rather than as a reference to environmental consequences. Many people think the Green Revolution was environmentally disastrous, and there are clearly some problems that need to be addressed. Nevertheless, we had such progress in food production that today, in a country like mine, where the population has more than tripled since 1947 (from 300 million to over a billion today), the government has so much grain that it is not sure where to store it. As much as 60 million tons of food grains are available in the stores (although there continues to be a large number of people going to bed hungry as they do not have the purchasing power, but that is another challenge that will not be addressed here).

The second reason I consider this a hopeful century is that, by and large, democratic institutions and culture are spreading across the world. Dictatorships are vanishing, and this is a good thing. When all is said and done, in democracies people have the right to say what they want to say, there is a free debate and the media is free. Whether we like what they say or not, the fact remains that everyone can discuss and debate. Democracy provides a mechanism for resolution of conflict, not through arms but through negotiation, through words and dialogues. In India, for example, one reason we collaborate with The Ohio State University (OSU) in the sustainable management of major soil types is that we feel confident that whatever scientific work we do can be spread largely because there are the democratic institutional structures at the local level. Every village has an elected government of its own called *Panchyat*. At least one third of each village governing council must be women, so there is gender balance, not a divide, with both sides working together. Therefore, there are opportunities through democratic institutions. On the contrary, in the last 20–30 years, many African countries have experienced famine that was not due to grain food shortage per se (although the Sahelian drought of the '80s did cause food shortages), but to civil wars and lack of peace and security in the region.

The third reason I consider this a century of hope is the possibility of reaching the heretofore unreachable. Modern information and communication technologies are bridging the digital divide. These are very important mechanisms for knowledge and skill empowerment of the poor. People can reach each other quickly, and there are excellent opportunities today for spreading new information and converting general knowledge into location-specific knowledge. Often, general knowledge is not needed in sustainable agriculture but rather location-specific knowledge in relation to the soils, microenvironment, etc. It is important to have methodologies by which this can be achieved. Wisdom lies in knowing that one does not know. Numerous opportunities await to enhance wisdom through development of user-controlled and demand-driven knowledge centers. Rural computer-aided knowledge centers for all age groups are also needed. These centers could help convert generic into location-specific information and advice; provide information related to health, livelihoods, weather and market; and enhance knowledge and skill empowerment.

In India, the last century can be divided into three phases. Phase one lasted from 1900–1950. Population was low, death rates were high, birth rates were high but infant mortality rates were also high and, at the time of independence in India, the average life span was 28 years. During this period, many illnesses that we now consider to be minor ailments were then great killers. Everything was a killer: malaria, smallpox (which has been nearly eradicated today), and numerous other diseases. This was the era prior to the discovery of antibiotics and the whole system of preventive and curative medicine. The growth rate of agriculture was 0.01% in food crops. In other words, during the British days, the growth rate in food supply was nil except in plantation crops and some of the commercial crops, which is why, in the early part of India's independence, wheat was imported as a cushion or many people would have died from hunger.

The second, or institution-building phase, lasted from 1950–1965. We are grateful for OSU's involvement at this time, particularly at the Punjab Agricultural University, which has been on the forefront of the Green Revolution movement. In the institution-building phase, arrangements were made to provide more irrigation, fertilizer factories were built, etc. However, the food deficit remained a problem even during the second phase (see Figure 1.1). Food security is a function of three factors: (1) availability, (2) access, and (3) absorption. Availability is a function of production, access is a function of purchasing power, and absorption a function of clean drinking water and environmental hygiene. Improvement has to be made in all three factors to enhance food security. In fact, in 1966, nearly 10 million tons of wheat was imported under the PL-480 program. Consequently, some started describing India as a country with "ship-to-mouth" existence.

The third phase, from 1966–2000, is the era of the Green Revolution. In 1968, Dr. William Gaud of the U.S. coined the term "Green Revolution" to indicate that, not only in the case of wheat, but in rice, corn, sorghum and many other crops, new opportunities had been opened up for a radical increase in growth rates. Formerly a small incremental pathway, evolution could now occur at revolutionary speed. Consider that wheat cultivation in India has a recorded history of over 4000 years. From those early days until 1950, total production had reached the level of 7 million tons. But between 1964 and 1968, another 7 million tons was added; in other words, 4000 years of wheat-production evolution was condensed into 4 years.

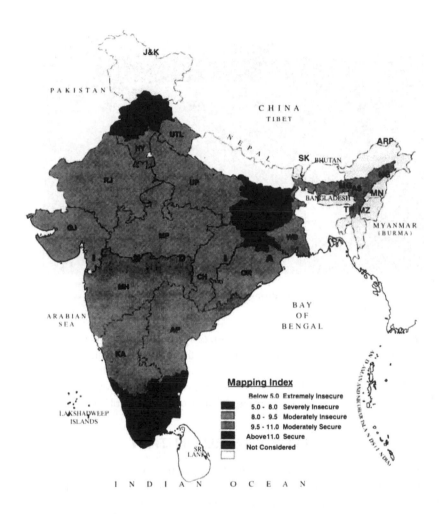

FIGURE 1.1 Food insecurity situation in India.

It is now clear that this revolution has its own problems. Social scientists say that the Green Revolution only makes the rich richer and the poor poorer, because inputs like seeds, fertilizer and water are needed for output; those who don't have the access or purchasing power for these inputs cannot benefit. Of these inputs, the availability of water is particularly important in India because of a large proportion of dry farming areas. When you don't have enough water, production is low unless water management is very good. Judicious water management is crucial to obtaining high yields. "Fertigation" and producing more yield or income per drop of water are important strategies. India receives most of its rainfall in just 100 hours out of 8760 hours in a year. If this water is not captured or stored (see Figure 1.2), there is no water for the rest of the year. Effectively captured and conserved, 100 mm of

FIGURE 1.2 Community water harvesting and cultivation of high-value, low-water-requirement crops (grain legumes).

rainfall falling on a 1-hectare plot can yield up to 1 million liters of water. Therefore, monsoon management is crucial. In addition, the Green Revolution also relied heavily on the use of pesticides. However, an excessive and indiscriminate use of pesticides can lead to the killing of pests' natural enemies, groundwater contamination, nitrate pollution and a whole series of environmental problems.

AN EVERGREEN REVOLUTION

The desire to solve these problems led to the development of the term "sustainable agriculture" during the last quarter of the 20th century. It refers to technology that is environmentally sustainable, economically viable and also socially acceptable. I coined the term "Evergreen Revolution" some years ago to indicate these kinds of sustainable advances in productivity, because the Green Revolution involves increased production through productivity improvement or yield per unit area. There are three basic steps toward achieving an Evergreen Revolution: (1) defending the gains already made, (2) extending the gains to additional areas and farming systems, and (3) achieving new gains in farming systems through intensification, diversification and value addition. Agricultural intensification, increasing yield per unit area, is an important strategy. For example, the average per capita arable land in India even today, with one billion people, is 0.15 hectare. The per capita arable land in China is even lower, less than 0.1 hectare. Obviously, with increasing urbanization and industrialization, land is going to go out of agriculture use. Therefore, there will be alternating demands on land and no option will exist except to produce more from diminishing land resources. This is what is called a vertical growth in productivity, in contrast to

FIGURE 1.3 Wheat production in India.

a horizontal expansion in area. The latter option is not open to us unless the remaining few forests are also to be lost. We have no option except to produce more from less land and less water, but produce it without the associated ecological or social concerns. This is what I defined as an "Evergreen Revolution," and that is why my book is called *The Century of Hope*. There is a prospect today for sustainable agriculture or an Evergreen Revolution based on productivity improvement per unit of water, per unit of land, and per unit of labor. At the same time, we should be able to increase the income of the farmer, because the smaller the holding, the greater the need for marketable surplus.

The Evergreen Revolution concept is especially relevant to production of wheat and rice in India. Wheat production in India now occupies the second position in the world (shown in experimental plots in Figure 1.3). However, the demand for wheat in India will increase by 40% between 2000 and 2020. There are opportunities to develop hybrid wheat, super-wheat with spikes that contain 50% more grains, wheat with high nutritional value (vitamin A, Fe and Zn contents), resistance to pests and improved physiological performance. New semi-dwarf varieties of wheat can produce 89 Kg of grains/ha/day. Similarly, hybrid rice has a vast yield potential (shown in Figure 1.4).

REACHING THE SMALL-SCALE FARMER

Advances in agriculture have been the most powerful instrument for poverty eradication in India because they touch the lives of so many people. In 1947, 80% of 300 million people in India were in farming; today, 70% of India's population of 1 billion still remain in farming. In other words, in absolute numbers, those who have to live by agriculture have increased enormously. If I am a farmer producing 1 ton of rice per hectare, then I have 200 kilograms to sell, but if I produce 5 tons of rice on the same land, then I have more than 4 tons to sell. The smaller the farm,

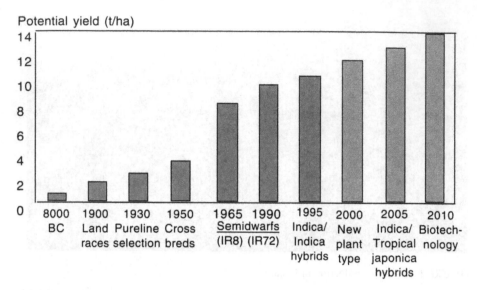

FIGURE 1.4 Progress in the yield potential of rice.

the greater the need for productivity improvement, largely because, unless there is cash flow, there is no marketable surplus. Small farmers require institutional structures to support them, like the soil management study between MSSRF (M.S. Swaminathan Research Foundation) and OSU. Success depends not only on the accumulation of scientific knowledge but also the ability to spread it around, which requires social engineering and the necessary mechanisms.

For instance, India is now the largest producer of milk in the world, having surpassed the U.S. We now produce 80 million tons of milk annually, while the U.S. produces only 72–73 million tons. The main difference is that milk in the U.S. is probably produced by only 200,000–300,000 farms, while India's 80 million tons of milk is produced by 50 million women farmers. How did they achieve the power of scale required both at the production site and the marketing site? In this particular case, the small producers formed into dairy cooperatives that had a single-window service system. This is a prime example of socially sustainable, economically viable and environmentally friendly small-scale agriculture. Enhancing the self-esteem of socially and economically underprivileged people and developing symbiotic linkages between knowledge providers and seekers (laboratory to land, and land to laboratory) are important strategies.

THE BIOVILLAGE

This term denotes a village where human development occupies a place of pride. *Bios* means life; *biovillage* implies human-centered development in which people are the decision makers. Their needs and feelings are ascertained through participatory rural surveys. The beneficial approach of development based on patronage gives way to an approach that regards rural people as producers, innovators and entrepreneurs. The enterprises are identified based on market studies and economic, environmental and social sustainability.

This concept is very relevant to eco-farming. In the 1st century BC, Varro, a Roman farmer, wrote, "Agriculture is a science which teaches us what crops should be planted in each kind of soil, and what operations are to be carried out, in order that the land may produce the highest yields in perpetuity." To achieve this, there is a specific three-step biovillage methodology: (1) microlevel planning, possibly based on geographic information system (GIS) mapping, (2) micro-enterprises based on markets, and (3) microcredit based on management by rural families.

There are numerous important applications of the concept to sustainable management of natural resources. Specific components include:

- Conservation of arable land
- Enhancement of soil quality
- Conservation and management of water
- Integrated gene management
- Integrated pest management
- Integrated nutrient management
- Minimizing post-harvest losses
- Development of integrated natural resources management committees at the local body level

Much of ecological farming requires a focused approach, whether it is watershed management, water conservation, saving water and sharing it, or integrated pest management (IPM). Writers have stated that IPM in the U.S. is not merely innovative technology but is also a question of social organization. If that is true in this country's larger farms, you can understand its significance for the small farms of India. Unless people can work together, new ecologically friendly technologies cannot be widely adopted. This is why the spread of democratic systems of governments at the grassroots level is an important and powerful ally in the movement for spreading eco-friendly and cost-effective technologies. We want to reduce the cost of production while increasing the income.

Apart from proprietary science, a separate world trade agreement on agriculture has been adopted for the first time since 1994. Previously, we had only bilateral agreements. The agreement is called AOA or Agreement on Agriculture. It is based on Ricardo's Principle of Comparative Advantage, which, in turn, was based on the observation that the differing fertility of land in different locales yielded unequal profits to the capital and labor applied to it. So, where can we produce most efficiently? Small-scale agriculture can have a lot of accountability, but today lacks the infrastructure, particularly the postharvest technology, sanitary and phytosanitary measures required by the western world.

In matters relating to quality, we should be concerned not only about exports but also about the food eaten at home. We should take the same precautions: *E-coli* and dysentery should become household words everywhere, and everyone should understand clearly what these terms mean. While we are working on the technological aspects of sustainable soil and water management, we should not forget the welfare of human beings. It is important also that the institutional structures and various methods by which people work together coalesce. In small-scale-farming conditions (whether in aquaculture, dairy or crop husbandry), it is very important to give farmers the power of scale; this makes ecologically friendly farming possible at the production site and provides more bargaining power at the marketing site. It also provides for the institution of some common facilities for sanitary and phytosanitary measures.

CONCLUSIONS

Achieving food security in India requires development and implementation of an integrated approach. The community food and water security system involves four components:

1. Gene bank or the *in situ* on-farm conservation of germ plasm
2. Seed bank or the formulation of *ex situ* seed bank as seed security reserve
3. Water bank or *in situ* conservation of rain, ground and surface waters
4. Grain bank or grain storage facilities where losses are minimal and reserves can be made available to cater to emergencies

This is an era of hope. Hope or despair is a state of mind. There are those people who are born optimists and those who are born pessimists. There is no use in being optimistic, though, without action. Therefore, I hope that this Century of Hope will give us the necessary impetus to work together and address the issues facing humankind. If we harness the power of partnership wisely, achieving a hunger-free world need not remain a dream.

REFERENCES

Mann, C. 1997. Reseeding the Green Revolution, *Science*, v. 277 (5329), p. 1038-1039 and 1041-1043.

Swaminathan, M.S. 2001. *Century of Hope: Harmony with Nature and Freedom from Hunger*. East-West Books, Chennai, India, 154 pp.

2 Natural Resources of India

Rattan Lal

CONTENTS

INTRODUCTION

Food grain production in India increased from 50 million tonnes (Mg = megagram = 1 metric ton) in 1947 to more than 200 million Mg in 2000. The Green Revolution — the use of high-yielding varieties along with intensive use of fertilizers on irrigated soils — enhanced agronomic production at a rate faster than that of the population growth. While these advances in production saved millions from starvation, some problems relevant to food security remain and new ones have emerged. Despite the large grain reserves, food is not accessible to a large proportion of the poor because of the lack of purchasing power. Further, expected food demand of 300 million tonnes of grains by the year 2050 will jeopardize natural resources already under great stress. The per capita availability of arable land and renewable fresh water are declining because of the increase in population. These resources are also being diminished by severe degradation of soil and pollution contamination of surface and groundwaters. Thus, there is an urgent need to develop strategies of sustainable management of natural resources while addressing the socioeconomic and political issues of equality, poverty, and postharvest losses due to lack of storage and processing facilities. There is little potential for further expansion of irrigation. Therefore, emphasis needs to be given to rain-fed agriculture. The Green Revolution strategies, as important a breakthrough as they were, need to be revisited in terms of the important issues pertaining to biophysical, socioeconomic and policy issues.

1-5667-0594-0/02/$0.00+$1.50
© 2002 by CRC Press LLC

TABLE 2.1
Dynamics of India's Population

Period	Population at the end of the period (millions)	Annual average growth rate (%/year)
1901–1911	252	0.56
1911–1921	251	–0.03
1921–1931	279	1.04
1931–1941	319	1.33
1941–1951	361	1.25
1951–1961	439	1.96
1961–1971	548	2.20
1971–1981	683	2.22
1981–1991	846	2.16
1991–2001	1001	1.85

Source: Adapted from Pachauri and Sridharan (1999; FAO (1998).

India is home to about 17% of the world population; its land area represents 2.9% of the world's total land mass. India's population increased from 252 million in 1900 to 1 billion in 2000, and is presently increasing at the rate of 1.85%/yr (Table 2.1). The country is endowed with a wide range of ecoregions, ranging from extreme heat to glaciers and from arid regions to those that receive more than 10 meters of rain every year. India has made outstanding progress in increased food-grain production, which has more than quadrupled over the five decades since independence. Currently, India has in excess of 50 million tons of food grains in reserves. Per capita dietary energy supply increased from 1980 cals in 1961 to 2267 cals in 1990 and 2415 cals in 1996 (Siamwalla, 2000). The present per capita food supply of about 2500 cals is adequate to meet the needs of its burgeoning population. Yet, more than 200 million people are undernourished, and infant mortality rates are among the highest in the world (Table 2.2). The malnutrition was 66% for children under age 5 for the period 1950–96 (Siamwalla, 2000). Poor composed 36% of the population in 1993 and 26% in 1999, while the literacy rate increased from 52% in 1991 to 65% in 2001 (*The Economist*, 2001).

Food security is a complex issue that is governed by a range of interacting biophysical, socioeconomic and policy variables. Food supply depends to a large extent on biophysical factors, but food availability is governed by complex socio-economic and policy considerations. In this chapter, food supply aspects related to resources such as soils, water availability and forest reserves are discussed.

LAND

India has diverse climates and ecoregions related to its large size. Rainfall averages range from less than 125 mm in the Thar Desert to 11,000 mm in Cherrapunji. Temperature, too, ranges widely, with a mean annual temperature of <4.5°C in Dras Kashmir to >45°C in Ganganagar, Rajasthan. India's climate is influenced by the

TABLE 2.2
Estimates of Hunger in India and Other Regions (Bread for the World Institute, 2000)

Country/region	Population (10⁶)			% population with access to safe water (1990–1997)	Mortality rate under 5 per 1000 live births			Per capita dietary energy supply (cals)	Under-nourished people 1995–97 (10⁶)
	1999	2025	Percent increase		1960	1997	Percent decrease		
India	998.1	1330.4	33.3	81	131	108	17.6	2495.6	204.4
South Asia	1340.3	1971.7	47.1	80	135	116	14.1	2448.8	296.6
Sub-Saharan Africa	596.7	1244.1	108.5	50	–	170	—	2182.8	179.6
Latin America & the Caribbean	511.3	696.7	36.3	77	53	41	22.6	2798.1	53.4
Developing countries	4793.2	6608.8	37.9	–	104	96	7.7	2650.0	791.4
World	5978.4	7823.7	30.9	72	94	87	7.4	2720.0	824.6

Source: From Bread of the World Institute, *Hunger 2000: A Program to End Hunger*, Silver Spring, MD, with permission.

Himalayan range in the north and by the Indian Ocean, Arabian Sea and Bay of Bengal, which surround the peninsula.

1. Rainfall: Depending on the geographic location, rainfall is highly site-specific and variable. Based on annual rainfall, India can be divided into the following regions: (a) the northeastern regions, neighboring areas and the west coast, which receives more than 2500 mm/yr; (b) the plains of the central and eastern upper peninsula, Bihar and West Bengal, which receive between 1250 and 1875 mm rainfall; (c) the region east of 79°E longitude and the west coast, which receive more than 1000 mm; (d) the northern plains between the northwest desert and the Brahmaputra Valley and the peninsula, excluding the coastal belt, which receive 500 to 750 mm rainfall; and (e) the northwestern region, which receives less than 250 mm of rainfall. About 70 to 80% of the rainfall occurs during the monsoon season from June to September.

TABLE 2.3
Land Use in India

Land use	Area (Mha)					
	1950	1960	1970	1980	1990	1998
Gross usable area	284	298	304	304	305	304
Not available for cultivation	48	51	45	40	41	—
Other cultivated land including fallow land	49	38	35	32	31	—
Fallow land	28	23	20	25	23	—
Total cropped area (gross)	132	153	166	173	185	—
Net area cropped	119	133	140	140	142	57
Net irrigated area	21	25	31	39	47	—
Cropping intensity	111	115	119	124	130	—

Source: From Ministry of Agriculture (1994) Annual Report, New Delhi, India; Pachauri and Sridharan (1999) *Looking Back to Think Ahead: Green India*, TERI, New Delhi, India; FAO (1998) Production Yearbook, Rome. With permission.

2. Land use: India has a large land area, much of which is suitable for cultivation. The gross cropped area, including land used to produce more than one crop per year, increased from 132 million hectares (Mha) in 1950 to 185 Mha in 1990 (Table 2.3). The corresponding net cropped area increased from 119 Mha in 1950 to 142 Mha in 1990. Net cropped area has stabilized around 140 Mha since 1970. The area under food grain in India changed little from 1977 to 1997 (Table 2.4). The net irrigated area increased substantially from 21 Mha in 1970 (17.6% of the net cropped area) to 47 Mha in 1990 (33.1% of the net cropped area). Irrigated land area in 1998 represented 57 Mha and contributed substantially to food grain production. Indeed, irrigation has played a major role in enhancing

TABLE 2.4
Area Under Food Grains in India

| | Area | | | |
| Particular | Mha | | % of total | |
	1977	1997	1977	1997
Food grains	122.6	125.5	67.2	58.9
Others	59.7	87.4	32.8	41.1

Source: From Kaosa-ard and Rerkasem (2000), *Growth and Sustainability of Agriculture in Asia,* Oxford University Press, New York, with permission.

TABLE 2.5
Forest Resources of India, 1995

Particulars	Area
Total land area	297.3 Mha
Total forest area	65.0 Mha
% of land under forest	21.9%
Per capita forest area in 1995	0.065 ha
Natural forest	50.4 Mha
Plantation	14.6 Mha

Source: From Kaosa-ard and Rerkasem (2000), *Growth and Sustainability of Agriculture in Asia,* Oxford University Press, New York, with permission.

food grain production. Total per capita land area, including irrigated area, is progressively declining due to population increases and its conversion to other land uses (Lal, 2000). The per capita arable land area in India is estimated to have decreased from 0.35 ha in 1960 to 0.07 ha in 2025 (Engelman and LeRoy, 1995).

3. Forests: In addition to agriculture, vast forest resources cover 21.9% of the total land area (Table 2.5). Natural forests cover 50.4 Mha and plantation forests cover 14.6 Mha. The quality of forest resources is highly variable. Further, there are differences between the recorded forest area and the actual forest area (Table 2.6). Dense forest with a crown density of >40% represents merely 60% of the total area under forest. The remaining 40% of the area with a low crown density has little biomass. In addition, protected areas represent about 15 Mha (Table 2.7) and include world heritage and wetlands areas.

4. Soils of India: The distribution of major soil types in India is shown in Table 2.8. The most productive soils, those of alluvial origin, are found in the flood plains of Indo-Gangetic and Brahmaputra basins and along

TABLE 2.6
Forest Resources of India

Category	Forest area (Mha)					
	1982	1989	1991	1993	1995	2000
Recorded forest area	75.1	75.9	77.0	77.0	76.5	75.0
Actual forest area	64.2	64.0	63.9	64.0	64.0	
(i) dense forest	361	37.9	38.5	38.6	38.6	
(ii) open forest	27.7	25.7	25.0	25.0	24.9	
(iii) mangroves	0.4	0.4	0.4	0.4	0.5	
(iv) scrub land	7.7	6.6	6.0	5.9	6.1	
(v) uninterpreted	1.2	0.4	1.9	0.0	0.0	
Nonforest area	255.7	257.8	256.9	258.8	258.7	

Dense forest = crown density > 40%
Open forest = crown density = 10-40%
Scrub land = crown density < 10%
Forest survey of India (1988, 1990, 1992, 1994)

Source: From Pachauri and Sridharan (1999), *Looking Back to Think Ahead: Green India,* TERI, New Delhi, India, with permission; FAO (2000).

TABLE 2.7
Protected Area in India

Particular	No.	Protected area (Mha)
National	344	14.3
International		
(i) world heritage	5	0.3
(ii) wetlands	6	0.2

Note: Number of malnourished children under 5 years of age in India was 76 million in 1993 and 59 million in 2010 (Rosegrant and Hazell, 2000).

Source: Kaosa-ard and Rerkasem (2000), *Growth and Sustainability of Agriculture in Asia,* Oxford University Press, New York, with permission

the east coast. These soils, comprising Inceptisols and Entisols, cover 76.5 Mha. They have been the basis for the Green Revolution. Vertisols in central India are also inherently fertile soils that cover 60.4 Mha. These are clay soils, have low infiltration rate, and develop large deep cracks on drying. Mollisols are highly fertile soils that cover only a small area of 1.8 Mha. Ultisols and Alfisols are highly weathered soils in the tropics and subtropics. Together they represent 117.7 Mha. Arid-

---

TABLE 2.8
Principal Soils of India (Personal Communication with H. Eswaran, NRCS)

Soil type	Area (Mha)
I. Non-soil	
Water bodies	4.6
Shifting sand	14.3
Rock	7.8
Others	2.1
Subtotal	28.8
II. Soil	
Gelisols	0.8
Vertisols	60.4
Aridisols	18.3
Ultisols	36.6
Mollisols	1.8
Alfisols	81.1
Inceptisols	51.7
Entisols	24.8
Subtotal	275.5
Total	304.3

---

isols, found in dry regions, can be cropped only with supplemental irrigation. Land areas under different land quality classes are found in Table 2.9. Good quality soils in classes I through III cover a land area of 110 Mha or 37% of the total land area and have few constraints related to crop production.

5. Water resources: India is also endowed with vast water resources. Annual internal renewable water resources are estimated to be 1850 Km^3. In addition, annual river flow from external resources is 235 Km^3 (Table 2.10). Because of the large population base, however, per capita water supply in India is low and declining. In fact, water scarcity will be a greater problem than land scarcity during the 21st century.

The per capita availability of renewable fresh water in India was 6008 m^3 in 1947, 5277 m^3 in 1955, 4237 m^3 in 1967, 3395 m^3 in 1977, 2737 m^3 in 1987, and 2263 m^3 in 1997 (Engelman and LeRoy, 1993; Pachauri and Sridharan, 1999). Data in Table 2.10 indicate temporal changes in per capita fresh water availability in India. Per capita water availability was 5,227 m^3 in 1955, 2451 m^3 in 1995 and 2085 m^3 in 2000. The projected population growth rate represents the medium projected U.N. population increase rate, and per capita available water resources will continue to decline to 1498 m^3 in 2025 and 1270 m^3 in 2050 (Table 2.11).

TABLE 2.9
Area in Different Land Quality Classes In India and its Population-Carrying Capacity at Low Input Lands

Land quality class	Land area (Mha)	Population carrying capacity (10^6)	Land characteristics
I	15.0	42	Few constraints to crop production
II	90.3	190	High temperature, low organic matter content, high shrink/swell potential
III	4.5	7	Seasonal wetness, short growing season due to low temperatures, minor root restriction
IV	8.5	8	Impeded drainage, crusting, compaction, high anion exchange capacity
V	103.7	62	Excessive leaching, calcareous/gypsiferous soils, aluminum toxicity, seasonal moisture stress
VI	6.0	2	Saline/alkaline soils, low moisture and nutrient status, acid sulphate soils, high nutrient fixation
VII	25.8	—	Shallow soils
VIII	4.7	—	Extended periods of low temperature, steeplands
IX	38.9	—	Extended periods of moisture stress
Total	297.3	310	

Source: From Beinroth et al. (2001), *Response to Land Degradation*, Science Publishers, Enfield, NH, with permission

TABLE 2.10
Water Resources of India.

Particulars	Value	Units
Annual interval renewable water resources	1,850	km^3
1998 per capita internal water resources	1,896	m^3
Annual river flow from external sources	235	km^3
Annual withdrawal of water volume	380	km^3
per capita withdrawal	612	m^3
proportion of internal resources	20.54	%
proportion of total resources	18.23	%

Source: From Kaosa-ard and Rerkasem (2000), *Growth and Sustainability of Agriculture in Asia*, Oxford University Press, New York, with permission

Despite abundant water resources, most of India's population experiences water scarcity due to the unequal distribution of rainfall in the region. Most rainfall is concentrated in three months between June and September. Consequently, both drought and floods are common throughout the country. Droughts are exacerbated by landscapes

TABLE 2.11
Annual Renewable Freshwater Availability in India

	Population (millions)				Per capita water availability (m³)			
Year	Actual	Low projection	Med. projection	High projection	Actual	Low projection	Med. projection	High projection
1955	395	—	—	—	5277	—	—	—
1995	850	—	—	—	2451	—	—	—
2000	1000	—	—	—	2085	—	—	—
2025	—	1286	1392	1501	—	1621	1498	1389
2050	—	1345	1639	1980	—	1549	1271	1053

Based on total annual renewable freshwater resources of 2085 km³

Source: Adapted from Engelman and LeRoy (1993), Sustaining water: Population and the future of renewable water supplies, population Action International, Washington, D.C.

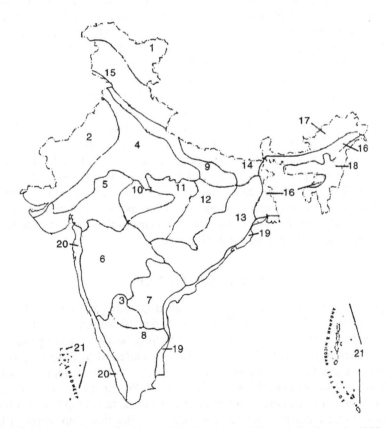

FIGURE 2.1. Agroecological regions of India (Adapted from Sehgal et al., 1990, ICAR, NBSS Publ. 24, Nagpur, India, with permission).

stripped of protective vegetal cover and by soils that are crusted and compacted and have low water-infiltration capacity. Most rainfall, therefore, is lost as runoff. Consequently, even high rainfall areas are often prone to drought stress.

The quality of surface and groundwater is poor. Most water resources are polluted, contaminated and unsuitable for consumption by people and domestic animals.

6. Agroecoregions of India: India can be divided into 21 ecoregions on the basis of rainfall and physiographic characteristics (Figure 2.1). Agriculturally important ecoregions in Figure 2.1 are 3, 4, 6, 7, 9, 14, 19, and 20. A brief description of these regions is given opposite, after Sehgal et al. (1990).

AGRICULTURAL PRODUCTION IN INDIA

Crop yields in India have increased considerably from the 1970s through the 1990s. Data in Table 2.12 indicate increased crop yields of 2.41 to 2.44%/yr for rice; of 3.10 to 4.26%/yr for wheat; and of 2.09 to 2.76%/yr for maize. Despite impressive gains, however, crop yields in India are below the world average (Table 2.13). The area under cereal production represents 14.3% of the world area, but total cereal production in India represents only 10.7% of the world production. Similarly, the area under rice cultivation in India is 28.1% of the total world area, but represents merely 21.7% of the world's total rice production. The area under sorghum cultivation in India is 25.2% of the total world area while the production is only 14.1% of the world's total sorghum production. The yield of soybeans in India is considerably lower. Area under soybean production in India represents 9% of the world's area, but produces only 3.9% of the world's total soybean production. Data in Tables 2.12 and 2.13 indicate a large potential for improving yields of grain and other crops in India through developing site-specific systems of soil, water, fertilizer and crop management. The demand for food grain production in India is likely to increase, not only because of the increase in population, but also because of increased demands for livestock products (See Table 2.14). Improvements in the livestock industry will also result in additional demand for food grains.

SOIL DEGRADATION

Soil degradation is a major cause of declining crop yields and low fertilizer- and water-use efficiencies in India (see Chapters 5 and 6 in this volume). Soil degradation results from water erosion, wind erosion, soil fertility decline, waterlogging, salinization and declining water table. The total land area affected by different processes of soil degradation is estimated to be about 59 Mha compared with 205 Mha in South Asia and 1965 Mha in the world (Table 2.15). Principal causes of soil degradation in India and elsewhere in South Asia include the non-adoption of soil conservation and management practices, extension of cultivation onto marginal lands

Eco-region #	Name	Description	Growing period (days)
1	Western Himalayas	Cold, arid, shallow skeletal soils	< 90
2	Western Plains & Kutch Peninsula	Hot, arid, saline soils	< 90
3	Deccan Plateau	Hot, arid, mixed red and black soils	< 90
4	Northern Plains & Central Highlands	Hot, semi-arid, alluvium-derived soils	90-150
5	Central Highlands & Kathiawar Peninsula	Hot, semi-arid, medium & deep black soils	90-150
6	Deccan Plateau	Hot, semi-arid, shallow & medium black soils	90-150
7	Deccan Plateau & Eastern Ghats	Hot, semi-arid, red & black soils	90-150
8	Eastern Ghats & Deccan Plateau	Hot, semi-arid, red loamy soils	90-150
9	Northern Plains	Hot, subhumid, alluvium-derived soils	50-180
10	Central Highlands	Hot, subhumid, medium & deep black soils	90-150
11	Deccan Plateau & Central Highlands	Hot, subhumid, mixed red & black soils	150-180
12	Eastern Plateau	Hot, subhumid, red & yellow soils	150-180
13	Eastern Plateau & Eastern Ghats	Hot, subhumid, red loamy soils	150-180
14	Eastern Plains	Hot, subhumid, alluvium-derived soils	180-210
15	Western Himalayas	Warm, subhumid, brown forest & podzolic soils	180-210(+)
16	Assam & Bengal Plains	Hot, humid, alluvium-derived soils	> 210
17	Eastern Himalayas	Warm, perhumid, brown & red hill soils	> 210
18	Northeastern Hills	Warm, perhumid, red & lateritic soils	> 210
19	Eastern Coastal Plains	Hot, sub-humid, alluvium-derived soils	150-210
20	Western Coastal Plains	Hot, humid-perhumid; red, lateritic & alluvium-derived soils	> 210
21	Islands of Andaman-Nicobar & Lakshadweep	Hot, perhumid, red loamy and sandy soils	> 210

FIGURE 2.1 (CONTINUED) Eco-Regions of India.

TABLE 2.12
Yield of Different Crops in India

Crop	Yield (Mg/ha)		Growth (%/yr)	
	1977	1997	1977-89	1987-97
Rice	1.86	2.87	2.41	2.44
Wheat	1.43	2.53	4.26	3.10
Maize	1.06	1.59	2.09	2.76
Coconuts	3.81	5.41	0.53	2.99
Rubber	0.80	1.45	1.41	4.48
Tea	1.51	1.84	0.01	2.00
Coffee	0.64	0.85	2.11	0.71
Sugarcane	53.4	66.5	1.24	0.95

Source: From Kaosa-ard and Rerkasem (2000), *Growth and Sustainability of Agriculture in Asia,* Oxford University Press, New York, with permission

TABLE 2.13
Food Grain Production in the World and India in 1998

Particular	World	India	% of the world
Population (billions)	6.0	1.0	16.7
Total area (Mha)	13387.0	382.7	2.9
Arable land (Mha)	1379.1	162.0	11.7
Irrigated land (Mha)	267.7	57.0	21.3
Total cereal area (Mha)	691.6	99.5	14.3
Total cereal production (m tons)	2054.4	219.4	10.7
Wheat area (Mha)	224.4	25.6	11.4
Wheat production (m tons)	588.8	66.0	11.2
Rice area (Mha)	150.3	42.3	28.1
Rice production (m tons)	563.25	122.2	21.7
Millet area (Mha)	37.6	13.3	35.3
Millet production (m tons)	29.2	10.5	35.9
Sorghum area (Mha)	44.4	11.2	25.2
Sorghum production (m tons)	63.5	9.0	14.1
Soybeans area (Mha)	70.7	6.4	9.0
Soybeans production (m tons)	158.3	6.1	3.9

Source: Recalculated from FAO (1998), Production Yearbook, Rome.

(e.g., steeply sloping, shallow soils), improper crop rotations, unbalanced fertilizer use, poor planning and improper management of canal irrigation and overpumping of groundwater (FAO, 1994).

Soil degradation is a biophysical process driven by socioeconomic and political forces. Among them are land shortage and declining per capita land area, land tenure

TABLE 2.14
Demand for Livestock Products in India

Particular	1993	2010
Per capita (kg)	4.3	5.8
Total demand (10^6 Mg)	3.8	6.8

Source: From Rosegrant and Hazell (2000), *Transforming the Rural Asian Economy: The Unfinished Revolution*, Oxford University Press, New York, with permission.

TABLE 2.15
Estimate of Land Area Affected by Soil Degradation

Process	India	South Asia Mha	World
Water erosion	32.8	81.8	1094
Wind erosion	10.8	59.0	549
Soil fertility decline	3.2	11.0	135
Water logging	3.1	4.6	?
Salinization	7.0	28.5	76
Lowering of the water table	2.0	19.6	?
Total	58.9	204.5	1965

Source: From FAO (1994), World Soil Resources Report 78, Rome; Oldeman (1994), *Soil Resilience and Sustainable Land Use*, CABI International, Wallingfor, U.K., with permission.

and tenancy, economic pressure and poverty. Depletion of the soil organic matter content of agricultural soils is also a widespread problem. The organic matter content of some soils is as low as 0.2%, because crop residues are either removed for use as fodder and fuel, heavily grazed or burnt. Animal waste, rather than being used as manure, is also used for household fuel.

WATER POLLUTION

A widespread problem of water pollution also exists. Principal sources of pollution are city sewage and industrial water discharges into rivers. Nonpoint-source pollution related to agricultural land uses also exists. Excessive and inappropriate application of fertilizers has led to increases in the nitrate content of well water, especially in Punjab, Haryana and Uttar Pradesh states. The nitrate contents in well water have ranged from 240 to 694 mg/l in Uttar Pradesh, from 419 to 1310 mg/l in Haryana, and from 265 to 567 mg/l in Punjab (Pachauri and Sridharan, 1 999). In addition to mineral fertilizers, manure and other organic residues are also important sources of nitrates in surface and groundwater. High contents of mercury, lead, manganese,

FIGURE 2.2 Seepage from an unlined canal is raising the water table.

DDT, phenolics and other compounds have also been observed in groundwater, and the concentration of these and other pollutants is increasing over time.

A problem of water imbalance also exists due to mismanagement of irrigation water. Waterlogging and salinity are severe problems in canal-irrigated areas with poor surface and subsurface drainage (Table 2.15). Excessive irrigation and seepage from canals (Figure 2.2)is causing groundwater levels to rise. In Bathinda, Punjab, the water table has been rising at the rate of 0.6 m/yr (FAO, 1990). Once waterlogging has occurred, soil salinity becomes a problem (Figure 2.3).Waterlogging can be addressed by judicious irrigation, by providing drainage or by reducing seepage losses. For flat topographies such as the Indo-Gangetic plains, disposal of drainage effluents is a major problem. In contrast to areas with canal irrigation, the water table is receding in areas irrigated by tube wells. For example, in the central region of Punjab, the water table is falling at the rate of 30 cm/yr. Once again, excessive irrigation, caused by subsidized water and electricity, has led to overexploitation of the groundwater resources.

AIR POLLUTION

Air is also a common resource that is prone to severe pollution. Air pollution in rural areas is caused by biomass burning (e.g., crop residue of rice and wheat) and the use of wood and dung or crop residue as a cooking fuel. Biomass fuels accounted

FIGURE 2.3 Waterlogging is followed by salinization.

for 74% of the household energy consumption in 1972, 66% in 1982 and 50% in 1989 (TERI, 1989). In 1978–79, 85 million households in rural areas and 19 million in urban areas used biomass fuels to meet their energy needs, especially to cook. At that time, the total annual consumption was 76 million Mg (Tera gram = 10^{12} g = 1 Tg) of wood, 16 Tg of crop residue, 22 Tg of dung cakes (NCER, 1985). Biofuel use in 2004–05 is estimated to be 300 to 330 Tg of wood, 192–221 Tg of crop residues and 90–104 Tg of dung cakes (Pachauri and Sridharan, 1999).

By contrast, air pollution in urban centers is primarily caused by automobiles, industry and thermal plants. Delhi is considered to be the fourth most polluted city in the world (Pachauri and Sridharan, 1999). Principal pollutants are particulate matter, sulfur dioxide, nitrogen oxides (NO_x), carbon monoxide, hydrocarbons, ozone and heavy metals such as lead and mercury. Pollutant emissions are estimated to be 1046 Mg/day in Delhi, 660 Mg/day in Mumbai, 305 Mg/day in Bangalore, 294 Mg/day in Calcutta and 226 Mg/day in Chenai (Pachauri and Sridharan, 1999). The Indian Ocean Experiment (INDOEX) reported high pollution levels over all of the northern Indian Ocean toward the Intertropical Convergence Zone at about 6° S (Leliveld et al., 2001). It was observed that agricultural burning, and especially biofuel use, enhanced carbon monoxide concentration, and that fossil fuel combustion and biomass burning caused a high aerosol loading. This extensive air quality degradation has global implications.

CONCLUSIONS

India is endowed with an abundance of natural resources. It has a wide range of climates and agroecoregions, soil types, rainfall regimes, and water resources. However, resource scarcities have resulted from rapid population increases during the 20th century. Population growth is expected to continue until the middle of the 21st century. Consequently, per capita arable land area and per capita renewable fresh water supply are progressively decreasing. Crop yields have increased substantially since the 1960s, but national average yields are still lower than their ecological potential. In some cases, crop yields are declining and incremental increases in yields per unit of fertilizer and other input are lower than they have been in the past. Inappropriate and indiscriminate use of chemical and organic fertilizers, pesticides and irrigation water have caused soil and environmental degradation, and water and air pollution. Accelerated soil erosion caused by water and wind results from India's lack of adoption of conservation-effective measures and the extension of agriculture onto marginal soils. Inappropriate use of irrigation is responsible for waterlogging and salinization in areas irrigated by canals, and excessive exploitation of groundwater in those irrigated by tubewells. Yet, India has a potential to enhance production and meet the demands of population increases. This will require restoration of degraded soils and ecosystems; improvement of irrigation water delivery systems; the return of crop residue and biosolids to the soil and adoption of sustainable systems of soil and water management.

REFERENCES

Beinroth, F.H., H. Eswaran and P.F. Reich. 2001. Land quality and food security in Asia. In: E.M. Bridges, I.D. Hannam, L.R. Oldeman, F.W.T. Penning de Vries, S.J. Scherr and S. Sombatpanit (Eds.) *Response to Land Degradation*, IBSRAM/ISRIC, Science Publishers, Enfield, NH, pp 83–97.

Bread for the World Institute. *Hunger 2000: A Program to End Hunger.* Silver Spring, MD, 161pp.

Economist. A survey of India's economy. *The Economist*, June 2001: 3-22.

Engelman, R. and P. LeRoy. 1993. Sustaining water: Population and the future of renewable water supplies. Population Action International, Washington, D.C.

Engelman, R. and P. LeRoy. 1995. Conserving land: population and sustainable food production. Population Action International, Washington, D.C.

FAO. 1990. Water and Sustainable Agricultural Development. A strategy for the implementation of the Mar del Plata Action Plan for the 1990s. FAO, Rome, Italy, 42 pp.

FAO. 1994. Land degradation in South Asia: Its severity, causes and effects on people. World Soil Resources Report 78, FAO, Rome, Italy, 100 pp.

FAO. 1998. Production yearbook. Rome, Italy.

Forest Survey of India. The State of Forest Report, Dehradun, India. 1988, 1990, 1991, 1994, 1995, 1996, 2000.

Kaosa-ard, M.S. and B. Rerkasem. 2000. Status of the natural resource base. In: *Growth and Sustainability of Agriculture in Asia*, Asian Development Bank, Oxford University Press, New York, Ch. 2, 303 pp.

Lal, R. 2000. Soil management in the developing countries. *Soil Sci.* 165:57-72.

Lelieveld, J., P.J. Crutzen, V. Ramanathan et al. 2000. The Indian Ocean Experiment. wide-
    spread air pollution from South and Southeast Asia. *Science* 291, 2000. pp 1031-1036.
Ministry of Agriculture. 1994. Annual Report, New Delhi, India.
Oldeman, L.R. 1994. The global extent of soil degradation. In: D.J. Greenland and I. Szabolcs
    (Eds.) *Soil Resilience and Sustainable Land Use*, CAB International, Wallingford,
    U.K., pp 99–118.
Pachauri, R.K. and P.V. Sridharan (Eds.). 1999. *Looking Back to Think Ahead: Green India
    2047*, TERI, New Delhi, India, 346 pp.
Rosegrant, M.W. and P.B.R. Hazell. 2000. *Transforming the Rural Asian Economy: The
    Unfinished Revolution*, Asian Development Bank, Oxford Univ. Press, New York, 512
    pp.
Sehgal, J.L., D.K. Mandal, C. Mandal and S. Vadiveul. 1990. Agroecological Regions of
    India. National Bureau of Soil Survey & Land Use Planning. ICAR, NBSS Publ. 24,
    Nagpur, India, 76 pp.
Siamwalla, A. 2000. *The Revolving Roles of State, Private and Local Actors in Rural Asia.*
    Asian Development Bank, Oxford Univ. Press, New York, 413 pp.
TERI. *TERI Energy Data, Directory and Yearbook 1989.* Tata Energy Research Institute, New
    Delhi, India, 281 pp.

3 Food Security: Is India at Risk?

Dina Umali-Deininger and Shahla Shapouri*

CONTENTS

INTRODUCTION

Recent developments indicate that India has made progress in terms of some key food security indicators. Food grain production grew by 2.7% per year over the last two decades, so that India at the national level achieved food grain self-sufficiency by the late 1990s. Indeed, the government held almost 60 million metric tons (mt) of food-grain (rice and wheat) stocks in 2001. The Food and Agriculture Organization (FAO) data (2001) indicate that the average per capita calorie available for consumption during 1996–98 had reached about 2,500 calories per day, an increase of 27% relative to 1980. Per capita incomes (GDP) grew at an even more extraordinary rate of about 5.5% per year during 1980–98 (constant 1995 price) leading to the expectation of significant improvements in food purchasing power and food security. These achievements, however, should not divert attention from the considerable remaining challenges, both current and future.

---

* Dina Umali-Deininger is lead agricultural economist at the World Bank and Shahla Shapouri is senior economist at the USDA-ERS. The findings, interpretations and conclusions expressed in this paper are those of the authors; they do not necessarily reflect the views of United States Department of Agriculture Economic Research Service and the World Bank and its executive directors or the countries they represent.

1-5667-0594-0/02/$0.00+$1.50
© 2002 by CRC Press LLC

With its population of 1 billion people, India's food security is of significance to global food security in many important respects. Ensuring adequate access to and utilization of food by about 17% of the world's population is a the tremendous challenge. Due to its size, the numbers of people who are potentially at risk also unavoidably become of global significance. Indeed, income poverty in India, a major factor contributing to food insecurity, is widespread and extremely high in absolute numbers by global standards. Although the figures are still subject to some debate, the government of India in 2000 declared that more than a quarter of the population (260 million people) is still living below the poverty line (Planning Commission, 2001). Per capita incomes are not only low (gross national income per capita in 2000 was $460) but income distribution is also highly skewed. The poorest 20% of the population receive about 9% of total income compared with the 39% received by the richest 20% of population (World Bank, 2000a). While India accounts for 20% of the world's children under age 5, it also accounts for about 62 million or 40% of the children who are malnourished (World Bank, 1998). Moreover, experience in the mid-1990s further illustrates how meeting unexpected local wheat production shortfalls and subsequent imports by India can push world wheat prices upward, affecting all other food import-dependent countries.

At the World Food Summit in November 1996, 186 countries committed themselves to reducing the number of undernourished people by half by 2015. The estimate of the number of hungry people* in 67 lower-income countries (excluding China) was 839 million out of a total population of 2.4 billion in 1995 and was expected to decline to about 774 million people by 2000 (Food Security Assessment, USDA-ERS, 2000). During the next decade, even though the number of people affected is expected to decline, the projected rate is slower than the years before. An important reason is the uncertainty about food availability in Africa, because of concerns for slowing agricultural output growth rate due to the spread of AIDS. Another reason is that in Asia, in particular India, the slow pace of poverty reduction depresses purchasing power and influences food access. Progress in improving food security in India has important ramifications at the global level because of the size of its population. In fact, eliminating hunger in India alone would cut the number of hungry people globally by half, thus achieving the goal of the World Food Summit.

The objectives of this chapter are to review the food security situation and prospects for 2010 in India, evaluate factors that contribute to food insecurity, examine India's food policies, and finally discuss policy options that can help improve the situation. In the next section, we assess India's current performance in ensuring household food security based on three indicators — status quo gap, nutrition gap and distribution gap, using the Economic Research Service (ERS) Food Security Assessment Model. Using the same model, we project India's prospects for achieving food security by 2010. In the subsequent sections, we examine how different factors, such as land quality, technology, water availability, and changing demand patterns, would influence the pace of progress in meeting the government's food security goals and then go on to describe the nature and scope of the government's food distribution policies. Finally, the last section outlines some key reform measures to ensure achievement of the government's longer-term food security goals.

---

* Defined as people consuming less than 2,100 calories per day.

ASSESSING INDIA'S FOOD SECURITY
PERFORMANCE: CURRENT AND FUTURE PROSPECTS

Food security is dependent on food availability (domestic supply), food access (ability to acquire food through production or purchase), and food utilization, which is affected by many factors such as education, health and access to safe water. Food insecurity can be either temporary or chronic, and overcoming each type requires a different set of strategies. The reasons for food insecurity are many: war, poverty, population growth, inadequate agricultural technology, inappropriate policies, environmental degradation, and poor education and health. It should be noted that, even among the prosperous countries, food insecurity persists in pockets of the population. For the affected populations, skewed purchasing power limits food access and causes food insecurity among the poor in these countries.

We assess the prospects of food security for India using the ERS Food Security Assessment Model (See Appendix for model details). Food in the model is defined to include grains, root crops and a category called "other," which includes all other commodities consumed, thus covering 100% of food consumption. All of these commodities are expressed in grain equivalent. The level of food security is assessed based on the gap between domestic food consumption (domestic production plus commercial imports minus exports and other nonfood use) and consumption targets. Although India has historically been receiving some food aid, this is not included in the projection of food availability.

We use three indicators to assess the food gap. These include:

1. Status quo gap: This gap represents the difference between projected food supplies and the food needed to maintain per capita consumption at the level of the most recent 3-year average (in this study 1997–99).
2. Nutrition gap: This gap represents the difference between projected food supplies and the food needed to support per capita nutritional standards at the national level (2,100 or 2,400 calorie consumption per capita per day).
3. Distribution gap: This is the amount of food needed to increase consumption in food-deficit income groups within a country to meet nutritional requirements.

The estimate of the status quo gap is an indicator of living standards. Maintaining per capita consumption at the same level implies no per capita income growth or changes in prices. In contrast, estimates of the nutritional and distribution gaps are long-term measures that reflect the well-being of the society. Reduction or elimination of these gaps requires growth in the purchasing power of consumers, i.e., income growth and a reduction in income inequality.

The estimates and projections of food availability in the next decade in India indicate that, on average, per capita food consumption will increase. The main assumptions underlying the model are listed in Table 3.1. As indicated, population growth projections are much lower than those of the last decade. The decline in population growth will reduce pressure on resources. The projected grain pro-

TABLE 3.1
ERS Food Security Model Assumptions

Variable	Annual growth,%	
	1989-99	2000-10
Population growth	1.80	1.35
Grain production growth	1.90	1.66
Area growth	-0.17	-0.09
Yield growth	2.09	1.75
Income growth	5.50	5.50
Export earnings growth	9.00	9.00
Other key assumptions		
Income elasticity for calories by income group	0.15	0.19
Change in stocks	No change	
Net foreign capital flow	constant at 1997-99 level	

Source: Authors' calculation

duction growth is lower than what was realized historically for reasons that will be discussed later. The economic and export growth rates are assumed to remain high, as experienced during the last decade. The growth in physical availability of food is projected to surpass the population growth, which means a continuation of growth in per capita food availability. This implies no status quo food gaps in the future.

The official Indian government per capita recommended daily allowance (RDA) for urban areas is 2,100 calories, while, for rural areas, it is higher: 2,400 calories. In terms of shares, the rural population accounts for 72% and the urban population 28%. Because the FSA model does not report urban and rural consumption separately, the two caloric standards can be viewed as the lower and upper boundary indicators of nutritional vulnerability in India.

Assuming an overall average nutritional standard of 2,100 calorie RDA, our analysis finds no nutritional gaps at the national level (Figure 3.1). This national level indicator, however, masks the impact of unequal food access. At the household level, we find that 20% of the population (200 million people) failed to meet the 2,100 caloric RDA in 2000. By 2010, despite improvements, about 10% or less of the population is projected to still face caloric deficiencies. When a 2,400-calorie RDA standard is applied to the model, a graver picture emerges. The average national nutritional gap is estimated at 15.8 million mt of food grain in 2000, which declines, but continues to remain sizeable at 4.5 million tons by 2010. Based on this standard, we estimate that 60% of the population (600 million people) consumed less than the 2,400 calories RDA in 2000, and will show no improvement by 2010 (Table 3.2). At a projected population of 1.16 billion people in 2010, as many as 116 million (2,100 calorie RDA) to 695 million (2,400 calorie RDA) people will still be subject to caloric deficiencies.

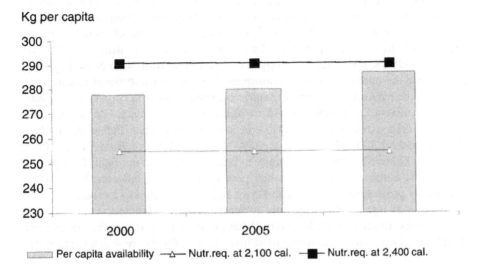

FIGURE 3.1 Per capita food availability (consumption) vs. requirement in India 2000–2010.

TABLE 3.2
Per Capita Caloric Consumption In India as a Percentage Share of Nutritional Requirement, 2000 and 2010

		Income Quintile			
RDA	Lowest (%)	2nd (%)	3rd (%)	4th (%)	Highest (%)
		2,100 calories			
2000	0.97	1.02	1.06	1.11	1.21
2010	1.00	1.06	1.10	1.14	1.25
		2,400 calories			
2000	0.85	0.89	0.93	0.97	1.06
2010	0.88	0.92	0.96	1.00	1.09

Source: Authors' calculation.

Under both scenarios, the distribution gap is projected to remain positive by 2010. As discussed earlier, the amount of food needed to increase food consumption for all income groups to the nutritionally required level is the distribution gap. Based on a 2,100 caloric standard, the gap is estimated at 1.6 million mt in 2000, which declines to less than 1 million mt in 2010. Based on a 2,400 caloric standard, the gap rises to 21.3 million mt in 2000, declining by about 25% to 16 million mt by 2010. Overall, although the indicators of nutritional vulnerability in India display some improvement, caloric deficiencies will not be eliminated by 2010 and will remain sizeable.

FAO's research suggests a much greater reduction in the number of hungry people in the past than ERS projections. According to the Food and Agriculture Organization of the United Nations (FAO) estimates, the rate of reduction of the number of hungry people was 26% in 1990/92 declining to 22% by 1995/97–1996/98. FAO's methodology is quite different from what is used by ERS. To estimate the number of people consuming less than the nutritional requirement, FAO uses the estimate of per capita calorie consumption of the country as its mean, while its variance is estimated based on household survey data. FAO's per capita minimum caloric requirement is also considerably lower than Indian standards at approximately 1,800 calories per day (FAO, *The Sixth World Food Survey*, 1996, Appendix 3, describes the methodology in detail). FAO does not publish its projections of the number of undernourished people by country, however, based on its commodity projections, food availability in India is projected to increase.

In sum, a large number of people suffer from caloric deficiencies in India and will continue to do so even by 2010. What possible instruments the government of India could pursue to eliminate food insecurity in the future is discussed in the next section.

MAJOR FACTORS INFLUENCING PROJECTIONS

India's projections of food security are based on assumptions regarding the performance of a number of factors such as productivity growth, technology use, and water availability. Therefore, any changes in performance of these factors will alter the projections. Because of the government's food self-sufficiency policies, imports currently play a small role in the domestic food supply. This means that the performance of the domestic agricultural sector will have a major influence on domestic food availability. The most important food crops are rice, followed by wheat, which together account for 78% of grain production in 1999–2000.* Domestic production of rice, wheat, and maize accounts for about 90% of food grain production, about 40% of gross cultivated area, and contributes 85% of the diet. Given the importance of domestic production in food security, in the following sections we briefly review factors that can change India's agricultural production performance in the future.

AGRICULTURAL LAND IS A LIMITING FACTOR

In India, similar to other Asian countries, population density is much higher than in most countries in other continents. According to the latest FAO report, there is no spare land available for agricultural expansion in South Asia (Agriculture: Toward 2015/30). About half of the suitable agricultural land in the region is already occupied by population settlements. Population growth alone will put further pressure on agricultural land and reduce the land available for food production. This also means that intensification of agricultural production and growth in crop yields, in particular rice and wheat, will play a major role in India's future food production growth.

---

* Food grains include rice, wheat, coarse cereals (e.g. sorghum, bajra, maize, etc.) and pulses (e.g. grain, peas, etc). Rice accounts for 42% and wheat 35% of total food-grain output.

Given the limited scope for land expansion, quality of land will be key to increasing yields. Land quality, as defined by soil quality, climate and rainfall, is a crucial factor in determining agricultural productivity. Cross-country analysis confirms that low cropland quality is significantly associated with low agricultural productivity. Rosen and Wiebe (2001) find that land quality not only affects yields directly, but also crop response to other inputs. The pace at which land for agriculture is lost — due to land degradation or expansion of urban areas — will therefore be a critical determinant of future production capacity in India.

It was estimated that 5.8 million hectares (ha) of irrigated land in 1991 were already degraded in India: 2.5 million ha were waterlogged, 3.1 million ha were affected by salinity, and 0.2 million by alkalinity (Ministry of Water Resources, 1991). This is equivalent to 20% of irrigated potential created or about 25% of the potential actually utilized. Intensive cultivation in India has also brought with it serious second-generation problems that threaten long-term agricultural growth. It contributed to environmental degradation in several ways. In many rice-growing regions in India, continuous monocultures along with inadequate soil conservation measures and unbalanced fertilizer use — in large part due to fertilizer price policies — have resulted in soil degradation. A study by Repetto in 1994 estimated that annual nutrient depletion due to topsoil removed by runoff is equal to all chemicals used in the country.

How much these estimates will be translated to losses in yields is not clearly known. While new technology has been successful in providing data on the existing quality of land, limited data are available on changes in land quality over time. Most studies are crop- and site-specific and cannot be generalized. Available data, however, indicate that land quality varies across India and is, on average, lower in low-income food-deficit regions than it is in high-income regions. This can have major implications on food security of the poor, who would be the least capable of coping with reductions in crop productivity and incomes. It also has important implications for policy makers, both in terms of exploring options for protection or improvement of land quality itself and in understanding the roles played by more conventional agricultural inputs in areas with differing land quality.

IMPORTANCE OF AGRICULTURAL PRODUCTIVITY GROWTH

Agricultural productivity growth is important for food security both through its impact on food availability as it contributes to output growth and to food access as it affects prices, farm incomes and the purchasing power of consumers. A major challenge for India will be not only sustaining, but also aiming to achieve higher yield growth to meet the rising food demand in the future.

The use of inputs, such as fertilizer, high-yielding varieties (HYVs), pesticides, surface irrigation and electricity- and diesel-powered tubewells, together contributed to the near doubling of yields between the 1970s and the 1980s. This has been referred to as the period of the Green Revolution in India. The growth in total factor productivity (TFP)* also accelerated during these two decades, spread-

---

* Total factor productivity is defined as the change in output that cannot be explained by changes in inputs, adequately adjusted for quality.

ing across all regions of India including the lagging agricultural regions of the eastern and southern states. Technological change, in fact, contributed one third of output growth, depending on the commodity and geographic coverage of the empirical studies (Desai, 1994: Dholakia and Dhokalia, 1993: Desai and Namboodiri, 1997; Kumar et al., 1998). Despite declining prices, this rapid technological change kept farming profitable, encouraging farmers to invest and use modern inputs. As on-farm productivity rose and demand for rural labor on- and off-farm rose, these pushed real rural wages up. Combined with declining food prices, these factors contributed to the significant reduction in poverty rates in India during this period. At the country level, it is estimated that the rise in real rural wages accounted for between 30 and 40% of the long-run impact of agricultural growth in reducing poverty (Ravallion and Datt, 1995).

Several studies, however, find that TFP in agriculture is declining or has become negative in the 1990s (Desai, 1994; Dhokalia, and Dhokalia, 1993; Kumar, et al., 1998; Rosegrant and Evenson 1994; Murgai, 1998; Fan, Hazell and Thorat, 1998). These studies indicate that, while output growth in the 1990s can be traced to more (private) investments and the application of more inputs and labor, their marginal productivity is now declining because of slower technological change. Unless redressed, declining TFP portends an eventual slowing of agricultural growth in the future. Indeed, the average annual growth rate of food-grain yields in India slowed from 2.7% during1980/81–1989/90 to 1.9% in the 1990/91–1998/99 (Figure 3.2). During the same period, the average annual yield growth for rice slowed from 3.6 to 1.9% and wheat declined slightly from 3.6 to 3.3%.

In the projections of food availability, the response of grain yield to a 1% increase in fertilizer use is 0.2. This, however, may overstate the reality. The recent estimation of the South Asian grain response to an increase in fertilizer use — after adjustments

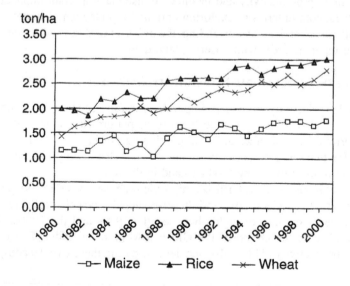

FIGURE 3.2 Yields of maize, rice and wheat, 1980–2000, mt per ha.

are made for land quality — is only 0.04 (Rosen and Wiebe, 2001). Fertilizer use has grown at the rate of 10% in the last two decades. The reason for such high growth is the government fertilizer subsidies that encourage application of fertilizer. Such an increase in fertilizer use is expected to lead to a reduction in the marginal response in crop output, but increase production costs. Growth in fertilizer use in the ERS's projection is 3% per year — the average growth rate of the last 3 years. The FAO projection of fertilizer use is much smaller, about 1.2% annually for 1995/97 to 2030. If the scenario of lower growth or lower yield and fertilizer response prevails, without any significant change in technology (i.e., use of high-yielding varieties), grain yields will be lower than what is projected in this chapter.

Continuing government subsidy of fertilizers, however, has become of major concern not only due to their rising fiscal costs, but also because they are adversely distorting farmer fertilizer-use behavior, with grave implications for future agricultural productivity growth. In 1999/2000, the fertilizer subsidy reached Rs 132.4 billion (US$2.9 billion), 0.02% of India's gross national product. Urea (nitrogen), because it receives a higher subsidy than phosphatic and potassic fertilizers, is the most utilized fertilizer and accounts for about 64% of fertilizer consumption in India. In 1999/2000, the average application ratios of N-P-K was 6.9:2.9:1 against a recommended N-P-K balance of 4:2:1. Such overuse of urea is not only inefficient, but, as many agricultural scientists conclude, leads to soil nutrient imbalances. The contamination of the groundwater, in particular, compounds the difficulty of managing micronutrient deficiencies that will have adverse implications for the productivity of agricultural lands in the longer term (World Bank, 1999a).

INCREASINGLY SCARCE WATER RESOURCES

Expansion of irrigation was one of the cornerstones of the success of the Green Revolution. Irrigation allowed intensive production and increased opportunities for diversification. Recognizing the importance of irrigation for food security and economic growth, the Indian government has invested almost Rs 920 billion (rupees) (nominal prices) in irrigation development from independence through the end of the Eighth Five-Year Plan (1992–97). Since the 1970s, gross irrigated area in India has nearly doubled, from 38.2 million ha in 1970/71 to 73.3 in 1996/97. This accounts for about 39% of gross cultivated area in the country.

While, in the future, the government plans continued investments to expand surface irrigation that can clearly help to sustain agricultural productivity growth, several major factors will make this increasingly difficult over the longer term. India has already developed almost 76% of the official estimate of ultimate gross irrigated potential of 113.5 million ha. The development of the remaining 24% will be difficult, as it will increasingly involve dam and canal construction in increasingly more difficult and environmentally fragile locations. Investment costs could also become prohibitive due to design, resettlement, and environmentally related issues (World Bank, 1999b). Rosegrant and Evensen (1994) estimated that the real costs of new irrigation more than doubled in the late 1960s and early 1970s in India. In view of

TABLE 3.3

Projected Utilization of Water in India

Sector	Base Year 1997, bcm*	Projections, bcm	
		2000	2025
Irrigation	501	630	770
Domestic	30	33	52
Industrial	20	27	120
Energy	20	30	71
Others	34		37
Total	605	720	1050

* bcm = billion cubic meters

Source: L.K. Joshi (2000)

the tight fiscal situation, obtaining the required resources to finance these investments in the contest for other competing fiscal demands will be a major challenge.

Various projections of water demand in India also point to the increasing competition for water resources among users, including agriculture, domestic, industrial, energy and other consumers. Rosegrant, Ringler and Gerpacio (1997) projected a 50% increase in water withdrawals between 1995 and 2020, including a 34% increase for agriculture and a 280% increase for domestic consumers and industry. Joshi (2000) similarly projected a 50% increase in irrigation water consumption by 2025. Of critical concern therefore, is the assessment that total domestic requirement for water will reach about 1,050 billion cubic meters (bcm) by 2025, which is nearly equal to total available water in India of 1,122 bcm (Table 3.3). To avert such water crisis in the longer term, improving water use efficiency, especially in the agricultural sector, will be critical.

Water and power policies and institutional weaknesses in state irrigation departments, however, are major culprits contributing to the inefficient use of water in agriculture. Price subsidies to canal irrigation and rural electric power for groundwater pumpsets contribute to the overuse of water. In turn, the unsustainable water-use practices of farmers are contributing to problems of deteriorating water quality, overexploitation of groundwater in some areas and salinization and water-logging in irrigated areas. Already, agricultural scientists and economists are raising concerns about the sustainability of rice–wheat cropping systems in the Indo-Gangetic region, most notably Punjab, Haryana and Western Uttar Pradesh, where the three-decade growth in cereal productivity jump-started the Green Revolution in India (Kumar et al., 1998, Chand and Haque, 1998). The unsustainable production practices, in large part fostered by the government's input policies (i.e. water, power and fertilizer) are also jeopardizing the country's food security in the future.

One result of the high subsidies and limited cost recovery for canal water and power is the deterioration in service delivery due to the inadequate resources for operations and maintenance and needed modernization investments. As illustrated in Figure 3.3, price subsidies to agricultural users of electricity for groundwater pumping is the major cause of the financial crisis in the state electric utilities (and many state

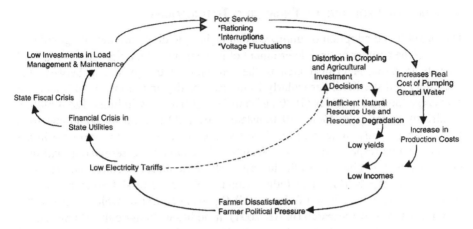

FIGURE 3.3 Vicious circle characterizing the power sector in India.

governments). The financial difficulties in turn lead to the inadequate supply of electricity and deterioration of service quality (frequent power outages and voltage fluctuations) to farmers, which have not only had adverse productivity impact, but have also distorted farmer cropping and investment decisions.* Moreover, poor quality of power supply increased consumer dissatisfaction and, in some cases, fostered unwillingness to pay, which further aggravated the low-cost recovery for the utility. These contributed to the perpetuation of the circle (Gulati, 1999; Aggarwal et al, 2001; World Bank, forthcoming 2001b). Dhawan (1998) estimates that the foregone agricultural value-added of electricity rationing at the country level could be as large as Rs9 per kWh (US $0.20) in 1996/97. An identical dilemma persists in most irrigation departments at the state level with respect to the delivery of canal water (World Bank, 1999a).

In sum, several factors contribute to the irrigated agriculture in India's performing below its potential. The inefficient and unreliable supplies of canal water and power to pump groundwater, combined with poorly functioning extension systems, have resulted in a large productivity gap in India's irrigated agriculture and lower farm incomes. It is estimated that a 10% improvement in the efficiency of water use alone would add some 14 million ha to the gross irrigated area (World Bank, 1999a). These improvements are achievable and could have a major impact on India's food security. Improving the performance of canal irrigation, as illustrated in the Bharda project in Karnataka for example, resulted in water savings of 22%, increased rice yields of 26%, and increased irrigated area of 18% (World Bank, 1999b).

---

* The poor quality of power supply also affects farmer behavior in several ways. Sharp voltage fluctuations, for instance, lead to frequent motor burnouts To ensure themselves against the risk of not having electricity when needed, farmers invest in "back-up" or "coping" strategies such as diesel pumps and tractors. These backup strategies further increase the effective costs of irrigation. The poor quality of supply distorts agricultural investment decisions in several ways. For instance, farmers tend to select robust motors that have thicker armature coil windings to reduce the frequency of motor burnouts, even though these motors have a lower overall efficiency. Farmers also tend to overinvest in the horsepower of the pump. From the farmer's viewpoint, a 10-hp motor operating under low voltage conditions is likely to perform as well as a 5-hp motor (World Bank forthcoming 2001b).

ADOPTION OF PRODUCTIVITY ENHANCING TECHNOLOGIES

The broad-based adoption of improved and higher yielding varieties of agricultural crops will be another critical determinant of long term productivity growth and food security in India. A critical area is the reduction in the yield gap between the laboratory and the field, particularly in the case of dryland crops. A 10% increase in average food-grain yields (1620 kg/ha in 1998/99) alone within the 125.4 million ha already under cultivation will translate to over 20 million mt additional food grains, significantly more than the food required to feed the hungry people in the country. In this respect, improved research and technology dissemination will play a critical role. India has one of the largest public agricultural research and extension complexes in the world.* The Indian Council of Agricultural Research complex alone has a manpower base of about 30,000 personnel, out of which nearly 7,000 are engaged in active research (ICAR 2001). In addition, 29 agricultural universities employ about 26,000 scientists for teaching, research and extension education; of these over 6,000 scientists are employed in the ICAR-supported coordinated projects.

Despite India's large investment in public research and extension, the quality of the agricultural research effort in the public system has weakened, while the agricultural extension system has virtually collapsed in the last two decades (Planning Commission, 2001). As the historical performance of the country indicates, strengthening the agricultural research and extension systems (both public and private) is essential to achieving rapid and sustained growth in agricultural productivity in the future.

The government of India's strategy, especially in the 1990s, has increasingly relied on input subsidies such as power, water, and fertilizer. These subsidies, along with increasing the minimum support prices for producers to promote increased agricultural production, have crowded out productivity-enhancing investments in rural infrastructure, irrigation, research and extension (Figure 3.4). Similarly, the deterioration of the state government finances has squeezed public investments in irrigation, roads, and technology upgrading. These public-expenditure patterns are not only fiscally costly but, to a large extent, also sacrifice long-term sustainable agricultural and economic growth. These short-term productivity gains are also likely to jeopardize future food security in India. The benefits of rebalancing expenditure priorities, therefore, are clearly going to be considerable.

CHANGES IN DEMAND CAN INFLUENCE PRODUCTION PATTERNS

India, despite its population increasing by about 450 million between 1970 to 2000, has so far managed to expand its food production capacity to respond to this growth. The growth in income also surpassed population growth and has led to changing consumption patterns. Indeed, average monthly per capita cereal** consumption has declined from 14.4 kg in 1987/88 to 12.8 kg in 1997 in rural areas and from 11.2

---

* The Indian Council of Agricultural Research operates through 46 central research institutes, 4 national bureaus, 10 project directorates, 27 national research centers, 90 all-India coordinated research projects, 261 krishi vigyan kendras and 8 trainers training tentres.
** Cereals include rice, wheat, sorghum, bajra, and maize.

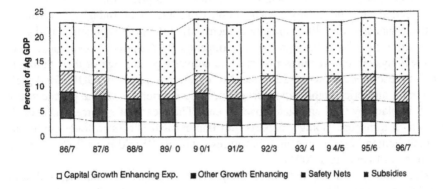

FIGURE 3.4 Trends in functional composition of agricultural expenditures (percentage of Ag GDP; 1986/87–1996/97).

kg to 10.3 kg in urban areas during the same period (NSSO 1998). At the same time, the consumption of dairy products, vegetables, fruits, meats and sugar has increased. If this pattern of consumption growth continues, it will have implications for the composition of future demand for food and consequently for the types of commodities produced in the country. For example, consumption of dairy products is expected to grow with income. Per capita consumption of milk grew by 4% during 1980–1999, and FAO projects a 3.1% annual growth in the next two decades. Despite the fact that a large share of the population in India is vegetarian, demand for meat is also expected to grow. Currently, per capita meat consumption in the country is very low, 4.5 kg in 1999, but demand for meat is highly income-elastic and the projected high-income growth can fuel the demand for meat. The FAO projects 3.4% annual growth in meat production in the next three decades. The FAO projection of poultry meat production growth is 7% per year from 1995/97 to 2015 for South Asia. The ERS's projections for milk and meat production is much lower than FAO, 3% per capita per year in the next decade. Nevertheless, in all these projections, the level of milk and meat production is expected to grow at rates higher than the 2.2% annual grain production growth predicted by both ERS and FAO. These trends mean increasing competition between use of grains for human consumption and animal feed. They also imply that, with no change in trade policy, such as allowing more imports of feeds, prices for lower-cost grains such as sorghum and millet, the foods of the poor, will rise, increasing their vulnerability to food insecurity.

GOVERNMENT FOOD DISTRIBUTION PROGRAM

To ensure the food security of its population, the government of India has several safety-net programs. The most important is to provide food grains to all consumers through a public food distribution system that sells subsidized rice and wheat. The government also has a buffer-stocking program whose mission is to stabilize domestic prices. The targeted public distribution system (TPDS) is the largest and most far-reaching of all safety nets in India. It aims to ensure access by consumers to essential commodities like rice and wheat at subsidized prices through private retail outlets called "fair-price shops." The Food Corporation of India, a government parastatal, manages the TPDS and government buffer-stocking programs. Food grain stocks are accumulated through the government's (GOI) price support operations and a levy system that requires rice mills to deliver a percentage of their output to FCI at a prescribed below-market price. Moreover, to support the GOI's food grain distribution and price stabilization program, trade restrictions on the private sector are put in place by GOI and state governments. The means of enforcement include controls on transport, storage, exports and imports and access to trade credit and risk management instruments such as futures contracts (World Bank 2001a, Umali-Deininger and Deininger 2001).

While this system may have been adequate in the past, when famines and large food deficits occurred frequently, it is currently the subject of widespread criticism as India has achieved national food self-sufficiency. Several studies have argued that these policies are undermining long-term food security by reducing the efficiency of markets and stifling their growth and modernization, thus contributing to rising physical losses, waste and costs (Radhakrishna et al. 1997, World Bank 2000, Planning Commission 2001, Umali-Deininger and Deininger 2001). Also, overwhelming evidence suggests that the public distribution system is hampered by poor targeting, rampant corruption and leakage of grains to the open market (Radhakrishna et al., 1997; World Bank, 2000b; Dev and Ranade, 1999; Kriesel and Zaidi, 1999). These programs are increasingly fiscally unsustainable — the central government food subsidy alone reached US$2.6 billion in 2000/01 or 0.6% of GDP (Ministry of Finance, 2001). Making these systems work to achieve India's food security goals more effectively will require broad-based policy reform, including measures to improve on the targeting and efficiency of delivery mechanisms for the TPDS, creating the enabling environment for increased private grain-sector efficiency and investments and improved efficiency and effectiveness of the Food Corporation of India.

SUMMARY AND CONCLUSION

Is India at risk of food insecurity? The answer continues to be yes. But at the same time, the dimensions and challenges for ensuring food security in India have clearly changed over the last decade. India has finally achieved its goal of food self sufficiency in the 1990s, with domestic public food-grain stocks mounting and expected to reach 70 million mt by the end of 2001. Indeed, by FAO's estimate, the national average per capita caloric availability in 1996–98 was a comfortable 2,500 calories.

Availability, however, has not translated to access by all. India is plagued with the paradox of hunger in the midst of plenty. Inequitable access to food, largely due to widespread poverty, leaves a large share of the population undernourished. Indeed, by our estimates, using the FSA model, as many as 200 million (2,100 caloric RDA) to 600 million people (2,400 caloric RDA) continue to suffer from caloric deficiencies in 2000. Projections to 2010 of absolute numbers of undernourished show little room for optimism. Under the assumption of 2,100 caloric RDA, the number of people suffering caloric deficiencies declines only slightly to 116 million people in 2010. Under the assumption of 2,400 caloric RDA, the situation deteriorates significantly to 695 million people consuming less than the RDA.

India achieved impressive agricultural and food production growth and significant reduction in poverty during the last decades, and it could continue to build on these successes in this century to achieve the goal of food security more rapidly for all. As noted in the previous section, this will require strong government commitment to actions in a number of fronts. These include reorienting government agricultural policies (i.e. price, production, trade and safety net policies) and expenditure priorities (i.e. subsidies vs. agricultural productivity and socially enhancing investments), and rebalancing the public- and private-sector roles in the rural sector to create the enabling environment for more rapid and sustained agricultural and overall economic growth and poverty reduction in India.

Notably, the government of India has taken action on a number of fronts. These include: (temporary) relaxation of some key domestic controls to private trade (e.g., storage, transport, credit), adjustments in the fertilizer subsidy, and improved targeting of food subsidies to the poor under the TPDS. In December 2000, the GOI introduced a new TPDS program called the *Antyodaya Anna Yoja*, specifically intended for the poorest of the poor and permitting the purchase of a greater allocation of food grains (25 kg) at a higher price subsidy. A large number of states have also implemented critical reform measures including irrigation-department restructuring integrated with increased user participation in irrigation system management and increased cost recovery to ensure sustainability of systems. Some states, under the umbrella of a broader power-sector reform program, are adjusting electricity tariffs to agricultural consumers matched with initiatives to improve the quality of supply.

These actions, hopefully, also portend faster progress by the GOI on the other required fronts. Faster progress in these areas will clearly help India avoid the potentially even graver and more widespread risks of food insecurity in the future.

REFERENCES

Aggarwal, R., Umali-Deininger, D. and Narayan, T. 2001. Impact of Electric Power Sector Reforms on Farm Incomes in India, Paper presented during the American Agricultural Economics Association Annual Meetings, Chicago, mimeo, August 5-8, 2001.

Chand, R. and Haque, T. 1998. Rice–wheat crop system in Indo-Gangetic region, issues concerning sustainability. *Economic and Political Weekly*: A108-112, New Delhi, India.

Desai, B.M. 1994. Contribution of institutional credit, self-finance, and technological change to agricultural growth in India. *Indian Journal of Agricultural Economics*, 49(3): 457-75, New Delhi, India.

Desai, B.M. and Namboodiri, N.V. 1997. Determinants of total factor productivity in Indian agriculture, *Economic and Political Weekly* 32(52):A165-A171, New Delhi, India.

Dev, S. M. and Ranade, A. 1997. Poverty and public policy: a mixed record. In: K.S. Parikh (Ed.), *India Development Report 1997*. New Delhi: Oxford University Press.

Dhawan, B.D. 1998. Latent threats to irrigated agriculture. In: B.M. Desai (ed.) *Agricultural Development Paradigm for the Ninth Plan under New Economic Environment*. New Delhi, India: Oxford and IBH Publishing.

Dholakia, R.H. and Dhokalia, B.H. 1993. Growth of total factor productivity in Indian agriculture. *Indian Economic Review* 38(1):25-40.

Fan, S., Hazell, P. and Thorat, S. 1997. Government spending, growth and poverty: an analysis of interlinkages in rural India, Environment and Production Technology Division Paper No. 33, International Food Policy Research Institute (IFPRI), Washington, D.C.

Food and Agriculture Organization (FAO), United Nations. 1996. *The Sixth World Food Survey*. Rome, Italy.

Food and Agriculture Organization, United Nations. *Agriculture: Toward 2015/30*, Technical Interim Report, Global Perspective Study Unit, Rome, Italy (April 2000).

Food and Agriculture Organization, United Nations. 2000. http://www.fao.org.

Gulati, A. 1999. Pricing of power and water for irrigation: issues related to efficiency and sustainability, Institute of Economic Growth, New Delhi, Draft Working Paper.

Indian Council of Agricultural Research. 2001. http://www.nic.in/icar.

Joshi, L.K. 2000. Irrigation and its management in India: need for paradigm shift. In Joshi L.K. and Hooja R. (Eds.) *Participatory Irrigation Management, Paradigm for the 21st Century*, Volume 1. New Delhi: Rawat Publications, pp.217-364.

Kriesel, S. and Zaidi, S. 1999. The targeted public distribution system in Uttar Pradesh–an evaluation. World Bank, Washington D.C. Draft mimeo.

Kumar, P., Joshi, P.K., Johansen, C. and Asokan, M. 1998. Sustainability of rice–wheat-based cropping systems in India. *Economic and Political Weekly:* A152-A158, New Delhi, India.

Ministry of Finance. 2001. *Economic Survey 2000/2001*. New Delhi: Ministry of Finance, Government of India.

Ministry of Water Resources. 1991. Waterlogging, Soil Salinity, and Alkalinity Problem Identification in Irrigated Areas with Suggested Remedial Measures, Report prepared by Ministry of Water Resources Working Group, Government of India.

Murgai, R. 1998. Diversity in economic growth and technical change, a district-wide disaggregation of the Punjab and Haryana growth experience: 1952-53 to 1990-91. University of California, Berkeley, mimeo.

National Sample Survey Organization, 1998. *Household Consumer Expenditure and Employment Situation in India 1997*. New Delhi: National Sample Survey Organization, Department of Statistics.

Planning Commission, 2001. Draft approach paper to the Tenth Five-Year Plan (2002-2007). New Delhi: Planning Commission, Government of India.

Radhakrishna, R., Subbarao, K., Indrakant, S., and Ravi, G. 1997. India's public distribution system: a national and international perspective. World Bank Discussion Paper No. 380. Washington D.C., World Bank.

Ravallion, M. and Datt, G. 1995. How important to India's poor is the urban-rural composition of growth, Policy Research Working Paper Series, No. 1399. Washington, D.C., World Bank.

Rosegrant, M.W. and Evenson, R.E. 1994. Total factor productivity and sources of long-term growth in Indian agriculture, EPTD Discussion Papers 7, International Food Policy Research Institute, Washington D.C.,1994.

Rosegrant, M.W, Ringler, C. and Gerpacio, R.V. 1997. Water and land resources and global food supply, paper presented at the 23rd International Conference of Agricultural Economists on Food Security, Diversification, and Resource Management: Refocusing the Role of Agriculture, Sacramento, California, August 10-16,1997.

Rosen, S. R. and Wiebe, K. 2001. Resource quality, agricultural productivity, and food security in developing countries, paper presented during the American Agricultural Economics Association Annual Meetings, Chicago, mimeo, August 5-8, 2001.

Umali-Deininger, D. and Deininger, K. 2001. Toward greater food security for India's poor: balancing government intervention and private competition, *Agricultural Economics*, in press.

United States Department of Agriculture-Economic Research Service. 2000. Food Security Assessment, International Agriculture and Trade Reports, GFA-12, Washington D.C.

World Bank.1998. Wasting Away, The Crisis of Malnutrition in India, Report No. 19887-IN. Washington, D.C.

World Bank. 1999a. India Towards Rural Development and Poverty Reduction, Report No. 18921-IN. Washington, D.C.

World Bank. 1999b. The Irrigation Sector, South Asia Rural Development Series. Washington, D.C.

World Bank. 2000a. World Development Report 2000/2001: Attacking Poverty. Washington, D.C.

World Bank. 2000b. India Foodgrain Marketing Policies: Reforming to Meet Food Security Needs, Volume I and II, Report No. 18329-IN. Washington, D.C.

World Bank. 2001a. India Improving Household Food and Nutrition Security, Achievements and the Challenges Ahead, Volume I and II, Report No. 20300-IN. Washington, D.C.

World Bank. 2001b. India Power Supply to Agriculture, Report No. 22171-IN. Washington, D.C., in press.

Appendix 3A:
Food Security Model

DEFINITION AND METHODOLOGY

The Food Security Assessment model was developed at the United States Department of Agriculture-Economic Research Service (USDA-ERS) for use in projecting food consumption and access, and food gaps in 67 low-income countries. Food security at a country level is evaluated based on the gap between projected domestic food consumption (produced domestically plus imported commercially minus exports and other nonfood use) and the consumption requirement. Although food aid is expected to be available during the projection period, it is not included in the projection of food consumption. It should be noted that, while the estimated results could provide a baseline for the food security situation in the selected countries, they are influenced by the assumptions and specifications of the model. Because the model is based on historical data, it implicitly assumes that the historical trend in key variables will continue in the future.

STRUCTURAL FRAMEWORK FOR PROJECTING FOOD CONSUMPTION IN THE AGGREGATE AND BY INCOME GROUP

Projection of food availability — The simulation framework used for projecting aggregate food availability is based on partial equilibrium recursive models of 67 lower-income countries. The country models are synthetic, meaning that the parameters used are either cross-country estimates or are estimated by other studies. Each country model includes three commodity groups — grains, root crops and "other." The production side of the grain and root crops is divided into yield and area response. Crop area is a function of 1-year lag return (real price times yield), while yield responds to input use. Commercial imports are assumed to be a function of domestic price, world commodity price and foreign exchange availability. Foreign exchange availability is a key determinant of commercial food imports and is the sum of the value of export earnings and net flow of credit. Foreign exchange availability is assumed to be equal to foreign exchange use, meaning that foreign exchange reserve is assumed constant during the projection period. Countries are assumed to be price takers in the international market, meaning world prices are exogenous in the model. However, producer prices are linked to the international market. The projections of consumption for the "other" commodities are simply based on a trend that follows the projected growth in supply of the food crops (grains plus root crops). Although this is a very simplistic approach, it represents an improvement from the previous assessments where the contribution to the diet of commodities such as meat and dairy products was overlooked. The plan is to enhance this aspect of the model in the future.

For the commodity group grains and root crops (c), food consumption (FC) is defined as domestic supply (DS) minus nonfood use (NF). n is country index and t is time index.

$$FC_{cnt} = DS_{cnt} - NF_{cnt} \tag{1}$$

Nonfood use is the sum of seed use (SD), feed use (FD), exports (EX), and other uses (OU).

$$NF_{cnt} = SD_{cnt} + FD_{cnt} + EX_{cnt} + OU_{cnt} \tag{2}$$

Domestic supply of a commodity group is the sum of domestic production (PR) plus commercial imports (CI) and changes in stocks $(CSTK)$.

$$DS_{cnt} = PR_{cnt} + CI_{cnt} + CSTK_{cnt} \tag{3}$$

Production is generally determined by the area and yield response functions:

$$PR_{cnt} = AR_{cnt} * YL_{cnt} \tag{4}$$

$$YL_{cnt} = f (LB_{cnt}, FR_{cnt} K_{cnt}, T_{cnt}) \tag{5}$$

$$RPY_{cnt} = YL_{cnt} * DP_{cnt} \tag{6}$$

$$RNPY_{cnt} = NYL_{cnt} * NDP_{cnt} \tag{7}$$

$$AR_{cnt} = f (AR_{cnt-1}, RPY_{cnt-1}, RNPY_{cnt-1}, Z_{cnt}) \tag{8}$$

where AR is area, YL is yield, LB is rural labor, FR is fertilizer use, K is indicator of capital use, T is the indicator of technology change, DP is real domestic price, RPY is yield times real price, NDP is real domestic substitute price, NYL is yield of substitute commodity, $RNPY$ is yield of substitute commodity times substitute price, and Z is exogenous policies.

The commercial import demand function is defined as:

$$CI_{cnt} = f (WPR_{ct}, NWPR_{ct}, FEX_{nt}, PR_{cnt}, M_{nt}) \tag{9}$$

where WPR is real world food price, $NWPR$ is real world substitute price, FEX is real foreign exchange availability, and M is import restriction policies.

The real domestic price is defined as:

$$DP_{cnt} = f (DP_{cnt-1}, DS_{cnt}, NDS_{cnt}, GD_{nt}, EXR_{nt}) \tag{10}$$

where *NDS* is supply of substitute commodity, *GD* is real income, and *EXR* is real exchange rate.

Projections of food consumption by income group — inadequate economic access, which is related to the level of income, is the most important cause of chronic undernutrition among developing countries. Estimates of food gaps at the aggregate or national level fail to take into account the distribution of food consumption among different income groups. Lack of consumption distribution data for the countries is the key factor preventing estimation of food consumption by income group. An attempt was made to fill this information gap by using an indirect method of projecting calorie consumption by different income groups based on income distribution data.* It should be noted that this approach ignores the consumption substitution of different food groups by income class. The procedure uses the concept of the income–consumption relationship and allocates the total projected amount of available food among different income groups in each country (income distributions are assumed constant during the projection period).

Assuming a declining consumption and income relationship (semilog functional form):

$$C = a + b \ln Y \qquad (11)$$

$$C = C_o/P \qquad (12)$$

$$P = P_1 + \ldots\ldots + P_i \qquad (13)$$

$$Y = Y_o/P \qquad (14)$$

$$i = 1 \text{ to } 5$$

where C and Y are known average per capita food consumption (all commodities in grain equivalent) and per capita income (all quintiles), C_o is total food consumption, P is the total population, i is income quintile, a is the intercept, b is the consumption income propensity, and b/C is consumption income elasticity (point estimate elasticity is calculated for individual countries). To estimate per capita consumption by income group, the parameter of b was estimated based on cross-country (67 low-income countries) data for per capita calorie consumption and income. The parameter a is estimated for each country based on the known data for average per capita calorie consumption and per capita income.

Endogenous variables:
    Production, area, yield, commercial import, domestic producer price, and
        food consumption.
Exogenous variables:
    Population — data are medium UN population projections as of 1998.

---

* The method is similar to that used by Shlomo Reutlinger and Marcelo Selowsky in *Malnutrition and Poverty*, World Bank, 1978.

World prices — data are USDA/baseline projections.

Stocks — USDA data, assumed constant during the projection period.

Seed use — USDA data; projections are based on area projections using constant base seed/area ratio.

Food exports — USDA data; projections are based on either the population growth rate or extrapolation of historical trends.

Inputs — fertilizer and capital projections are, in general, an extrapolation of historical growth data from FAO.

Agricultural labor — projections are based on UN population projections, accounting for urbanization growth.

Food aid — historical data from FAO, no food aid assumed during the projection period.

Gross Domestic Product — World Bank data.

Merchandise and service imports and exports — World Bank data.

Net foreign credit — is assumed constant during the projection period.

Value of exports — projections are based on World Bank (*Global Economic Prospects and the Developing Countries,* various issues), IMF (*World Economic Outlook,* various issues), or an extrapolation of historical growth.

Export deflator or terms of trade — World Bank (*Commodity Markets — Projection of Inflation Indices for Developed Countries*).

Income — projected based on World Bank report (*Global Economic Prospects and the Developing Countries,* various issues) or extrapolation of historical growth.

Income distribution — World Bank data. Income distributions are assumed constant during the projection period.

4 Fertilizer Needs to Enhance Production — Challenges Facing India

*Amit H. Roy**

CONTENTS

INTRODUCTION

Agriculture, an important sector in the economy of India, accounts for approximately 30% of the gross domestic product and employs about 60% of the labor force. Sustainable economic development, food security and social and political stability are, therefore, intricately linked to the sustainable growth of the agricultural sector. Forty years ago, the world regarded India as an economic basket case because of the low productivity of agriculture and a high rate of population growth.

---

* President and CEO, IFDC — An International Center for Soil Fertility and Agricultural Development, Muscle Shoals, Alabama

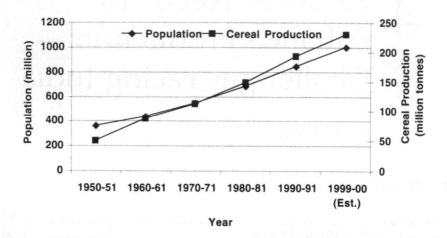

FIGURE 4.1 Growth in population and cereal production — India.

But, with the advent of the Green Revolution technologies, the last 35 years have witnessed a phenomenal growth in production and productivity of Indian agriculture (Figure 4.1). During this period, the cereal production more than doubled to 230 million metric tons (mt). In comparison, the global cereal production increased from about 1,200 million mt in 1970 to about 2,100 million mt in 2000, with developing countries accounting for nearly 70% of this increase (Figure 4.2). Today, India manages the world's largest public grain stock and has even become a minor grain exporter.

In the past, this production increase in developing countries was brought about by growth in both area cultivated and crop yields per hectare; however, since the 1960s, increases in yields per unit of land area have played a dominant role and

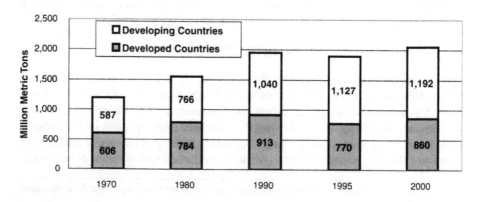

FIGURE 4.2 Cereal production in developed and developing countries, 1961-2000.

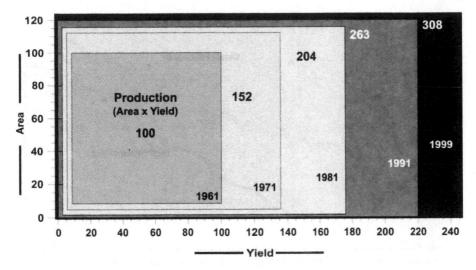

FIGURE 4.3 Cereal production in developing countries, 1961-99.

contributed to more than 80% of the growth in cereal production (Figure 4.3). Fertilizer accounted for 55–57% of the rise in average yields per hectare and 30–31% of the total increase in production (Pinstrup-Andersen, 1976). Consequently, cereal production and fertilizer use are closely associated in developing countries (Figure 4.4), where cereal production increased to 1,200 million mt in 2000 from a base of 400 million mt in 1961. During this period, fertilizer use increased from about 10 million mt nutrients to the present level of 91 million mt nutrients; India accounted for 20% (18.7 million mt) of this use level (Figure 4.5).

Since independence in 1947, India has accorded a high priority to the agriculture sector and pursued a goal of meeting its fertilizer demand through domestic production. As a result, domestic production capacity has progressively increased to

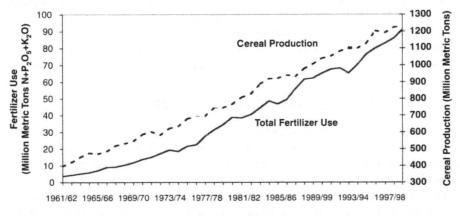

FIGURE 4.4 Developing countries: total cereal production and total fertilizer use, 1961/62 – 1999/00.

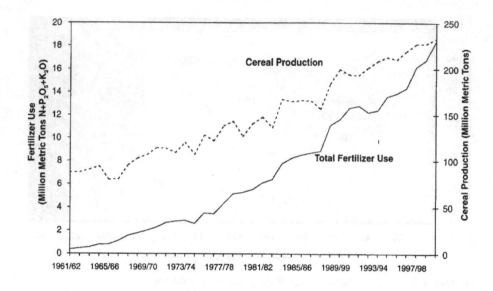

FIGURE 4.5 Growth in cereal production and fertilizer use — India.

the current level of nearly 15.0 million mt of nutrients based on both indigenous and imported raw materials (Awasthi, 2000). To boost agricultural productivity, India continues to subsidize inputs. The current subsidy on fertilizer alone is about rupees (Rs) 140 billion (approximately US$3.0 billion).

In spite of great progress, India faces many challenges in feeding its expanding population, which passed 1 billion in 2000 and is expected to be around 1.33 billion by 2025 (United Nations, 1998). Food production has to increase by more than 50% by then in the face of shrinking per capita arable land and water resources and environmental constraints. Agricultural intensification is essential, requiring more plant nutrients — particularly fertilizers. However, some of these plant nutrients can be derived through improvements in the efficiency of fertilizer use, recycling of plant nutrients and adoption of improved biotechnology.

This chapter traces the history of development of the Indian fertilizer sector, challenges facing the country and the projected requirement of fertilizers to meet the food and fiber demands of the expanding population for the coming decades.

BRIEF HISTORY OF THE FERTILIZER SECTOR

At the time of independence in 1947, India's per capita income was a meager $50 in nominal terms, average daily per capita grain consumption was only 400 g and the agricultural sector was characterized by low productivity of land and labor. Yet, agriculture provided employment for almost 75% of India's population, which was expanding very rapidly. To meet the increasing demand for food for the growing population as well as to increase incomes of farmers, the government focused on strengthening the agricultural base of the economy (Lele et al., 1994). Despite progress in development of irrigation and other rural infrastructure, the dependence on uncertain monsoon seasons continued to trouble India. This necessitated large imports of food during the late 1950s and early 1960s.

THE EARLY YEARS

In the early 1950s, Indian farmers used traditional methods of production with little awareness of the role of fertilizers in increasing land productivity. The introduction of Green Revolution technology, with its emphasis on high-yielding varieties (HYV) of seeds and the appropriate plant nutrients, plant protection chemicals and water, highlighted the crucial role of fertilizers in increasing land productivity. Consequently, the consumption of fertilizers increased from 69,000 mt nutrients in 1950 to 1.1 million mt in 1966/67 and domestic production constituted about 42% of the consumption. The subsequent years saw a rapid rise in fertilizer consumption, reaching a level of about 18.7 million mt in 2000.

In the early years, the government concentrated on ensuring distribution of imported and domestically produced fertilizers to all parts of the country at an affordable price. The need to focus on domestic production of fertilizers began to be seen as an essential component of food security because of recurring food shortages and chronic foreign exchange scarcity. Development of the fertilizer sector was initially envisaged to be in the public sector. In 1966, the government liberalized policies to allow private investment, including foreign investment, for distribution of fertilizers. With better domestic availability and the spread of extension education, consumption increased threefold during the period 1966/67 to 1976/77. The level of imports declined from 50% in 1968 to 35% in 1976/77.

The increase in fertilizer prices following the oil crisis of 1973 and the continued foreign exchange shortages gave impetus to increasing domestic fertilizer production. Discovery of natural gas offshore in "Bombay High" provided the needed raw materials for producing nitrogenous fertilizers. Both private sector and cooperative sectors invested heavily in the gas-based plants, starting in the late 1960s and continuing into the mid-1990s.

While many *ad hoc* measures by the government were directed to foster domestic investment in the first four 5-year plans, one of the most important policy instruments to encourage domestic production was the Retention Price Scheme (RPS) introduced in 1977 on the basis of the recommendation of the Marathe Committee. The sixth, seventh and eighth plans continued with the focus on indigenous production. The eighth plan categorically stated, "Government of India's policy regarding fertilizer sector ... has been the achievement of a maximum degree of self-sufficiency in nitrogen production based on the utilization of our own feedstock, leaving only marginal quantities to be met through imports." This focus on domestic production for phosphatic fertilizers was somewhat weaker because India had only a few economically exploitable phosphate deposits. As a result, more than 90% of the country's phosphate requirements are imported either as an intermediate (e.g., phosphoric acid) or finished product (diammonium phosphate [DAP]). In the case of potassic fertilizers, 100% of the country's requirements were imported.

The emphasis in the fertilizer sector so far has been guided by the objective of augmenting domestic fertilizer production to achieve self-sufficiency in food.

However, in 1991, there was a major paradigm shift in the general economic policy framework. This shift resulted in a more open foreign trade regime accompanied by deregulation of economic activities within a framework of macroeconomic stability.

CONSEQUENCES OF POLICY OPTIONS

The Indian fertilizer market is the second largest in the world at 18.7 million mt nutrients in 1999/00 after China (36.7 million mt nutrients). Collectively, China and India accounted for about 39% of the world's fertilizer consumption of 141.4 million mt nutrients in 1999/00. This level of consumption in India was achieved through several policy actions undertaken by the government. The most significant action was setting up the Marathe Committee in 1976, which recommended, as stated earlier, the RPS for nitrogen producers. The objectives of RPS were to: (1) stimulate fertilizer use and crop yields by keeping retail fertilizer prices low and (2) keep the domestic producers viable given the sharp increase in raw material costs. The RPS for phosphatic fertilizers was implemented in 1979. In concert with the implementation of RPS, the government also set the maximum sales price (MSP) for fertilizers in the country. The difference between the retention price and the plant revenue realized from selling at the MSP along with the fixed transportation costs to distribution points are paid to the producer as subsidy.

While this pricing policy had the desired effect of increasing food production, it came at a price to the government exchequer. The subsidy bill increased from Rs 5 billion in 1980/81 to Rs 44 billion in 1990/91. The economic reform of 1991 forced the government to reduce the subsidy bill resulting in an increase of 40% in retail prices. This price increase was later rolled back to 30% and small farmers were exempted, but this exemption was impossible to administer (Gregory et al., 2000).

Since that time, the government, faced with internal and external pressures, during the 1990s took an extraordinary set of *ad hoc* measures that are summarized in Table 4.1. These measures created extensive uncertainty within the sector, nutrient price differentials and trade distortions due to an Rs 3,400 subsidy difference between domestic and imported DAP in April–June 2000.

Because of many of the policy interventions, the Indian fertilizer subsidy has still increased exponentially in recent years (Figure 4.6). In 1999, urea subsidies accounted for two thirds of the total subsidy of about Rs 140 billion. The factors that have caused this explosion appear to be:

- Subsidies are paid on more tonnage each year.
- The governmentis reluctant to increase retail prices in general and of urea in particular.
- The steady depreciation of the rupee against the U.S. dollar has significantly increased the local cost of imported raw materials, intermediates and final products.

TABLE 4.1
Indian Fertilizer Subsidy Policy Changes 1991 to 2000

1991	Controlled selling prices for all fertilizers increased 40% (later reduced to 30%). Small farmers were exempted. Low-analysis fertilizers were decontrolled.
1992	Phosphate and potash fertilizers were decontrolled. Controlled urea price was lowered 10%. Low-analysis fertilizers were brought back under government control.
1992	*Ad hoc* subsidies introduced to cover all decontrolled fertilizers except single superphosphate (SSP) and other measures implemented to reduce shock of decontrol.
1993	Convertibility of rupee and unified exchange rate introduced. *Ad hoc* subsidies reintroduced for DAP, but only indigenous product eligible. *Ad hoc* subsidies introduced for SSP.
1994	*Ad hoc* subsidies continued at the same levels as 1993. Retail urea price increased 20%. Low analysis fertilizers decontrolled again.
1996	Subsidy on indigenous DAP increased to Rs 3,000. Subsidy on imported DAP reintroduced at Rs 1,500.
1997	Hanumantha Rao Committee (HRC) set up to explore reforms of the RPS. Empowered Committee set up to determine the subsidies for decontrolled fertilizers. DAP subsidies increased Rs 750, but Rs 1,500 differential between indigenous and imported DAP still existed.
1998	Empowered Committee disbanded.
1999	Interministerial group (IMG) set up to make subsidy recommendations. IMG makes initial or provisional estimates at start of the fiscal year. Producers are paid 80% of the initial subsidy. Each quarter a final subsidy is determined based on exchange rates and cost of imported materials during the previous quarter. Import duties of 5.5% imposed on all fertilizer imports.
2000	March 2000 provisional subsidies lowered to Rs 3,900 for indigenous DAP and Rs 950 for imported DAP. Changes nearly triple differential to Rs 2,950. Provisional subsidies for the 2000/01 fertilizer year set at Rs 2,800 for indigenous DAP and Rs 950 for imported DAP. First quarter subsidies of Rs 4,450 on indigenous DAP and Rs 1,050 on imported DAP increase differential to a record Rs 3,400 or $75 mt.

The Indian policies have kept the MSP of nitrogen low relative to phosphate and potash, particularly since the economic liberalization in 1991. As a result, nitrogen use has increased sharply relative to phosphate and potash. The recommended target nutrient ratio (N:P:K) in India is 4:2:1. However, following the liberalization, the N:K ratio has surged to almost 10:1 (Figure 4.7). Such distortions add considerably to the subsidy bill. A recent study indicates that balanced fertilizer use in India could produce an additional 160 million mt rice and 25 million mt wheat over current production if state fertilizer recommendations were universally applied (Stauffer et al., 2000).

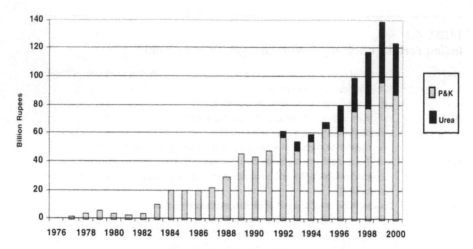

FIGURE 4.6 Fertilizer subsidy cost — India.

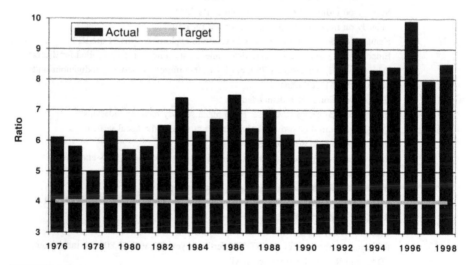

FIGURE 4.7 N:K use ratio — India.

The economic cost of the fertilizer sector subsidy policies has been under intense scrutiny and debate. As a result, the government of India constituted a high-powered review committee under the chairmanship of Professor C. H. Hanumantha Rao to study the fertilizer pricing policy and recommend options to meet domestic demand under the WTO regulations (Anon., 1998). The committee recommended abolishing RPS in favor of a ceiling farm-gate price of fertilizers along with the adoption of uniform normative price for determining the subsidy. In July 2000, the Ministry of Chemicals and Fertilizers issued a draft long-term fertilizer policy based on the Hanumantha Rao Committee report, which advocated adoption of a single retention price for the entire industry while providing some cost reimbursements for high-cost plants for a limited period . In September 2000, the Expenditure Reforms Commission of the Finance Ministry recommended gradual phasing-out of fertilizer subsidies by 2006 with a 7% annual real price increase in the price of urea and a fixed quantity of subsidized fertilizer per farmer.

SOURCES OF CHANGE AND FUTURE CHALLENGES FACING INDIA

Absolute growth in Indian population is expected to reach 1.33 billion by 2025 — an increase of more than 300 million over the current level. This growth in population, coupled with increased income, urbanization and changes in dietary habits, will require more food, feed and agricultural products. This is an immense challenge in the face of land and water constraints. However, emerging technologies and practices may alleviate some of these constraints.

Present estimates indicate that India's cereal production will have to increase by 50% by 2025. This increase will require an additional 15 million mt of nutrients. Adequacy of requisite natural resources is the most obvious concern when facing a substantial increase in future demands.

LAND CONSTRAINT

Most of the food for the Indian population in the next decades will be produced in today's arable lands. Arable land per person in India has declined from 0.25 hectares (ha) in 1980 to 0.19 ha in 1995 and is projected to be only 0.12 ha in 2025 (Engelman and LeRoy, 1995). The need for agricultural intensification will become more important as arable land scarcity increases due to the population growth in India. Agricultural intensification will require widespread adoption of improved yield-enhancing technologies that will be more demanding of plant nutrients. Most of this demand will have to be met through increased use of fertilizers and the rest could be through improvements in the efficiency of fertilizer use, recycling of plant nutrients and adoption of improved biotechnology.

WATER CONSTRAINT

The quantity of fresh water that is continually renewed through the global water cycle is a finite natural resource. In India, average water availability per person/year was 5,831 cubic meters (m^3) in 1950, 2,244 m^3 in 1995 and is expected to be only 1,567 m^3 in 2025 (Engelman and LeRoy, 1993). Thus, investments in technologies to increase the efficiency and productivity of water use will be crucial to increasing agricultural productivity in India.

Future agricultural productivity in India will be more dependent on technologies that can help ameliorate the constraints associated with land and water scarcities while maintaining agricultural output at high levels. Improvements in biotechnology, crop management and input use technologies, including more efficient integrated nutrient and pest management, will be needed to meet land and water scarcities.

IMPROVEMENTS IN BIOTECHNOLOGY

Biotechnology is one of the most promising emerging technologies for improving agricultural productivity in India. Improvements in biotechnology will include efforts to increase the productivity of land and water resources by increasing the efficiency in use of plant nutrients and by improving the biological control of pests and diseases of crops and the uptake of micronutrients necessary for human nutrition. The initial emphasis of biotechnology has been on increased productivity through the genetic modification of soybean, cotton and maize seeds to have crops with increased insect and herbicide resistance. There is also considerable research activity in developing hybrids of rice, maize, sorghum and wheat that are tolerant of high levels of solubilized aluminum in acid soils. Another area receiving considerable attention is genetic modification of input traits to improve the quality of crop outputs such as increased nutritional value, virus resistance and handling characteristics.

The biotechnologies currently being introduced by the commercial sector are expected to reduce the use of herbicides and pesticides, but not to significantly affect the use of fertilizers in the near future. However, if these genetically engineered seeds are successful, they will increase the possibility for future genetically modified input traits that could reduce the demand for some fertilizers. For example, the development of varieties with increased nutrient use efficiency or that incorporate the ability to biologically fix nitrogen could significantly impact fertilizer consumption.

TECHNOLOGIES FOR MORE EFFICIENT USE OF WATER AND NUTRIENTS

In the foreseeable future, fertilizer use is more likely to be affected by precision agriculture. Precision agriculture technologies involve variable-rate fertilizer application based on crop needs and the available nutrient reserve present in the soil, thereby increasing nutrient use efficiencies and minimizing nutrient losses. Experiences in the United States and Europe suggest that this technology can help Indian farmers achieve the right balance between economic food production and environmental stewardship (Giese, 1997).

An additional technology, which has the potential to increase both nutrient and water use efficiency while helping to maintain the environmental integrity of the natural resources (soil and water), is fertigation. This technology is having an increasingly important role in meeting the need for increased agricultural productivity in semiarid regions where a constraint for water availability is often accompanied by low soil fertility, land scarcity and concern for maintenance of water quality. The major disadvantages to fertigation, particularly drip irrigation, are the initial high investment, more expensive inputs and the skilled labor required for maintaining the system. At present, fertigation is best suited for high-value crops, but in the future, water scarcity may be the key driving force for the increased use of this technology. Fertigation is expected to increase in importance in India for cultivated lands subject to water scarcity.

Adoption and adaptation of minimum tillage technologies to promote soil conservation, as well as the development of technologies incorporating more intensive recycling of nutrients are also very much needed. The latter will be of particular relevance to India because projected population growth will result in the increased need for recycling of urban and industrial wastes. Technologies to recycle some of these wastes as "clean" sources of nutrients for crop production in an environmentally sound way would address two important problems — soil fertility maintenance and waste management — and could be highly beneficial to all segments of society (Hedley and Sharpley, 1998).

RESOURCE CONSERVATION AND ENVIRONMENTAL CONCERNS

There is no question that, in the last three decades, Green Revolution technologies (hybrid seeds, fertilizers, etc.) allowed India to avoid the famine predicted 25–30 years ago. The Green Revolution also had a largely ignored positive impact on the environment. Without the adoption of these technologies, feeding current populations at present nutritional levels would have required 60% more land (ODI, 1994; Gill, 1995), which would have demanded cultivation of ecologically fragile areas and would have resulted in widespread, severe natural-resource degradation.

Today, however, there are many questions as to whether present agricultural intensification practices can be sustained without considerable damage to natural resources. These questions are particularly relevant when one considers the need to feed a still-growing population while simultaneously moving the agricultural economy in the direction of exploiting comparative advantages in a global market.

In India, intensive rice and wheat cultivation has produced a number of effects that are contributing to degradation of natural resources, most importantly, nutrient mining of the soil. For example, soil nutrient (NPK) losses are estimated to range from less than 20 kg/ha per year in Punjab, Uttar Pradesh and Chhattisgarh to more than 80 kg/ha per year in Maharashtra, Tamil Nadu and other states (Figure 4.8). This depletion is mainly affected by distortion in NPK balance, causing a large depletion of potash (Figure 4.9). This depletion of soil nutrients has definite negative implications for the long-term sustainability of agriculture in India. The "mining" and depletion of nutrients in the soil occurs when the outflow of nutrients from farmers' fields, due to off-take in harvest of crop outputs and through losses, is not replaced by inflows through recycling, the application of external sources of nutrients and the use of biological means. The depletion of soil nutrients and the loss of good soil physical properties due to the loss of soil organic matter and other factors lead to soil fertility decline and land degradation. Ultimately, the cumulative effect of this process will lead to the loss of land resources, environmental damage and increased poverty, hunger and malnutrition.

Other problems (micronutrient deficiencies, soil toxicities, soil erosion, sedimentation, soil salinization, mining of groundwater and deforestation) resulting from agricultural intensification combined with global environmental concerns (climate change, loss of biodiversity) pose additional challenges for sustainable development. The Green Revolution provided time to reduce population growth without pervasive famine. A new set of technologies must now be developed and adopted to address this new set of problems.

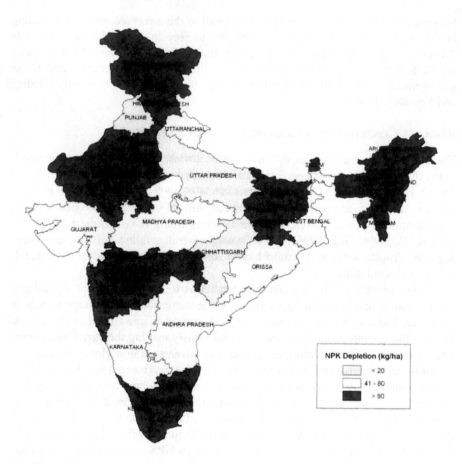

NPK Depletion (kg/ha)

     < 20

     41 - 80

     > 80

FIGURE 4.8 Annual soil NPK depletion.

WORLD TRADE AGREEMENT IMPLICATIONS

The agricultural and fertilizer sectors in India are facing the challenges and impact resulting from the commitments under the Uruguay Round Agreement on Agriculture (URAA) of the General Agreement on Tariffs and Trade (GATT) (now World Trade Organization [WTO]). Under the WTO agreement, the government of India was to have removed quantitative restrictions (QR) on various items including fertilizers by April 1, 2001. As a result of the removal of QR on urea, a certain percentage of the indigenous capacity, especially that based on naphtha, will have to be closed. The government is presently rationalizing closure of such plants. In the case of DAP, the government of India is bound to the level of 5% customs duty to protect domestic plants. If the degree of protection were to be enhanced, protracted negotiations with the initial rights holders according to WTO procedure would be required. It is possible that some domestic production capacities may have to be closed and demand met through joint ventures or imports.

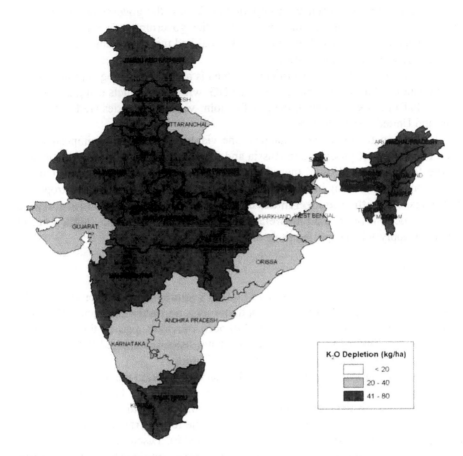

FIGURE 4.9 Annual soil K_2O depletion.

GROWTH IN FERTILIZER DEMAND

NITROGEN

As mentioned earlier, India has offshore natural gas reserves that partially provide feed for domestic nitrogen plants. In 2000, nitrogen production in India reached 13.6 million mt of nutrients (mainly urea) ranking third behind China and the United States. In addition to natural gas, several plants use fuel oil and naphtha, which are three to four times more expensive than natural gas. The reason for the use of this higher-cost feedstock is the nonavailability of adequate amounts of indigenous natural gas. The current domestic gas supply is 65 standard cubic meters per day(m^3/d) as against a demand of 110 m^3/d. It is highly unlikely that the government will be able to increase domestic gas supplies to existing plants or allocate to new facilities. To supplement this shortfall between supply and demand of natural gas, the government is considering import of liquefied natural gas (LNG), but the very high infrastructure costs required may force a reassessment of this option.

To reduce the dependency on imported feedstock, the government is exploring the coal-based technology concept. Further, the government is also considering unconventional resources of natural gas like coal bed methane, natural gas hydrates and underground coal gasification.

A third option that the government of India is actively pursuing is joint ventures with countries that possess natural gas or LNG, with the products being exclusively for the Indian market. Some of the possible joint-venture countries are Russia, Iraq, Iran and United Arab Emirates.

In addition to the above strategies, one viable option involves importation of urea from low-cost producers in Russia and the Gulf States. Following the breakup of the former Soviet Union, there has been a precipitous drop in fertilizer use in those countries, resulting in a significant surplus capacity — particularly nitrogenous fertilizers. While the use level in those countries has shown signs of recovery, the excess capacity is projected for several years. In addition to meeting India's demand, this oversupply will keep the price competitive.

PHOSPHATES

At present, India imports about 33% of its DAP requirements while producing 66% indigenously through imports of phosphate rock and phosphoric acid. The joint-venture plants in Senegal, Jordan and Morocco provide a part of this phosphoric acid. During 1999/2000, indigenous phosphate deposits supplied 0.4 million mt P_2O_5, or 8.5% of the phosphate requirement of India.

Among the known deposits, only the Jhamarkotra rock, in the State of Rajasthan, is commercially mined. It is unlikely that this mining will be expanded or any existing deposit brought to commercial operation. As a result, to meet increasing demand, India will have to import phosphates in the form of raw rock, intermediates (phosphoric acid) or finished products (mainly DAP). Discussions are under way with traditional phosphate producers in North Africa and the Middle East for new projects or expansion of existing joint-venture capacities.

POTASH

As stated earlier, India has no known potash deposits; as a result, the entire requirement of the country will continue to be imported from Canada, Germany, Jordan, Israel and Russia. However, studies are under way to assess the feasibility of commercializing a deposit recently discovered in Thailand.

CONCLUSION

India has achieved tremendous increases in agricultural production and productivity in the last 35 years. The adoption of high-yielding varieties of wheat and rice in particular, in conjunction with increased use of fertilizers, has more than doubled cereal production. However, in the next 25 years, India's food requirements are expected to grow by at least 50%, while the resource base of arable land and water will shrink. The challenge for India is to increase agricultural production by increasing the productivity of agriculture while maintaining the long-term intrinsic productivity of the natural resource base. Improvements in biotechnology, technologies to increase water and nutrient use efficiency, crop and natural resource management technologies and input use technologies (viz., precision agriculture, fertigation) will be required to achieve these goals.

The widespread adoption of high-yielding varieties, agriculture intensification practices and government policies are adversely affecting fertility of agricultural lands due to "mining" of nutrients. Therefore, prevention of such mining of soil nutrients is a necessary condition for the sustainable development of agriculture in India. Investments are needed to monitor the nutrient and fertility status of soils and to design and implement measures to convert nutrient depletion and imbalances and other potential adverse consequences of agricultural intensification and adoption of new yield-enhancing technologies.

The continued population growth, urbanization and changes in dietary habits will require the cereal production in India to double by 2025. This will result in a substantial growth in demand for fertilizers. Given that India has limited natural resources needed for fertilizer production, the future supply will be increasingly dependent on international trade within the guidelines of the WTO agreements.

REFERENCES

Anon. 1998. Fertilizer Pricing Policy, Report of the High Powered Review Committee, Department of Fertilizers, Ministry of Chemicals and Fertilizers, Government of India.

Awasthi, V. 2000. Long Term Policy for Fertilizer Sector in India — Implications for Industry and Farmers, Paper presented at the Fertilizer Ammoniation of India Seminar on Agricultural Subsidies — Global Dimensions, December 11-13, New Delhi, India.

Engelman, R. and P. LeRoy. 1995. Conserving Land: Population and Sustainable Food Production, Population and Environment Program, Population Action International.

Engelman, R. and P. LeRoy. 1993. Sustaining Water: Population and the Future of Renewable Water Supplies, Population and Environment Program, Population Action International.

FAO. 2001. Food and Agriculture Organization of the United Nations. Available at <http://faostat.fao.org/default.htm>.

Giese, A. 1997. Environmental responsibility on the farm — the impact on the fertilizer and agriculture industry, in *Proceedings of an International Workshop: Environmental Challenges of Fertilizer Production — An Examination of Progress and Pitfalls*, IFDC, Muscle Shoals, Alabama, September 17-19.

Gill, G. 1995. Major Natural Resource Management Concerns in South Asia. Food, Agriculture and the Environment, Discussion Paper 8, International Food Policy Research Institute, Washington, D.C.

Gregory, D., A. Roy and B. Bumb. 2000. Agricultural Subsidies in Developing Countries With Particular Focus on Fertilizers, Paper presented at the Fertilizer Association of India Seminar on Agricultural Subsidies — Global Dimensions, New Delhi, India, December 11-13.

Hedley, M. and A. Sharpley. 1998. Strategies for Global Nutrient Cycling, Paper presented at the 11th Annual Workshop of the Fertilizer and Lime Research Center, Massey University, Palmerston North, New Zealand, February 11-12.

Lele, U. and B. Bumb. 1994. *The Evolving Role of the World Bank: The Food Crisis in South Asia — The Case of India*. The World Bank, Washington, D.C.

ODI (Overseas Development Institute). 1994. The CGIAR: What Future for International Agricultural Research? Brief Paper Series, London: Overseas Development Institute.

Pinstrup-Andersen, P. 1976. Preliminary estimates of the contribution of fertilizers to cereal production in developing countries, *J. Econ.* II.

Stauffer, M. and G. Sulewski. 2000. Asia's Potential for Fertilizer Use, Paper presented at The Fertilizer Institute Outlook 2001 Conference, Alexandria, Virginia, November 13-14.

United Nations. 1998. *World Population Prospects: The 1998 Revisions*.

5 Economic Impacts of Agricultural Soil Degradation in Asia

Sara J. Scherr

CONTENTS

INTRODUCTION

Agricultural expansion and intensification in the developing countries of Asia have been accompanied by considerable soil degradation. Whether this loss of natural capital actually threatens food security and agricultural development in the short- or long term or justifies greater public investment to combat soil degradation, however, has been a matter of considerable debate. The objective of this chapter is to examine

1-5667-0594-0/02/$0.00+$1.50
© 2002 by CRC Press LLC

the question in light of the evidence from research on the agricultural productivity-related economic impacts of soil degradation, most published in the past decade. Environmental and off-site effects, while very important, are not addressed.

AGRICULTURAL DEVELOPMENT AND SOIL DEGRADATION

The period since World War II has seen remarkable growth in rural population, agricultural land area and agricultural productivity in the developing world. Although the rural population growth rate declined from 2.2% in 1960–65 to 1% in 1990–95, the absolute number of rural dwellers grew almost 40%, from 2.0 to 2.8 billion. Total growth rates for agricultural production in developing countries of Asia (4.1%/yr in East Asia between 1970–1988, 3.1%/yr in South Asia, the Near East and North Africa), compared with 2.6%/yr in Latin America and the Caribbean and 1.8%/yr in Subsaharan Africa), have surpassed historical growth rates in the industrialized countries (1.2%/yr), though not on a per capita basis. This growth came in part from extensive clearing of land for arable crops since the mid-1960s, plus much more for pasture and perennial crops. However, yield increases on land already in production contributed far more to production. For example, over 90% of growth in developing country cereal production between 1961 and 1990 came from yield growth. Arable land per capita declined from just under 0.5 ha in 1950 to just under 0.3 ha in 1990 (Scherr 1997; 1999).

It should be unsurprising that rural population increase, area expansion and intensification on such a scale would be associated with some degradation of soil resources. Furthermore, tropical and subtropical soils are typically more sensitive to degradation and subjected to more severe climatic conditions than the main Temperate Zone producing areas. Large numbers of people in Asia reside in areas with poor-quality soils that require careful management of organic matter, micro-organisms and physical structure even to maintain production. Of all global agricultural soils free from significant constraints,* only 15% lie within the tropics. While, globally, 45% of all agricultural lands are found on slopes above 8%, in the tropics, that proportion is 78% (Wood, Sebastian and Scherr 2000). An econometric analysis of agricultural productivity for 110 countries from 1961–1997 shows that good soils and climate are associated with an increase of about 13% in output per worker relative to poor soils. In Asia, that increase is 34% (Wiebe, et al. 2000). Assessments of vulnerability and risk of soil degradation, based on soil types and land pressure, suggest that a high proportion of the people of Asia live in areas with moderate to high vulnerability (Beinroth, Eswaran and Reich 2001). The nature of soil degradation has varied in different pathways of agricultural intensification, reflecting their different resource endowments. Five broad pathways can be identified. *Irrigated lands* have expanded and become highly productive beneficiaries of Green Revolution technologies. However, poor water management has led to widespread salinization and waterlogging, while more

---

* Major constraints included in FAO's Fertility Capability Classification include poor drainage, low cation exchange capacity, aluminum toxicity, acidity, high phosphorus-fixation, vertisol clays, low potassium reserves, alkalinity, salinity, nitric properties, shallow or gravelly soils, organic soils and low moisture-holding capacity.

subtle nutrient management problems associated with multiple cropping have slowed down yield increases in recent years. In *high-quality rainfed lands*, cropping intensity has also increased greatly. In these lands, excessive clearing and poor management of natural vegetation and inappropriate machinery use have sometimes led to soil compaction and exposed soils to erosion. Substitution of organic inputs with chemical fertilizers has led to declining organic matter and acidification on vulnerable soils.

In *densely populated marginal areas*, low quality and degradation-prone soils that were traditionally managed through moderate-to-long-fallow systems are now used for intensive crop production. Soil erosion from poor soil cover and nutrient depletion from inadequate management of organic matter and fertilizers, are common results. In *extensively managed marginal lands*, degradation has been caused principally by the land-clearing process itself, by nutrient depletion due to declining fallow periods and by widespread burning to control weeds and pests and provide ash for plant nutrition. *Urban agriculture*, whose importance accelerated dramatically in the 1980s, has contributed to food security, income generation and recycling of urban wastes, but also to some environmental problems. Soil contamination with urban pollutants may pose a health hazard to consumers and reduce production, while insecure land access and tenure discourage sustainable grazing and soil management practices (Scherr 1999).

ESTIMATES OF THE PHYSICAL EXTENT AND SEVERITY OF SOIL DEGRADATION

While there is an enormous amount of empirical data on physical soil degradation worldwide, most is plot- or site-specific and quite difficult to aggregate and interpret on a national, regional or global basis. The Global Assessment of Soil Degradation (GLASOD), based on a formal survey of local soil experts, was the first worldwide comparative analysis of soil degradation. The continental-scale study concluded that 747 million ha in Asia, of a total of 2,787 million ha, had been degraded since World War II — 27% of used land. This includes 38% of all cropland, 20% of permanent pasture and 27% of forest and woodland. Of total used land, 73% was not degraded, 11% was lightly degraded and 16% had experienced a significant decline in productivity. The primary causes of degradation were water erosion (441 million ha) and wind erosion (222 million ha), chemical degradation (74 million ha) and physical degradation (8 million ha); the latter two mainly affect cropland (Oldeman, et al. 1992).

Available regional studies (Table 5.1) provide estimates of soil quality change of similar magnitude, with large areas undegraded, but some subregions and countries having a high incidence of degradation. They suggest that the extent and severity of degradation is much worse in tropical than temperate regions; greater in drylands (South Asia) and hilly and mountainous agricultural areas (such as the Philippines, Thailand) and most serious for annual cropland (relative to other agricultural uses). A high proportion of irrigated farmlands — Asia's "breadbaskets" and "rice bowls" — suffer from salinization. Organic matter (OM) content — probably the single factor that best reflects productive potential — seems to be declining on a high proportion of land in many regions. Nutrient depletion is associated with areas of intensive crop production with low input use.

TABLE 5.1
Regional Estimates of the Physical Extent and Severity of Soil Degradation in Asia

Area	Source	Method	Main findings
Asia	Oldeman et al. (1994)	Qualitative, systematic survey of regional soil experts; sampling for continental results (GLASOD)	Evidence of soil degradation on 38% of agricultural lands, 20% of permanent pastures, and 27% of woodland and forests. 16% of all used land was moderately, strongly or extremely degraded following World War II
Asia-Pacific	FAO (1992)	Literature review for 13 countries	31% of arable and permanent cropland soils were degraded
Asian drylands	Dregne and Chou (1992)	Literature review	Land (not just soil) degradation affected 35% of irrigated lands, 56% of dry rain-fed lands, 76% of rangelands; 39% were severely affected.
South Asia	Young (1993)	Adjustment of GLASOD with national soil data	50% of all agricultural land had been moderately or strongly degraded.
South and SE Asia	Van Lynden and Oldeman (1997)	Qualitative survey of regional soil experts; sample for national results	Agriculturally induced degradation 27% of all land since World War II
China	Lindert et al. (1995), Lindert (1996)	Analysis of soil survey data 1940s–1990s	Declining organic matter and nitrogen in North China; little overall change in nutrient status in South China
Java	Lindert (1997)	Analysis of soil survey data 1940s–1990s	Declines in organic matter and nitrogen; increases in total phosphorus and potassium

These estimates have convinced many soil scientists and environmentalists that soil degradation poses a significant threat to current, and certainly future, food security. Their perspective is reinforced by widespread anecdotal evidence of yield decline and out-migration from degraded areas and by the historical evidence of economic collapse associated with uncontrolled soil degradation (Hillel 1991). To combat soil degradation, major international policy initiatives are being proposed, such as the Convention to Combat Desertification; incorporation of land rehabilitation into the Global Environment Facility; increased allocation of international research resources to soil management (e.g., the Soil Water, Nutrient Management program of the CGIAR-supported Future Harvest Centers); and increased allocation of bilateral and multi-lateral aid programs toward land quality improvement (e.g.,

UNCED 1992). Programs for soil protection and rehabilitation have expanded in many Asian countries through government, NGO and farm organization initiatives (Bridges, et al. 2001).

IS SOIL DEGRADATION ECONOMICALLY SIGNIFICANT?

However, such programs draw upon resources that are scarce, given stagnating or declining national agricultural investment budgets. Some have been unconvinced that such a reallocation of resources to combat soil degradation is economically justifiable. Pierre Crosson, for example, after reviewing the limited available literature on the impacts of soil degradation at the beginning of the 1990s, concluded that reducing, even eliminating, the present rate of natural resource degradation and fully restoring the productivity of presently degraded resources will make only a small contribution [relative to changes in production technology] to the global (especially LDC) increases in food production needed to meet the demand scenario for 2030 (1994:32). Even where data indicating high rates of physical degradation are considered to be reasonably reliable (often not the case), for several reasons, combating soil degradation may not be considered a high-priority public investment.

First, soil degradation statistics can be misleading. Degradation may not be an economically important problem at the farm level. Otherwise, farmers — who rationally would wish to preserve the resource base of their livelihood — would be taking the necessary steps to protect or rehabilitate their soils (and participate actively in soil conservation programs); when the problem becomes sufficiently serious they can be expected to do so (Enters 1998; Scherr, et al. 1996). Microeconomic evidence from some areas does support this view (Templeton and Scherr 1999). But it is clear that farmers may underinvest in soil protection even when they recognize it as a serious problem, due to incapacity to mobilize the necessary resources, inadequate knowledge, weak economic incentives, inadequate institutional support, tenure insecurity and other disincentives for long-term investment, and lack of available and locally adapted conservation technology (Scherr 2000). In some places where private costs of soil degradation are insufficient to trigger a response, the aggregate impacts of reduced agricultural productivity on economic growth, employment and agricultural markets may generate serious social costs worthy of policy attention. On some types of soil, degradation may be irreversible.

A second argument accepts that the economic costs of degradation are real and that key factors constrain farmers' ability to respond effectively without policy intervention, but asserts that the impact of degradation can largely be compensated for by increased use of other factors or through advances in production technology (e.g., fertilizer or improved seed). Returns to public investment in these alternative activities (e.g., fertilizer distribution or agricultural research) are observed or presumed to be superior to those of soil-improving investments (Crosson 1995). The validity of this argument hinges critically on the degree of complementarity between soil quality and other inputs or investments, as well as the absolute costs of degradation relative to the gains from productivity improvements.

A third argument against prioritizing soil-improvement efforts also accepts that there are serious economic impacts from soil degradation, but concludes that production on some of these degrading soils is not sustainable or competitive over the long term. Rather than invest scarce resources in improving poor quality soils, it is argued that it would be cheaper and more efficient to supply agricultural products from the country's better, less degradation-prone soils or imports. This position was asserted in much of the literature from the 1970s and 1980s arguing against investment in marginal lands (Nelson et al. 1997). While this argument may be valid in some degrading regions, in many others, sharp geographical shifts in production would entail unacceptable social and economic costs. The capacity to substitute imports depends upon export potential, and many countries with severe degradation problems have agriculture-dependent economies. This position also ignores the promising development of more sustainable production and soil management systems now being developed for tropical environments (Bridges, et al. 2001), as well as the potential threats to long-term food security posed by dependence on concentrated production in a few areas or foreign imports.

All of these arguments are essentially empirical. Each is correct for some places and not for others. Until recently, however, there was little quantitative evidence of the economic importance of soil degradation that would permit comparison with other types of agricultural problems, alternative investment opportunities or the costs of good soil husbandry and rehabilitation. Fortunately, the past decade has seen a marked increase in relevant studies.

OVERVIEW OF LITERATURE AND METHODS

This chapter draws from a review of more than 20 studies of the economic impacts of soil degradation in developing countries of Asia. Most address the impacts of soil erosion, nutrient depletion or salinization.

SOURCES OF DATA ON SOIL PRODUCTIVITY CHANGE AND EFFECTS

Economic impact studies require biophysical data on changes in soil quality over time and on the impact of such changes on productivity. Such studies are plagued by chronic technical data problems relating to soil quality change, the relation of soil quality change to productivity and the capacity to compensate for or reverse degradation through input use. Since large longitudinal datasets are rarely available, researchers must estimate production impacts. For this purpose, many use the Universal Soil Loss Equation (USLE) model, which integrates effects of erosion rates, rainfall erosivity, soil erodibility, slope length and steepness, farming practice and soil conservation measures. But, like most methods for soil quality assessment, the USLE was developed for use at plot scale and it generally overestimates erosion, especially when aggregated to higher scale. Thus, researchers trying to assess quality change at higher scales have also used approximation approaches, including expert consultation (using standard criteria), comparative evaluation of published studies from multiple sites, extrapolation of field experiment and survey results or estimates constructed from an examination of secondary data on land use change, etc. Existing data are not ideal for use in economic analysis, as few are based on actual historical time series data on yield and production cost and even fewer control for other variables in attributing yield or cost change to soil quality change (such as soil health, pests or weed problems).

ECONOMIC EVALUATION METHODS

Studies of the productivity-related economic effects of soil degradation can be divided into three periods. Those published in the late 1970s and 1980s were intended mainly to draw public attention to the issue, simply calculating gross aggregate effects of soil erosion on agricultural lands without conservation and the value of gross economic losses. Analyses published in the early 1990s were more systematically designed and reflective of field experience. They relied mainly on secondary data, literature reviews and surveys of regional soil experts and used fairly simple economic analyses. Since the mid-1990s, a third generation of studies has been undertaken, using more sophisticated data collection and analytical methods.

Eight types of economic analysis were found in the literature reviewed. Qualitative impact assessments included surveys of either farmers or soil experts. Aggregate gross valuation of the economic losses due to degradation, usually the cost of replacing lost nutrients or lost production, was used in many of the earlier studies. Comprehensive evaluations, based on data disaggregated by soil type, farming system, ecozone or crop, were used to calculate continental soil nutrient balances or erosion measures and national environmental accounts.

More sophisticated modeling approaches have increasingly been used, although their advantages are sometimes outweighed by the poor quality of the underlying data and oversimplified assumptions. Biophysical models of degradation–yield relationships, often with an economic module to permit valuation of outcomes, were constructed using either primary data from field trials or surveys or secondary data from literature. Cost–benefit models were used for *ex-ante* prediction of soil degradation and associated economic losses over an extended time period. Mathematical models, such as household, plot or district-level regression models and production functions, were developed to disentangle the effects of soil degradation from other factors. Global-scale partial equilibrium models were constructed to evaluate the impacts of soil quality on output and, in one case, consumption and prices. Over 30 different indicators of economic impact, often measured with quite different methods, were used, making comparative assessment problematic.

EVIDENCE OF THE ECONOMIC IMPACTS
OF SOIL DEGRADATION

Economists and policymakers typically consider soil quality not as a policy objective itself, but as an input into achieving other policy objectives. Four types of agricultural productivity-related economic impact from soil degradation are of particular interest:

1. Aggregate supply, stability or price of agricultural output, i.e., if lands with degrading soils are a significant source of market supply for national consumers or export markets and alternative sources of supply are not available or economic
2. Agricultural income or economic growth, i.e., if soil degradation reduces agricultural income and its multiplier effects on an economically significant scale, through lower production or higher costs and alternative sources of economic growth are limited or expensive to develop
3. Consumption by poor farm households, i.e., if lands with degrading soils are a critical source of food security for subsistence or semisubsistence producers with few alternative livelihood options
4. National wealth, i.e., if degradation reduces the long-term productive capacity of soil resources deemed to be of future economic or environmental significance, threatening the resource base and food security of future generations

AGRICULTURAL SUPPLY

More researchers have evaluated the agricultural supply effects of soil degradation than of other economic impacts. Unfortunately, most studies are limited to simple analyses of loss in the annual productive potential of land relative to its nondegraded state. Impacts of declining production from degraded sites on overall domestic market supply, trade, consumer and producer prices are not calculated. It is thus impossible to determine how much of the lost supply might be made up by producers elsewhere responding to higher prices. Only a few studies distinguish the impacts on production due to declining productive capacity of soils from the impacts due to farmers' decisions to reduce use of variable inputs on more degraded sites. And, as noted above, the soil degradation–productivity relationships underlying these economic studies are, in many cases, unconfirmed empirically or modeled in a fairly crude way. Even with these caveats in mind, however, the magnitude of productivity impact suggested in the more than 20 studies reviewed is troubling, given expected future growth in agricultural demand and import constraints.

Major findings from regional and national studies for Asia are summarized in Table 5.2. While the global GLASOD study estimates that 84% of the huge continent of Asia is nondegraded or only lightly degraded, important areas were identified with declining soil productivity. Degradation is occurring both in some of the most intensively cultivated, irrigated surplus production areas of South Asia and in dryland and mountain areas undergoing intensification in South Asia and montane Southeast Asia. In both regions, rural poverty is already very high. Cumulative productivity losses in Asian cropland, based on GLASOD data, are roughly estimated to be 12.8%.* While this loss is similar to that of South America, only half that of Africa and a third that

---

* These figures were calculated by multiplying the GLASOD areas with different soil degradation categories by a coefficient of yield loss. In the case of crop land, the coefficients were 15% loss for "light" soil degradation, 35% for "moderate," 75% for "strong," and 100% for "extreme" degradation. In the case of pasture land, the corresponding coefficients were 5% for light, 18% moderate, 50% strong. For combined crop and pasture land, two different sets of coefficients were used: 5% for light, 18% for moderate, 50% for strong; and 15, 35 and 75% respectively.

TABLE 5.2
Effects of Soil Degradation on Agricultural Supply in Asia

Area	Source	Method	Principal findings
Asia	Dregne and Chou (1992); Dregne (1992)	Literature review	Over a third of irrigated land and over half of dry rain-fed lands had experienced a 10% loss in productive potential. On 8% of irrigated and 10% of rain-fed drylands, there had been at least a 25% loss. Over half rangelands had experienced over 50% loss. Well-confirmed permanent soil productivity loss of at least 20% from human-induced water erosion in 8 countries, and presumptive evidence in 5 others. Wind erosion had little productivity effect.
Asia	Oldeman (1998)	Standard coefficients of yield decline applied to GLASOD data	Average productivity loss was 13% in cropland and 5–9% in pastures.
South and SE Asia	Van Lynden and Oldeman (1997)	Qualitative, systematic survey of regional soil experts, sampling for national results (ASSOD)	Moderate or worse productivity decline was found on 7.5% of total land from erosion-induced topsoil loss; 6.5% terrain deformation; 2.5% fertility decline; 1.0% salinization; 1.3% waterlogging. Serious fertility decline or salinization affected at least 15% of arable land. Major irreversible productivity loss was found only in small areas. Moderate or worse impacts were found on a tenth of all land, but incidence much higher in Pakistan, the Philippines and Thailand (over 30%), and India (20%). In Bangladesh, China, Cambodia, Malaysia and Vietnam 5–12% of lands were badly degraded.
India (Nat'l)	Seghal and Abrol (1994)	Synthesis from soil surveys, expert survey and experimental data	Degradation was insignificant on 36% of land area, low (less than 15% yield loss) on 5%; moderate (15–33% loss) on 11%, high (33–67% loss) on 43% and 5% so degraded that soils are unusable.

(continued)

TABLE 5.2 (CONTINUED)
Effects of Soil Degradation on Agricultural Supply in Asia

Area	Source	Method	Principal findings
India (Uttar Pradesh)	Joshi and Jha (1991)	Household and plot survey in four villages in Uttar Pradesh, 1985–86	Over 10 years, paddy yields on soils affected by salinization declined 61%, wheat by 68%. Average yield of high-yielding varieties on alkaline plots decreased by 51% and local varieties by 45%; under waterlogged conditions, the figures were 41% and 26%. Alkalinity accounted for 72% of the difference in gross income between normal and salt-affected plots; the rest was attributed to reduced input use on degraded soils.
India (4 sites)	Cassman et al. (1995)	Long-term experiments on intensive irrigated farming systems	Negative soil productivity trend without farmyard manure in annual double-crop irrigated rice systems in the warm, subhumid tropics; with manure, the trend was flat. In the warm, subhumid tropics, irrigated rice-wheat productivity trends were rising for rice, flat for wheat. A warm semi-arid subtropics maize-wheat system had flat productivity trend for maize, positive for wheat.
Bangladesh	Pagiola (1995a)	Trend analysis from farm, experimental data	In intensive cropping systems, time trends for yields declined 0.36 tons/yr 1975–85, most likely due to deteriorating nutrient balance and organic matter decline.
Pakistan (Punjab)	Ali and Byerlee (2001)	Cost function analysis of productivity changes, using secondary district-level data, 1966–94	Crop total factor productivity was relatively high in the wheat-cotton and wheat-mungbean systems, modest in one system (wheat-mixed) and negative in the wheat-rice system. Soil and water degradation lowered productivity growth by about 58% in the province. In the wheat-rice system, resource degradation offset most of the benefits of improved production technology.
China (Nat'l)	Huang and Rozelle (1994); Rozelle et al. (1997)	Grain yield functions for 1975–90, w/pooled secondary data	Elasticities of grain yield calculated for 23 provinces were −0.146 for soil erosion, −0.003 for salinization, and −0.276 for multiple cropping intensity. In the early 1990s, degradation cost the country the equivalent of nearly 30% of yearly grain imports.

of Central America, it is still 50 to 60% higher than comparable productivity losses in Europe and North America.

Higher quality data from the later ASSOD expert survey showed higher degradation in Asia than did GLASOD, with especially high rates of productivity loss documented for India, Pakistan, the Philippines and Thailand. Fertility decline and salinization notably affected supply potential in Asian cropland.

One study used a global agricultural model (IMPACT) to simulate the effects of a 10% decline, relative to the baseline scenario, in crop productivity in the developing countries after 25 years. A second scenario assumed the same rate of degradation, but also further reduction of crop yield growth in Pakistan by 50% (reflecting declining area due to salinization), a further 5% decline in growth of rice yields and a 21% decline in other crop yield growth in China. The study concluded that the first scenario would result in world prices higher by 17–30% in 2020, particularly for maize, rice, roots, tubers and wheat. The second scenario does not further affect world prices, but does result in higher wheat imports, especially in Pakistan and China. These are large effects, although the authors argue that the impact of inadequate conventional agricultural research and investment would be even larger (Agcaoili, Perez and Rosegrant 1995).

ECONOMIC GROWTH

Regional and national studies have produced a wide range of estimates of the magnitude of economic losses from soil degradation in Asia, reported as a proportion of the AGDP (Table 5.3). Economic losses are calculated in a fairly simple manner, in most cases, either in terms of the financial value of lost crop yields or the cost of purchasing fertilizer to replace nutrients lost through erosion or depletion. The estimates are unexpectedly high — 3–7%/year in a regional study and national studies of India, Java and Pakistan; less than 1% in China. Calculating the discounted future stream of losses from soil degradation raises the cost to a figure equivalent to 36–44% of AGDP in Java (Repetto, et al. 1989). It is hard to evaluate whether the figures are overestimates or underestimates. On the one hand, few take into account market and price impacts and responses that would tend to dampen the effects of degradation; on the other hand, only two reflect the economic multiplier effects of that lost income through the larger economy and most include only the effects of soil degradation on the principal staple crops.

CONSUMPTION BY POOR FARM HOUSEHOLDS

There has been no global spatial mapping of the relationship between poverty and soil quality or soil degradation. However, a number of factors lead us to assume that soil degradation has a particularly negative impact on the rural poor. Studies in Asia in the 1980s found that the rural poor were more dependent on agriculture than the nonpoor. The poor depend more on annual crops that are typically more degrading; they rely more on common-property lands, which tend to suffer more degradation than privately managed land. Where the principal assets of the poor are low-productivity or degrading lands and their mobility to seek more remunerative livelihood

TABLE 5.3
Magnitude of Economic Losses from Soil Degradation in Asia, as %AGDP*

Study region	Authors	Types of degradation	Annual loss (or GAIL) as %AGDP	Discounted future loss as %AGDP
South and Southeast Asia	Young (1993)	Soil erosion, fertility decline, salinization and waterlogging	7	—
China	Huang and Rozelle (1994); Huang et al. (1996)	Soil erosion, salinization, fertility decline	<1	—
India	Young (1993)	Soil erosion, fertility decline, salinization and waterlogging	5	—
Indonesia (Java)	Magrath and Arens (1989)	Soil erosion	3	—
	Repetto et al. (1989)	Soil erosion	4	40 (CLFP)
Pakistan	Young (1993)	Soil erosion, salinization	5	—

*Annual loss: the lost value for that year due to soil degradation. AAVA: Annual agricultural value added. CLFP: Capitalized loss of future productivity (the value of the stream of future loses due to a particular year's soil degradation). GAIL: Gross annual immediate loss (the lost value for gross cropland output in a single year due to land degradation in the previous year).

options is restricted by economic, political or social conditions, they may fall into a poverty "trap," without sufficient assets to undertake the land husbandry and investment necessary to maintain or increase productivity (Malik 1998:18–20). The poor tend to be "pushed" to marginal lands, either by political forces, expulsion of squatters from higher quality land during modernization or by land markets in which they cannot afford to compete for higher quality land. Because the poor use fewer inputs, they rely more on intrinsic soil quality.

There has been extensive debate over the role of the poor in *causing* soil degradation. Some studies have shown that poverty may exacerbate degradation where subsistence food, feed and fuel needs can be met only through overexploitation of natural vegetation (with resulting exposure of soil to accelerated degradation) and consumption of organic residues from farming and livestock-keeping that would otherwise contribute to soil replenishment. But there is also evidence that the poor have higher incentives to conserve soil than do the more prosperous (Pagiola 1995b; Scherr 2000). Few studies have quantified impacts of soil degradation on the poor.

The IMPACT model simulations discussed above projected that accelerated soil degradation would reduce food security for the poor not only from contraction in production, but also from reduced demand due to higher prices. In the first scenario, with general productivity growth decline in developing countries, child malnutrition rates remain unchanged, while in the baseline scenario, malnutrition declined by nearly 3.6%. In the second scenario, child malnutrition actually increased 0.6%, with the major wheat- and rice-producing and consuming countries, especially in Asia, exhibiting the largest increments in malnutrition (Agcaoili, Perez and Rosegrant 1995). An econometric analysis of degradation impacts at district level in China found them systematically more serious in poorer and more densely populated areas (Rozelle, Huang and Zhang 1997).

NATIONAL WEALTH

No studies were found on the impacts of degradation on Asian national wealth in financial terms. Rather, in various studies, wealth effects were assumed to result from land abandoned and lost to agriculture irreversibly due to soil degradation and from declines in productive potential on other land. GLASOD data indicate that only a few percent of all agricultural land in Asia had been permanently lost to agriculture since the mid-1940s. ASSOD reported that 7.5% was so degraded that losses could not be compensated for even with high levels of management and was unproductive under low management (van Lynden and Oldeman 1997). Without proactive efforts, millions of hectares of irrigated land in Asia may go out of production due to salinization.

Of greater concern is the decline in soil quality on land that remains in production. Over a third of all the developing world's arable and perennial cropland is currently located in just three Asian countries — India (22%), China and Indonesia. In these countries and others with large farmland areas, such as Pakistan, Myanmar and Thailand, food supply and rural poverty concerns are probably more pressing reasons for careful attention to protection of land assets than is preservation of long-term soil wealth. However, 11 Asian developing countries have only 1–10 million ha of cropland, and in nearly all, land pressure is already high (0.16 to 0.30 ha/capita) or very high (under 0.15/capita) (Table 5.4). For these countries, conserving farmland quality must be a strategic long-term food security consideration (Scherr 1999).

CONCLUSIONS

Notwithstanding the data limitations, some tentative conclusions can be drawn about the economic and policy importance of soil degradation in Asian developing countries.

CONTINUED AGRICULTURAL SOIL DEGRADATION IS LIKELY TO RAISE PRICES AND IMPORT DEMAND IN MAJOR ASIAN REGIONAL MARKETS

Evidence suggests that soil quality on about 16% of Asia's total agricultural land has been seriously degraded since the middle of the 20th century — about the average for the developing world and just above the world average. Thus, soil quality on most used land is stable or only lightly degraded. Soil degradation does not present a "global" threat in terms of supplying international markets. For the major suppliers to these markets, other factors (increased land in production and under irrigation, increased productivity though new varieties and inputs and improved marketing systems) have probably compensated for — or at least masked — some of the degradation-induced

TABLE 5.4
Arable Land Resources of South and Southeast Asia, 1994

Extent of arable land (million hectares)	Very high (under 0.15)	High (0.16-0.30)	Medium (0.31-0.45)	Low (over 0.45)
Very extensive (over 30)	China Indonesia	India	—	—
Extensive (10.0-29.9)	—	Myanmar Pakistan	Thailand	—
Moderate (5.0-9.9)	Bangladesh Philippines Vietnam	—	Malaysia	—
Limited (1.0-4.9)	North Korea South Korea Sri Lanka Nepal	—	—	—
Very limited (Under 1)	Papua New Guinea	Cambodia Laos	—	—

Source: FAO (1995)

productivity loss. Elements tempering any global effects are the considerable global capacity for supply substitution from non-degrading lands, the dominance of temperate producers in international wheat and maize markets and the modest share — 10% — of food consumption that is traded on international markets (McCalla, 2000). However, accelerated soil degradation in Asia, especially in the larger, economically fast-growing countries, would likely raise international food prices and regional import demand.

SOIL DEGRADATION POSES A MAJOR ECONOMIC THREAT IN MANY SUBREGIONS OF ASIA

Although soil degradation in Asian cropland appears to be widespread and the pace has almost certainly accelerated in the past 50 years, productivity effects have so far been geographically limited. In those regions where degradation is prominent, there appear to be large impacts on food consumption by the rural poor, agricultural income and economic growth and. in some cases, national wealth. At particular economic risk are the many subregions with degradation-prone soils (particularly in drylands and hillside regions), inadequately managed irrigation (particularly in South Asia), and rapidly intensifying production without the economic incentives or the technology for good resource husbandry (densely populated marginal lands in many parts of Asia). Some of these agricultural regions have limited alternative livelihood options, sources of food supply or nonagricultural development potentials. Thus, while posing particular problems for the poor, soil degradation is also likely to have far-reaching impacts on economic development in many Asian countries and sub-regions, in both the short and long term.

RECOMMENDATIONS FOR RESEARCH

Recent years have seen the beginning of serious efforts by soil scientists to quantify the extent and severity of soil degradation and the relationship of soil quality to productivity, joined with serious efforts by economists to link those changes to policy-relevant outcomes. Nonetheless, available data has serious limitations as a guide for informing policy priorities.

The empirical basis for drawing policy recommendations remains weak. The data on which Tables 5.1, 5.2 and 5.3 are based rely on variable, often subjective methods to assess the impact of soil degradation on productivity. To facilitate cross-site analysis, this suggests the need for greater standardization of methods used. Soil quality change and consequent productivity impacts on different types of crops remain poorly understood for many types of tropical soils.

The geographic coverage of economic studies is limited; for example, no sub-national or national-scale economic studies were found for East Asia and the Pacific or West Asia. For no country does a critical mass of economic data exist that is sufficient to formulate well-targeted policy interventions. Few studies have been done of the economic impacts of change in soil physical properties, such as compaction or acidification or the economic impacts of soil degradation on grazing lands. Few studies evaluate the *net* effect of soil degradation on supply, taking into account price effects, substitution of supply by other producing areas or other secondary impacts. Only a handful of studies evaluated the impact of soil degradation on food consumption or nutrition of poor farmers. There were no studies on the impacts of degradation on national wealth; none even assessed the relative economic importance of those lands currently suffering productivity loss or going out of production. Few studies were found, at subnational, national or regional scales, of the extent and economic impacts of farmers' management and investments to *improve* soil quality, which we know anecdotally is occurring.

Even if these empirical gaps are filled and the economic impacts of soil degradation are more accurately documented, this may not be sufficient to inform priorities for policy action. Additional contextual information is required. First, so long as soil degradation is reversible at an economically acceptable cost and other investment opportunities are more attractive, prevention is not always preferable — or even cheaper — to cure. For example, farmers may cease to undertake soil-protecting investments during periods of prolonged low food prices, but resume those practices when prices rise. Farmers may practice soil nutrient mining over some period to accumulate alternative forms of more economically valuable capital, and subsequently use those resources to rebuild soils. Unfortunately, for many soil types, we simply do not know much about the impacts of degradation, the costs of rehabilitation or the thresholds for soil quality below which future investment in restoration will be uneconomic.

Even if the economic impacts of degradation are high, it may not always be necessary to intervene. Future structural changes in the economy, trade, infrastructure, climate and human settlements may shift the geographic pattern of agricultural production and farmer incentives and capacity for good soil husbandry. Areas that are currently major food suppliers may be marginalized and the relative value of their soil resources reduced even without degradation. Empirically based analysis of such trends is conspicuously absent. The impact of inputs (for example, fertilizer or irrigation) on productivity loss for different types and severity of soil degradation is poorly understood.

The economic and noneconomic impacts of soil degradation on environmental services, such as species habitat, hydrological function, water quality and global carbon cycles (Pagiola, 2001) were not addressed in this chapter, nor were the effects of soil degradation on downstream economic activities (Enters, 1998) taken into account. These are often considerable, raising the social costs of soil degradation in all five agricultural pathways. They must clearly be taken into account, together with productivity-related impacts, in setting policy priorities and strategies. At the same time, care must be taken not to blame farmers for environmental impacts (e.g., sedimentation) that recent research shows may be due to other factors (Bridges, et al. 2001). Recent advances in hydrology, geology, ecology and other fields are changing our understanding of agriculture–environment interactions and illustrating ways that improved farm management and landscape design can contribute to watersheds and biodiversity (McNeely and Scherr 2001).

PRIORITY RESEARCH CHALLENGES

There remains, thus, a substantial research challenge to answer the critical questions for setting policy priorities related to soil degradation. Economic analyses have typically considered soil simply to be an inert substrate to which external inputs are applied to produce agricultural outputs; this view is clearly mistaken. Soil quality — topsoil depth and function, chemical and physical properties — is itself a critical factor of production and should be evaluated explicitly in agricultural economic and policy analysis.

Further conceptual work is needed, particularly for the assessment of changes in soil wealth. Methodologies for data collection and analysis of soil quality effects on agricultural supply needs and economic growth need further development. Economists need to work more closely with soil scientists, geographers and landscape ecologists to develop cost-effective soil quality indicators and sampling strategies for collection of socioeconomic, soils and production data, at various scales. Standard indicators must be developed to assess economic impacts of degradation — for all types of impacts — to permit meaningful comparative analysis across space and time.

To set action priorities, more accurate information is needed on the actual geographic patterns of serious soil degradation — and soil improvement and their impacts on agricultural supply, economic growth, rural poverty and soil wealth. Studies should distinguish farming areas with different soil types, agroclimatic zones, land use intensity, market environment and type of producer. Such studies have been made feasible by new remote-sensing technology, sampling designs that provide spatially explicit socioeconomic survey data and geographic information systems. The focus of analysis should be subnational, where soil quality change and impacts can be meaningfully measured and interpreted and policies implemented.

Spatial variations in the economic importance of soil quality change should be expected and should motivate targeted policy intervention.

REFERENCES

Agcaoili, M., N. Perez and M. Rosegrant. 1995. Impact of Resource Degradation on Global Food Balances. Paper prepared for the workshop on Land Degradation in the Developing World: Implications for Food, Agriculture and Environment to the Year 2020, April 4-6, Annapolis, Maryland; Washington, D.C.: International Food Policy Research Institute.

Ali, M. and D. Byerlee. 2001. Agricultural productivity growth and resource degradation in Pakistan's Punjab. In: E.M. Bridges, F.W.T. Penning De Vries, I.D. Hannam, R.L. Oldeman, S.J. Scherr and Samran Sombatpanit, Eds. *Responses to Land Degradation.* Oxford and IBH Publishing, New Delhi, pp. 186-199.

Beinroth, F.H., H. Eswaran and P.F. Reich. 2001. Land quality and food security in Asia. In E.M. Bridges, F.W.T. Penning De Vries, I.D. Hannam, R.L. Oldeman, S.J. Scherr and Samran Sombatpanit, Eds. *Responses to Land Degradation.* Oxford and IBH Publishing, New Delhi, pp. 83-97.

Bridges, E.M., I.D. Hannam, L.R. Oldeman, F.W.T. Penning de Vries, S.J. Scherr, Samran Sombatpanit. 2001. *Response to Land Degradation.* Oxford and IBH Publishing, New Delhi, 507 pp.

Cassman, K., R. Steiner, A.E. Johnson. 1995. Long-term experiments and productivity indexes to evaluate the sustainability of cropping systems. In: V. Barret, R. Payne and R. Steiner, Eds. *Agricultural Sustainability, Environment and Statistical Considerations.* John Wiley and Sons, Chichester: 231-244.

Crosson, P.R. 1995. Soil Erosion and its On-Farm Productivity Consequences: What do We Know? Resources for the Future Discussion Paper 95-29. Resources for the Future. Washington, D.C.

Crosson, P. 1994. Degradation of Resources as a Threat to Sustainable Agriculture. Paper presented for the First World Congress of Professionals in Agronomy, Santiago, Chile, September 5-8.

Dregne, H. E. 1992. Erosion and soil productivity in Asia. *Journal of Soil and Water Conservation.* 47(1):8-13.

Dregne, Harold E. and Nan-Ting Chou. 1992. Global desertifications, dimensions and costs. In: H.E. Dregne (Ed.) *Degradation and Restoration of Arid Lands,* Texas Tech University, Lubbock, Texas: 249-82.

Enters, T. 1998. Methods for the economic assessment of the on- and off-site impacts of soil erosion. Issues in Sustainable Land Management No.2. International Board for Soil Research and Management. IBSRAM, Bangkok.

FAO. 1995. *FAO Production Yearbook.* Food and Agriculture Organization of the United Nations, Rome, Italy.

FAO. 1992. Regional Strategies for Arresting Land Degradation. Twenty-First FAO Regional Conference for Asia and the Pacific, February 10-14, New Delhi, India.

Hillel, D.J. 1991. *Out of the Earth: Civilization and the Life of the Soil.* The Free Press, New York.

Huang, J. and S.Rozelle. 1994. Environmental stress and grain yields in China. *American Journal of Agricultural Economics* 77: 246-256.

Huang, J. and S. Rzelle 1996, or Huang et al. 1996.

Joshi, P.K. and D. Jha. 1991. Farm-Level Effects of Soil Degradation in Sharda Sahayak Irrigation Project. Working Papers on Future Growth in Indian Agriculture, No. 1, Central Soil Salinity Research Institute, ICAR and International Food Policy Research Institute. September.

Lindert, P., J. Lu and W. Wanli. 1995. Soil Trends in China since the 1930s. Agricultural History Center Working Paper Series No. 79. University of California, Davis. August.

Lindert, P. 1996. The Bad Earth? China's Agricultural Soils Since the 1930s. Working Paper Series No. 83. Agricultural History Center, University of California, Davis, December.

Lindert, P. 1997. A Half-Century of Soil Change in Indonesia. Working Paper Series No. 90. Agricultural History Center, University of California, Davis, June.

Magrath, W.B. and P. Arens. 1989. The Costs of Soil Erosion on Java: a Natural Resource Accounting Approach. Environment Department Working Paper No. 18. The World Bank, Washington, D.C. August.

Malik, S.J. 1998. Rural poverty and land degradation: What does the available literature suggest for priority setting for the Consultative Group on International Agricultural Research? Report prepared for the Technical Advisory Committee of the CGIAR. February. Draft.

McCalla, A. F. 2000. Agriculture in the 21st Century. CIMMYT Economics Program, Fourth Distinguished Economist Lecture, March 2000. International Maize and Wheat Improvement Center, Mexico, D.F.

McNeely, J. and S.J. Scherr. 2001. Common Ground, Common Future: How Ecoagriculture Can Help Feed the World and Conserve Wild Biodiversity. IUCN and Future Harvest: Washington, D.C.

Nelson, M., R. Dudal, H. Gregersen, N. Jodha, D. Nyamai, J.-P. Groenewold, F. Torres, A. Kassam. 1997. Report of the Study on CGIAR Research Priorities for Marginal Lands. Technical Advisory Committee, CGIAR, FAO, Rome, March.

Oldeman, L.R., R.T.A. Hakkeling and W.G. Sombroek. 1992. World Map of the Status of Human-induced Soil Degradation: An Explanatory note. Wageningen, International Soil Reference and Information Centre, Nairobi, United Nations Environment Programme. 27 pp + 3 maps. Revised edition.

Oldeman, L.R. 1994. The global extent of soil degradation. In: D.J. Greenland and T.Szaboles (Eds). Soil Resilience and Sustainable Land Use. Commonwealth Agricultural Bureau International, Wallingford, U.K., pp 99-118.

Oldeman, L.R. 1998. Soil degradation: A threat to food security? Report 98/01. International Soil Reference and Information Centre, Wageningen.

Pagiola, S. 1995a. Environmental and Natural Resource Degradation in Intensive Agriculture in Bangladesh. Environmental Economics Series Paper No. 15. Environment Department. The World Bank, Washington, D.C.

Pagiola, S. 1995b. The Effects of Subsistence Requirements on Sustainable Land Use Practices. Paper presented at the Annual Meeting of the American Agricultural Economics Association, Indianapolis, August 6-9.

Pagiola, S. 2001. The global environmental impacts of agricultural land degradation in developing countries. In: Bridges, E.M., I.D. Hannam, L.R. Oldeman, F.W.T. Penning de Vries, S.J. Scherr, Samran Sombatpanit. Eds. Responses to Land Degradation. Oxford and IBH Publishing, New Delhi,

Repetto, R.W. Magrath, M. Welk, C. Beer and F. Rossini. 1989. Wasting Assets. World Resources Institute, Washington, DC.

Rozelle, S., J. Huang and L. Zhang. 1997. Poverty, population and environmental degradation in China. Food Policy 22(3): 229-251.

Rozelle, S., G. Veeck and J. Huang. 1997. The impact of environmental degradation on grain production in China's provinces. *Economic Geography,* 73 (June): 44-66.

Scherr, S.J. 2000. A downward spiral? Research evidence on the relationship between poverty and natural resource degradation. *Food Policy,* 25: 479-498.

Scherr, S.J. 1999. Soil Degradation: A Threat to Developing Country Food Security in 2020? Food, Agriculture and the Environment Discussion Paper, International Food Policy Research Institute, Washington, D.C.

Scherr, S.J. 1997. People and environment: What is the relationship between exploitation of natural resources and population growth in the South? *Forum for Development Studies* 1:33-58.

Scherr, S.J., G. Bergeron, J. Pender and B. Barbier. 1996. Policies for sustainable development in fragile lands. EPTD Paper. International Food Policy Research Institute, Washington, D.C.

Scherr, S.J. and S. Yadav. 1995. Land Degradation in the Developing World: Implications for Food, Agriculture and the Environment to 2020. Food, Agriculture and Environment Discussion Paper 14. International Food Policy Research Institute, Washington, D.C.

Seghal, J. and I.P. Abrol. 1994. *Soil Degradation in India: Status and Impact.* New Delhi, India, Oxford and IBH.

Templeton, S. and S.J. Scherr. 1999. Effects of demographic and related microeconomic change on land quality in hills and mountains of developing countries. *World Development* 27(6):903-918.

UNCED. 1992. *Agenda 21.* United Nations Commission on Environment and Development, Rome, Italy.

van Lynden, G. and L.R. Oldeman. 1997. *Soil Degradation in South and Southeast Asia.* Wageningen, International Soil Reference and Information Centre. For UNEP.

Wiebe, K., M. Soule, C. Narrod and V. Breneman. 2000. Resource quality and agricultural productivity: A multi-country comparison. Contributed Paper for presentation at the 24th International Conference of Agricultural Economists, 13-19 August 2000, Berlin.

Wood, S., K. Sebastian and S.J. Scherr. 2000. *Pilot Analysis of Global Ecosystems: Agroecosystems.* International Food Policy Research Institute and the World Resources Institute: Washington, D.C.

Young, A. 1993. Land Degradation in South Asia: Its Severity, Causes and Effects Upon the People. Final Report Prepared for Submission to the Economic and Social Council of the United Nations (ECOSOC). FAO, UNDP and UNEP, Rome, Italy.

6 Soil Degradation as a Threat to Food Security

R.P. Narwal, B.R.Singh and R.S. Antil

CONTENTS

INTRODUCTION

Food security is the very basis for economic, social and cultural development and for political stability of a country. It is projected that, between 1997 and 2020, a growing and urbanizing population with rising incomes will increase global demand for cereals by 35%, amounting to 2497 million tonnes. However, the growth in cereal yields is slowing down in both developed and developing countries and is projected to further decrease in coming decades. The net cereal imports by developing countries are forecast to almost double by 2020, with maximum increase expected in East and South Asia. India, home to about 20% of the world's population and more than 15% of the world's livestock, has only 10% of the land resources of the world to meet their basic requirements. With the present rate of increase in human and animal population, India will face a great challenge and tremendous task in sustaining food security for its population in coming years.

TABLE 6.1
Area, Food-Grain Production and Population in India

Year	Area (million ha)	Production (M Mg)	Population (million)	Land:people ratio
1950–51	97	51	361	0.34
1960–61	116	82	439	0.32
1970–71	124	108	548	0.28
1980–81	127	130	683	0.24
1990–91	128	176	846	0.20
1999–2000	—	206	987	0.16
2000–2001(T)	—	212	1006	0.14

Source: Extracted and modified from FAI (2000)

One of the finest Indian success stories of the post-independent era is the Green Revolution of the 1960s, which transformed the country from "begging bowl" to "breadbasket." Presently, food production growth in India is keeping pace with that of its population. But the question remains whether this momentum can be sustained. While the population has increased threefold since independence, the production of food grains has quadrupled from 51 million mega grams (M Mg = million tonnes) in 1950–51 to 206 MMg in 1999–2000 (Table 6.1). The land area in India is limited and shrinking, whereas human and animal populations are increasing, leading to a fast decline in per capita arable land area. The demographic projections indicate that per capita land availability declined from 0.34 ha in 1950 to 0.14 ha in 2000 and will decrease to 0.10 ha in 2025 (FAI, 2000). This implies that the increase in food production has to come by vertical expansion of productivity per unit area.

India is likely to reach a population of 1225 million of people and 600 million livestock by 2015, necessitating 275 MMg of food grains, 1000 MMg of green fodder and 235 million m³ of fuel wood compared with the present production of 206 MMg of food grains, 513 MMg of green fodder and 40 million m³ of fuel wood. Kanwar and Katyal (1997) estimated that India may need 301 MMg of food grains by 2025, if the present trend of population growth continues. This scenario shows that food-grain production must be increased at the rate of 5–6 MMg yr^{-1} to keep pace with the population growth. If the present rate of population growth continues unabated, India's population will exceed that of China by 2050, but its agricultural production will remain far behind. Thus, it is a matter of great concern that, unless population growth is drastically curtailed, India will face a serious problem of food security, nutritional quality, social security and environmental safety. Brown and Kane (1994), of the World Watch Institute, estimated that India may be compelled to import 45 MMg of food grains by 2025, as its production will fall short of its needs. According to the World Watch Institute report entitled Beyond Malthus (Brown,1998)," India's population will increase from 1 billion in 2000 to 1.535 billion in 2050. The per capita availability of arable land will decrease to <0.10 ha and shortage of water will be accentuated." The report warns that "the resulting cutbacks in irrigation could reduce India's harvest by 25%."

LAND RESOURCES

The population explosion has nullified the benefits of the Green Revolution. Thus, the problems of food security, poverty and overpopulation may become even more acute and serious in the future. The problems are further compounded by the concern for the sustainability of high-production systems with regard to the adverse effects on soil and environment. Because the major share of human food comes from the land, sustainable management of soil, water and other natural resources has become a challenging task for scientists, administrators and planners for meeting the future food requirements of a growing population.

Out of 329 M ha of total geographical area in India, as much as 142.5 M ha is cropped. Such a high proportion of land area under agriculture prohibits any further expansion of agricultural area without even further jeopardizing the environment. Furthermore, the competition from industry, urbanization and civic use is growing and the availability of prime lands for agriculture is decreasing. In fact, agriculture is being extended into more marginal lands and fragile environments. Nearly 87% of the cropped area is under food grains and only 13% is available for other crops (commercial crops, horticulture and plantation crops). The problem is further aggravated by the rapid march of various soil-degradative processes that have already affected, to varying degrees, 221 M ha (Table 6.2) of the total geographical area (Singh, 1999).

Soil and water are the two most important natural resources constituting the backbone of almost all life-supporting systems. The exploitative and inappropriate

TABLE 6.2
Degraded Land Areas in India

Type of Degradation	M ha	Percent to Total
Total geographical area	329.0	
Water and wind erosion	162.4	49.4
Area degraded by special problems	58.2	17.7
a) Water-logged area	11.6	3.5
b) Alkali soils	4.5	1.4
c) Saline soils including coastal sandy area	5.5	1.7
d) Acid soils (pH < 5.5)	25.0	7.6
e) Ravine, gullies and torrents	6.7	1.2
f) Shifting cultivation	4.9	1.5
Total problem area	220.6	67.0
Annual loss of nutrients (MMt)	5.4–8.4	

Source: Extracted from Singh (1999)

land use is self defeating and results in soil degradation and decline in soil productivity. Buringh (1989) estimated that about 15–30% decline in world food production over a 25-year period may be caused by soil degradation. The impact of the Green Revolution on long-term sustained production is already waning. Intensive cultivation of land without enhancing soil fertility and improving soil structure may exacerbate desertification. Irrigation without drainage facilities can increase alkalinity or salinity. Indiscriminate use of insecticides, fungicides and herbicides may cause adverse changes in flora and fauna and reduction in soil biodiversity. Land application of industrial effluent may destroy good agricultural lands. Therefore, prevention of soil degradation through erosion management and restoration of productivity of degraded soils is absolutely essential to preserving the finite land resources. Only vertical expansion of food production can enhance agricultural production and achieve food security.

SOIL DEGRADATION

Soil degradation implies loss of natural fertility and regenerative capacity for the production of food crops and other raw materials. The most widespread phenomena are the loss of organic matter and the essential nutrients needed for biomass production caused by improper soil management practices coupled with accelerated soil erosion by wind and water. Also, concern is increasing about the decline of soil quality in large areas of India due to salinization, acidification and contamination with toxic heavy metals and organic pollutants. Soil degradation is defined as a process, which lowers the current or potential capability of soil to produce (quantitative and or qualitative) goods and services (FAO,1978).

The three principal types of soil degradation are physical, chemical and biological. Physical degradation includes compaction and hard setting, laterization and desertification, and erosion and depletion of nutrients by water and wind. Chemical degradation encompasses nutrient imbalances, acidification, sodification and toxic compounds. Decline in soil organic matter, reduction in macro- and microfauna and increase in soil-borne pathogens lead to biological degradation in soils (Lal and Stewart, 1992).

PHYSICAL DEGRADATION

Soil Erosion

Soil erosion, one of the most important soil degradative processes, has affected, in some cases irreversibly, about 430 M ha of land area covering 30% of the present cultivated area in different parts of the world (Oldeman, 1994). On a global scale, soil loss through erosion alone is assessed at 2500 MMg every year and 0.2 to 0.3 million ha of irrigated land are being lost by salinization and waterlogging every year (Oldeman, 1994). In general, soil erosion is more severe in hilly areas than in undulating areas. Erosion influences several soil properties that regulate soil quality and determine crop yield, e.g., topsoil thickness, soil organic carbon content, nutrient status, soil texture and structure and available-water-holding capacity.

The loss of topsoil resulting in reduced productivity is a serious degradation problem in the Indian subcontinent. The erosion due to water and wind occurs over large areas. Overpopulation, harsh climate, overexploitation and unwise use of soil resources, deforestation and nutrient imbalances have left the ecosystem extremely vulnerable to soil erosion and erosion-induced land degradation.

Soil Erosion by Water

Erosion by water is the most serious degradation problem in the Indian context. Analysis of the existing soil loss data indicate that soil erosion takes place at an average rate of 16.4 Mg ha^{-1} yr^{-1} totaling 5,334 MMg yr^{-1}. About 29% of the total eroded soil is lost permanently to the sea. In addition, nearly 10% is deposited in reservoirs, resulting in the reduction of their storage capacity by 1–2% annually. The remaining 61% of the eroded soil is redistributed over the landscape (Deb, 1995).

Singh et al. (1992) prepared an iso-erosion map of India based on 21 observed and 64 estimated data points spread over different land resource regions of the country. The annual water erosion rate ranged from <5 Mg ha^{-1}(for dense forests, snow-clad cold deserts and arid regions of western Rajasthan) to >80 Mg ha^{-1} in the Shivalic hills. Ravines along the banks of the rivers Yamuna, Chambal, Mahi, Tapti and Krishna and the shifting cultivation regions of Orissa and the northeastern states revealed soil losses exceeding 40 Mg ha^{-1} yr^{-1}. The annual erosion rate in the Western Ghats coastal regions ranged from 20 to 30 Mg ha^{-1} yr^{-1}.

The soils mainly supporting rain-fed agriculture are subjected to severe sheet and rill erosion with an annual soil loss of 20 to >100 Mg ha^{-1} yr^{-1} (Dhruvanarayana and Ram Babu, 1983). Even under normal cultivation (for sorghum or cotton) on land slopes of 1–3%, annual losses at the rate of 13.6 Mg ha^{-1} yr^{-1} have been observed (NBSS staff, 1987). The red soils, covering about 70 mha^{-1}, are another major soil group subject to severe water erosion problems. The red soils with low water intake capacity, where crusting is a serious problem, suffer from rapid surface runoff and erosion. The northeastern states of India have severe water erosion problems because of the prevalent practices of shifting cultivation (*jhuming*). In the past, the practice of jhuming had a long fallow cycle of 20–30 years. But due to population pressures, the cycles have been reduced to 3–6 years, thus aggravating erosion and degradation problems.

Wind Erosion

Wind erosion is a serious problem in the arid and semi-arid regions of Rajasthan, Haryana, Gujarat, Punjab and coastal areas and in the cold desert regions of Leh in extreme northwestern India. Removal of natural vegetation cover through excessive grazing and the extension of agriculture to marginal areas is the major human-induced factor leading to accelerated wind erosion. In India, wind erosion is moderate to severe in the arid and semi-arid regions of the northwest, covering an area of 28, 600 km^2, of which 68% is covered by sand dunes and sandy plains (Gupta, 1990). However, active wind erosion is observed in the extreme western sectors of the country.

CHEMICAL DEGRADATION

Chemical degradation of soils can occur through a number of processes, that is, the loss of nutrients or organic matter and the accumulation of salts or pollutants.

Nutrient Depletion and Fertilizer Use

Among the soil groups, alfisols, ultisols and oxisols are prone to chemical degradation due to nutrient depletion. Alfisols are relatively less prone, as their base saturation is generally more than 60%. Oxisols and ultisols, on the other hand, are more prone to chemical degradation, particularly because they are marginal soils with low reserves of nutrients and poor nutrient retention capacity.

Soils of India are more hungry than thirsty. About 70% of soils are low in organic carbon (<1%) with widespread micronutrient deficiencies throughout the country. Soil is an exhaustible storehouse of plant nutrients. Nutrients depleted through low-input agriculture are not being replenished with natural and cultural methods. In a densely populated country like India, one with a long history of civilization, nutrient reserves have been exploited for millennia. Prior to the Green Revolution (before 1965) the population pressure on soil resources to produce food was far less than in 2000. In 1951–52, food-grain production was merely 52 MMg and the fertilizer consumption only 0.07MMg, whereas the food-grain production during 1999–2000 increased to 206 MMg and fertilizer consumption to about 18 M Mg. In spite of a more than 250-fold increase in fertilizer consumption during the past 50 years, the gap between crop removal of nutrients and their restoration through fertilizers, manure and other sources has widened. This has resulted in undesirable mining of soils for plant nutrients, leading to imbalance of nutrient availability to crops. For the present level of production, the estimated NPK removal is about 28 MMg, whereas their return through fertilizer is only 18 MMg. Thus, a net negative balance of about 10 MMg is estimated (Figure 6.1). Organic manures and biofertilizers

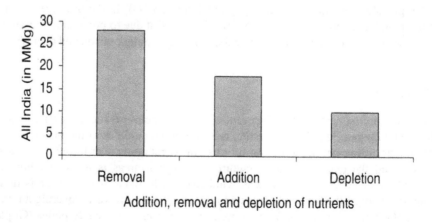

FIGURE 6.1 Nutrient mining in India during 2000–2001. Source: Data calculated from FAI (2000).

contribute to about 4 MMg, which means that about 6 MMg of these nutrients have to be replenished by the soil itself. Thus, we are presently mining the soils at the rate of 6 MMg annually. This is a serious soil-quality hazard and requires urgent attention from all concerned. If the production must be doubled in the next 25 years to feed the increased population, the nutrient removal would be more than double the present level (28 MMg) to about 56 MMg, because the nutrient requirement for incremental production would be higher. Thus, the gap between nutrient supply and removal would further escalate to more than 12 MMg from the present level (1999–2000) of about 6 M Mg, presuming that the contribution of organic and biofertilizer sources will also be doubled as compared with that of the present level (4 MMg). This implies that the soil-quality problem, which is already a serious threat to soil productivity, would be further aggravated.

When the high yielding varieties (HYV) of wheat were introduced in the mid-1960s, high yield could only be obtained by the application of nitrogenous fertilizers. Soon the soils were depleted of available phosphorus and phosphoric fertilizers had to be applied, along with nitrogenous fertilizers to sustain high yields. Overexploitation of soils with multiple cropping and use of high doses of fertilizers and other agrochemicals with high rates of chemical purity have resulted in deficiencies of macro, secondary and micronutrients. Analysis of the soil samples from 103 districts of irrigated ecosystems of the Indo-Gangetic plains and southern states (Andhra Pradesh and Tamil Nadu) indicated that most of the soils were low in N and P fertility (Table 6.3). The magnitude of P deficiency was apparently greater in soils of Uttar Pradesh and Andhra Pradesh than that of Punjab. Except for two districts of Tamil Nadu, none of the districts in other states could be categorized as being under high P fertility status. Potassium fertility status rated medium to high in most of the districts in all states. For micronutrients, 251,660 samples were collected from all states and analyzed for Zn, Cu, Fe and Mn. The results showed that, on an average for all of India, 49% of soils were deficient in Zn, followed by 12% in Fe, 5% in Mn and 3% in Cu (Singh, 2001). More than 50% of soils in Maharastra, Karnatka, Haryana, Tamil Nadu Andhra Pradesh, Orissa and Bihar were deficient in Zn. The magnitude of micronutrient deficiency varied widely among soil types and agro-ecological zones.

Antil et al. (2001) assessed organic carbon levels in soils of Haryana. The organic carbon status was low in 80% of soils, medium in 18% and high in 2%. In 1996, organic carbon status was low in 92%, medium in 8% and high in only a few soils (Table 6.4). There was a major shift in available P during these years, as most soils containing high and medium levels in 1980 changed to low and medium levels in 1996. For the state on the whole, 70% of soils were low, 25% medium and 5% high in P during 1996. So far, K is not a limiting nutrient for crop production in the state. But there was reduction in K content of soils from 1980 to 1996 (Table 6.4). The K content in 1996 was low in 5%, medium in 33% and high in 62% of soils. In 1967, a long-term field experiment was started on the use of farmyard manure and N fertilizer on a coarse, loamy, typic ustocrept soil at CCS Haryana Agricultural University, Hisar, India, using a pearl millet–wheat cropping sequence. Neither farmyard manure nor N was added to the control treatment. Consequently, there was a severe depletion of organic C and available N, P and K contents in soil (Table 6.5).

TABLE 6.3

Statewide Distribution of N, P and K Fertility Classes in Irrigated Ecosystems of India

State	Total Districts Sampled	N-fertility			P-fertility			K-fertility		
		Low	Medium	High	Low	Medium	High	Low	Medium	High
Andhra Pradesh	22	18	4	Nil	18	4	Nil	1	4	17
Tamil Nadu	14	11	3	Nil	6	6	2	2	12 (86)	Nil
Uttar Pradesh	55	—	—	—	41	14	Nil	16	34	5
Punjab	12	—	—	—	4	8	Nil	Nil	7	5

— indicates that N was not estimated.

Source: Extracted from FAI (1992), Ghosh and Hasan (1979), Tandon (1987)

TABLE 6.4
Soil Fertility Status of Soils of Haryana

Nutrients	Percentage of Soil Samples		
	Low	Medium	High
A. 1980			
Organic Carbon	80	18	2
Available P	25	40	35
Available K	—	18	82
B. 1996			
Organic carbon	91.5	7.9	0.6
Available P	70.3	25.4	4.3
Available K	5.3	33.2	61.5

Source: Extracted and modified from Antil et al. (2001)

TABLE 6.5
Nutrient Depletion Due to Continuous Cropping in Soil Without Addition of Manure or Fertilizer under Pearl Millet–Wheat Cropping System

Properties	1967	1979	1992
Organic C (%)	0.47	0.36	0.21
Available N (kg ha^{-1})	200	150	104
Available P (kg ha^{-1})	26	15	7
Available K (kg ha^{-1})	498	354	196

Source: Extracted and modified from Ruhal and Shukla (1979) and Gupta et al. (1992)

Although fertilizers have played a major role in India's Green Revolution, their use on a per hectare basis in India is still much lower than in its neighboring countries in Asia (Figure 6.2). Lower consumption of fertilizer in India than in China has resulted in lower yield levels of most crops (Figure 6.3). Besides lower consumption of NPK, a most disturbing feature in fertilizer consumption is an apparent imbalance in the use of N, P_2O_5 and K_2O (nutrient consumption ratio). The nutrient consumption ratio for N, P_2O_5 and K_2O in 1998–99 was 9:3:1 for India, 5:2:1 for eastern regions, 37:9:1 for northern, 4:2:1 for southern and 10:5:1 for western regions. These comparisons show that the northern zone had a much wider ratio than other zones. These observations suggest that it will be difficult to sustain higher yield levels and avoid detrimental effects on soil quality in India without increasing fertilizer input at ideal nutrient consumption ratio. Simply balancing the nutrient consumption ratio on an all-India basis is not enough. The strategy is to improve the consumption ratio on

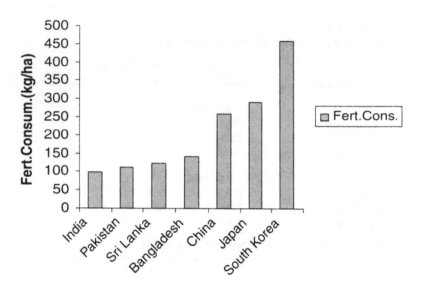

FIGURE 6.2 Fertilizer consumption in India as compared to other countries in Asia. Data from FAI (2000).

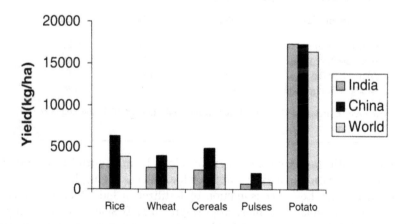

FIGURE 6.3 Yield levels of major crops in India as compared to China and the world average. Data drawn from FAI (2000).

crop and cropping system basis for different ecoregions. Kumar (1998) made projections on area requirements, total production and yield levels of major crops to meet the demands of increased population in 2020 (Table 6.6). He estimated that 294 MMg of food grains would be required to meet the needs of the growing population. Sekhon (1997) estimated that the fertilizer nutrient needs for irrigated

TABLE 6.6

Projections on Area, Production and Yield of Crops to Meet Demands of Food Grains and other Crops in 2020

	1999–2000			2020		
Crop	Area (M ha)	Production (M Mg)	Yield (kg/ha)	Area (M ha)	Production (M Mg)	Yield (kg/ha)
Rice	44.6	85.9	1928	42.2	221	2895
Wheat	27.4	70.8	2583	26.2	103	2918
Cereals	101.5	188.2	1854	99.1	266	2682
Pulses	23.8	14.8	822	21.7	28	1282
Foodgrains	125.4	203.0	1620	121.	294	2447

Source: Extracted from Kumar (1998) and FAI (2000)

crops in India, based on a response of 10 kg food grain per kg of nutrient fertilizer will range from a minimum of 17.5 MMg to a maximum of 30 MMg of N, P_2O_5 and K_2O to achieve the food-grain production of 237 or 359 M Mg, respectively.

Salinization and Alkalinization

Excessive salt concentration in soil is a major threat to large areas of irrigated agricultural land worldwide. Substantial areas of productive lands are affected by salinity and alkalinity. In India, 10 Mha of land are affected by salinity and alkalinity (Table 6.2). These soils predominately occur in the irrigated areas in arid and semi-arid regions and also along the coastal belt. In an attempt to feed the increasing population of the country, large areas have been brought under irrigation. Increase in irrigated land area from about 20 M ha in 1950 to 53 M ha in 2000 has played a pivotal role in boosting agricultural production. A large fraction of the irrigation is by surface water through canals, which, in the absence of adequate drainage, has resulted in a rise in the groundwater table. In four northern and western states (e.g., Punjab, Haryana, Rajasthan and Gujrat) saline, sodic and saline-sodic waters compose 20, 37 and 43% of irrigation waters. Inadequate drainage, use of poor-quality water for irrigation and inappropriate management of soil and irrigation have contributed to high levels of soil salinity and alkalinity and adversely affected agricultural production. In almost all cases, the groundwater table, which was 20 to 30 meters deep prior to the introduction of irrigation, now reaches only a few meters below the surface. When the groundwater table reaches within 2 meters of the surface, it contributes significantly to evaporation from the soil surface, which leads to soil salinization. Thus, efforts to bring more lands under irrigation have contributed to large areas of land affected by salinization.

Soil salinization is a major constraint to agricultural production on irrigated land in India and elsewhere in arid and semi-arid regions. The area of salt-affected land in some of the worst-affected states of India is about 4.3 M ha or 6.3% of the total arable land. Abrol and Bhumbla (1971) estimated about 7 M ha of salt-affected soils, of which 2.5 M ha was under alkali soils in the Indo-Gangetic plain. It was also stated that nearly 50% of the canal-irrigated areas are suffering from salinization or alkalinization due to inadequate drainage, inefficient use of available water resources and other sociopolitical reasons. Salinity caused by the rise in the groundwater table is widely observed in Uttar Pradesh, Haryana, Rajasthan, Maharastra and Karnatka.

Pollution by Toxic Substances

Soil contamination by land application of industrial effluents and indiscriminate use of chemicals and pesticides has adversely affected soil quality and crop productivity. The accumulation of toxic substances of industrial and urban origin is increasingly contributing to soil degradation.

In some intensively cultivated areas of the Indo-Gangetic plains (especially Punjab, Haryana and western parts of Uttar Pradesh), farmers use large quantities of fertilizers, pesticides and herbicides. Contaminants in these fertilizers (e.g., Cd in P fertilizers) and the residues of agrochemicals may accumulate in soils. There are also problems of leaching of nutrients (e.g., nitrate and other chemicals) to groundwater. In Punjab, where heavy doses of N-fertilizers are used, the nitrate content of groundwater ranges from 12.1 to 17.8 mg NO_3 L^{-1}, which is above the safe limit of 10 mg NO_3 L^{-1} set by the World Health Organization for drinking water. Very high levels of nitrate content in groundwater (from 156 to 530 mg NO_3 L^{-1}) have also been measured in Tamil Nadu, Uttar Pradesh and Rajisthan. These high concentrations have been caused by either natural geological deposits of high nitrate content or leaching from septic tanks.

Industrial and domestic effluents are indiscriminately discharged on prime agricultural lands in the vicinity of cities. Such effluents are rich in heavy metals (i.e., lead, nickel, cadmium, zinc, chromium, arsenic) and harmful pathogens. Land applications of industrial effluents contaminate soil with heavy metals and organic pollutants (Narwal et al., 1992). Continuous use of such effluents for long periods has contaminated soils with heavy metals (Table 6.7).

TABLE 6.7
Heavy Metal Concentration in Soils (mg kg⁻¹) Irrigated with Industrial Effluent in Haryana

Metals	Source of irrigation	
	Tube-well irrigation	Effluent irrigation
Zn	91	12188
Cu	14	1199
Pb	20	280
Cd	0.8	5.4
Ni	26	6000

Source: Extracted and modified from Narwal et al. (1992)

SOIL PHYSICAL DEGRADATION

Problems of physical degradation of soils are generally related to a decline in the organic matter content. Soils with low organic matter content are prone to crusting and high water runoff. Agricultural intensification based on use of heavy farm machinery exacerbates soil physical and chemical constraints such as formation of traffic pan, soil compaction and crusting. Puddling for rice cultivation in rice–wheat rotation is necessary to create ponds but it also adversely affects soil structure. The use of heavy machinery results in the development of a hard compacted layer of subsoil that hinders root proliferation, water penetration and free exchange of gases, resulting in lower yields of wheat that follows rice. Subsoil compaction is serious even in light-textured soils cultivated for other crops (e.g., cotton). This dense layer hinders root development and accentuates waterlogging and anaerobiosis.

Waterlogging and Anaerobiosis

The term waterlogging refers to a condition of short- or long-term inundation caused by changes in hydrological regime, landscape, silting up of river beds, increased sedimentation and reduced capacity of the drainage systems. Repeated flooding is yet another cause of waterlogging in coastal and flood-plain areas of major rivers. Problems of short- or long-duration flooding in India have been increasing rapidly over the years. This is largely attributed to deforestation in catchment areas, destruction of surface vegetation, changes in land use, urbanization and other development activities. The irrigated areas, which have contributed significantly in increasing foodgrain production (e.g. in Haryana, Rajasthan, Uttar Pradesh, Madhya Pradesh, Karnatka and Gujarat), are now facing a serious problem of rise in the groundwater table and soil salinization. Such problems are most severe in areas with canal irrigation. Some typical examples of waterlogging in canal-irrigated areas are given in Table 6.8. The National Bureau of Soil Survey and Land Use Planning, Nagpur,

TABLE 6.8
Area and Annual Increase in Waterlogging and Soil Salinity under Some Irrigation Projects

Irrigation Project	States	Waterlogging		Soil salinity	
		Area 1000 ha	Annual increase 1000 ha	Area 1000 ha	Annual increase 1000 ha
Sriramsagar	Andhra Pradesh	60	10.0	1	0.2
Gandak	Bihar, Uttar Pradesh	211	3.5	400	36.4
Mahi Kadana	Gujarat, Rajasthan	82	3.9	36	1.7
Chambal	Madhya Pradesh	99	7.6	40	3.1
Sharda Sahayak	Uttar Pradesh	303	5.7	50	0.9
Ramaganga	Uttar Pradesh	195	27.9	352	50.3

Source: Adopted and modified from Yadav (1996)

India has estimated that about 5.7 M ha or 1.74% of the geographical area is affected by waterlogging (NBSS Staff,1987).

SUMMARY AND CONCLUSIONS

Soil degradation is creating a scenario in which it is increasingly difficult to manage soil and water resources efficiently and economically. The important issues of national food security, nutritional quality, environmental safety, enhancing soil productivity and leaving a good heritage for future generations depend on our ability to curtail soil degradation. Because 98% of human food comes from land, management of soil and water resources and biodiversity are priorities for scientists, administrators, planners and land managers. The main problems of soil degradation relating to food security in India are summarized below:

- Inadequate and imbalanced use of fertilizers, as evident from wide N: P: K ratios, has resulted in increasing deficiencies of P and K and in declining yields.
- The gap between the input (fertilizer nutrient) and output (nutrient removal) is increasing, leading to nutrient mining of soils. Unless steps are taken to bridge this gap, the overmining of the nutrient reserves of the soil will further lead to decline in yields and increase deficiencies of macro- and micronutrients. Therefore, to sustain crop production, we must stop extra mining of soil nutrients.
- The need for agricultural inputs has increased considerably, which has accentuated the cost of production. Crop yields are low, especially when inputs are below the recommended rates. Whether production increases obtained through the chemical inputs can be sustained is debatable. Groundwater resources are also being depleted.
- About 30 MMg nutrients, through mineral fertilizers and organic sources, will be required to meet the food demands of India by 2025. In addition, many thousands of tonnes of micronutrients (Zn, Cu, S) will also be needed.
- The average yields of main crops and fertilizer application rates in India are much lower than those in China and the developed world.
- Increasing crop yields from existing lands will need balanced and integrated nutrient input, efficient management of soil and water resources and use of good crop management practices to produce synergistic effects. Integrated nutrient management is undoubtedly the key to achieving food security.
- Severe water erosion removes considerable quantities of plant nutrients and the topsoil. Effective erosion management needs a strategy for soil conservation and for arresting further depletion of productive topsoil.
- Intensive agriculture coupled with surface irrigation has caused serious waterlogging and the rapid march of salinization and alkalinization.
- Land application of industrial and domestic effluents in the vicinity of cities and towns results in the accumulation of heavy metals and toxic

substances. Because most of such contaminated soils are used for growing vegetables for human consumption, serious human health hazards can arise if proper measures are not taken to avoid soil contamination.

- Accumulation of nitrate and other chemicals in groundwater is a severe health hazard.
- Enhancing soil quality and crop productivity without jeopardizing environmental quality is a major challenge.
- Soils of India are low in nutrient reserves and require supplemental doses of inputs to enhance soil fertility. Soil is a living entity that nourishes crops but also needs to be nourished. "Soil is like a bank. You cannot withdraw from it more than what you have deposited in it. Nature permits no overdrafts. Hence not only the fertility of the soil is to be conserved but it also needs to be enriched," said Charan Singh, the late prime minister of India.

RESEARCH NEEDS AND PERSPECTIVES

Some important missing links and important areas of research to counter impediments to achieving food security in India are as follows:

- Preventing, combating and reversing soil degradation and improving the quality of soil and water through efficient and scientific soil and water management.
- Creating mass awareness about soil and water resources, their potential, problems, constraints and management options.
- Monitoring soil quality and changes in productivity and developing a warning system to indicate the dangers of decline in soil quality.
- Developing interdisciplinary teams for interaction among agroscientists, engineers, farmers and various central and state agencies to solve the problem of soil degradation.
- Identifying balanced plant nutrient (NPKS) ratio for different crops and cropping systems in principal agroclimatic regions of the country based on nutrient uptake, nutrient-use efficiency, indigenous soil nutrient-supplying capacity and availability of other inputs of agriculture.
- Developing precision farming approaches in fertilizer use.
- Creating integrated plant nutrient systems (IPNS) to enhance efficiency of fertilizer use.
- Using biofertilizers organic manures, crop residues, legumes in rotation and green manuring to improve soil productivity, efficiency of microbial activity and fertilizer use.
- Diversifying the rice–wheat system, which is prevalent in the Indo-Gangetic plains. the system is causing excessive mining of nutrients and groundwater, decline of soil quality and increase in incidence of pests, diseases and weeds. Identifying and evaluating efficient alternate cropping

systems for different situations, especially for higher productivity zones and problem is a priority issue.
* Linking agricultural sustainability with resource degradation.
* Developing appropriate systems of disposal of solid and liquid effluents from industries and enforcing regulatory measures for environmental protection.

REFERENCES

Abrol, I.P. and D.R. Bhumbla. 1971. Saline and alkali soils in India — their occurrence and management. Paper presented at FAO/UNDP seminar, Soil Fertility Research, FAO World Soil Resources Report No. 41:42-51.

Antil, R.S., V. Kumar, R.P. Narwal and M.S. Kuhad. 2001. Nutrient Removal and Balance in Soils of Haryana. Bulletin, Department of Soil Science, CCS Haryana Agricultural University, Hisar, India. pp. 7-12.

Brown, L. and H. Kane. 1994. Full House Reassessing the Earth's Population Carrying Capacity. The World Watch Environment Alert Series, WW Norton and Co. New York, p 261.

Brown, L. 1998. Full House Reassessing the Earth's Population Carrying Capacity. The World Watch Environment Alert Series, WW Norton and Co. New York.

Buringh, P. 1989. Availability of agricultural land for crop and livestock population. In: D. Pimental and C.W. Hall (Eds.) *Food and Natural Resources*, Academic Press, San Diego, pp. 69-83.

Deb, D.L. 1995. *Natural Resource Management for Sustainable Agriculture and Environment.* Angkor Publishers (P) Ltd. New Delhi, India.

Dhruvanarayana, V.V. and R. Babu. 1983. Estimation of soil erosion in India. *J. Irrigation and Drainage Engg.* 109:419-434.

FAI. 1992. Fertilizer and Agriculture Statistics (southern region), 1991-92. The Fertilizer Association of India, New Delhi. pp. 105-106.

FAI. 2000. Fertilizer Statistics, 1999-2000. The Fertilizer Association of India, New Delhi. Part III, pp. 9 and 126; Part IV, pp. 20.

FAI. 2001. Nutrient Mining — Emerging Scenario. *Fertilizer News* 46: 11-12.

FAO. 1978. Methodology for assessing soil degradation. FAO/UNEP, Expert Consultation, FAO, Rome, Italy. pp. 84.

Ghosh, A.B. and R. Hasan. 1979. Phosphorus fertility status of soils of India. In: Phosphorus in Soils, Crops and Fertilizers, Bulletin 12, New Delhi, India. *J. Ind. Soc. Soil Sci.*, pp. 1-8.

Gupta, J.P. 1990. Sand dunes and their stabilization. In: Abrol, I.P. and Dhruvanarayana, V.V. (Eds.)*Technologies for Wasteland Development*, ICAR Publication, New Delhi, Pp. 59-68.

Gupta, A.P., R.P. Narwal, R.S. Antil and S. Dev. 1992. Sustaining soil fertility with organic C, N, P and K by using farmyard manure and fertilizer-N in a semiarid zone: A long-term study. *Arid Soil Research and Rehabilitation.* 6:243-251.

Kanwar, J.S. and J.C. Katyal. 1997. *Proc. Symp. Plant Nutrient Needs, Supply, Efficiency and Policy Issues*, 2000-2025, National Academy Agricultural Sciences, New Delhi, India. pp.326.

Kumar, D. 1998. Food Demand and Supply Projection for India. Agricultural Economics Policy Paper 98-01, IARI, New Delhi. Pp1-141.

Lal, R. and B.A. Stewart. 1992. Need for land restoration. *Advances in Soil Sci.* 17:1-9. Springer Verlag, New York.

Narwal, R.P., R.S. Antil and A.P. Gupta. 1992. Soil pollution through Industrial effluent and its management. *J. Environ. Contam.*, 1: 265-272.

NBSS Staff. 1987. Benchmark soils of India. Mondha Series, Characteristics, Classification and Interpretation for Land Use Planning, Bulletin No. 15, NBSS&LUP, Nagpur, India, pp. 24.

Oldeman, L.R. 1994. The global extent of soil degradation In: D.J. Greenland and I. Szabolcs (Eds.) *Soil Resilience and Sustainable Land Use.* CAB International. Wallingford, UK pp. 99-118.

Ruhal, D.S. and U.C. Shukla. 1979. Effect of continuous application of farmyard manure and N on organic carbon and available N, P and K content in soil. *Indian J. Agri. Chem.* 12:11-18.

Sekhon, G.S. 1997. Plant Nutrients. In : Kanwar, J.S. and Katyal, J.C. (Eds.), *Proc. Symp. Plant Nutrient Needs, Supply, Efficiency and Policy Issues: 2000-2025,* National Academy Agricultural Sciences, 1997, pp. 78-90.

Singh, G., R. Babu, L.S. Bhusan and I.P. Abrol. 1992. Soil erosion rates in India. *J. Soil Water Conserv.* 47: 97-99.

Singh, G.B. 1999. Natural resource management for sustainable agriculture. In: Singh, G.B. and Sharma, B.R (Eds.) *50 Years of Natural Resource Management Research,* ICAR, New Delhi, pp. 4-11.

Singh, M.V. 2001. Evaluation of current micronutrient stocks in different agro-ecological zones of India for sustainable crop production. *Fertilizer News* 46(2):25-42.

Tandon, H.L.S. 1987. Phosphorus research and agriculture production in India, New Delhi: Fertilizer Development and Consultation Organization, pp 160.

Yadav, J.S.P. 1996. Extent, nature, intensity and causes of land degradation in India. In: Soil Management in Relation to Land Degradation and Environment, Bulletin, New Delhi, India. *Indian Soc. Soil Sci.*, pp. 1-26.

7 Importance of Biotechnology in Global Food Security*

*Prem P. Jauhar** and Gurdev S. Khush***

CONTENTS

* Note: Mention of a trademark or proprietary product does not constitute a guarantee or warranty of the product by the USDA or imply approval to the exclusion of other products that also may be suitable.
** United States Department of Agriculture — Agricultural Research Service, Northern Crop Science Laboratory, Fargo, North Dakota
*** International Rice Research Institute, Manila, The Philippines

INTRODUCTION

Conventional plant breeding practiced over the centuries has produced crop cultivars that sustain humankind today. Genetic improvement of crops has mainly been achieved through sexual hybridization of crop species with land races and related species, resulting in cultivars of food, fiber, oilseed and other crops with high yields and other superior agronomic traits. Thus, largely through exploitation of hybrid vigor, grain yields of maize (*Zea mays* L.), pearl millet (*Pennisetum glaucum* [L.] R. Brown), and sorghum (*Sorghum bicolor* [L.] Moench) registered a phenomenal increase in the period from around 1965 to 1990 (Khush and Baenziger, 1996; Jauhar and Hanna, 1998). More importantly, improved high-yielding varieties of wheat (*Triticum aestivum* L.) and rice (*Oryza sativa* L.) developed in the 1960s and 1970s launched the Green Revolution in Asia.

Global food security is threatened by several factors including diseases — fungal, bacterial, and viral. Severe malnutrition among the masses poses another serious problem, which must be addressed using all scientific tools at our disposal. Conventional breeding, although slow, sometimes combined with classical cytogenetic techniques, has been the main method of crop improvement. Since the late 1980s, the advent of the novel tools of biotechnology* has facilitated direct gene transfer into crop plants. These tools, collectively termed "genetic engineering," help mobilize specific genes for value-added traits into otherwise superior crop cultivars. The process involves the insertion of a well-characterized gene(s) into regenerable embryogenic cells followed by recovery of fully fertile plants with the inserted gene(s) integrated into their genome. The tools of modern biotechnology help asexually engineer new traits that are otherwise very difficult or impossible to introduce by traditional breeding. Thus, the new technology allows access to an unlimited gene pool for genetic enrichment of crop plants. Combination of modern biotechnology and conventional breeding will help sustain global food supply. The importance of biotechnology in global food security is discussed in this chapter.

MODERN BIOTECHNOLOGY: INTERFACE WITH CONVENTIONAL PLANT BREEDING

Plant breeding deals with the generation, manipulation and combination of genetic variability into plant forms most useful to humankind. The art of plant breeding was developed several thousand years ago, long before the principles of genetics became known. Working under a myriad of cultural contexts, the early plant breeders, or perhaps selectionists, turned the relatively useless weedy species into useful crop cultivars that sustain us today. The advent of the principles of genetics and cytogenetics at the turn of the 20th century catalyzed the growth of plant breeding, making it a science-based technology that helped to raise the yields of major crops considerably. The period from 1960 to 1980 witnessed a dramatic increase in crop yields, particularly in cereal grains, leading to the Green Revolution (Khush, 1999; Figure 7.1). The availability of cytoplasmic male sterility (CMS) in maize, pearl millet, and other crops

---

* The terms "biotechnology" and "genetic engineering" are used interchangeably in this chapter.

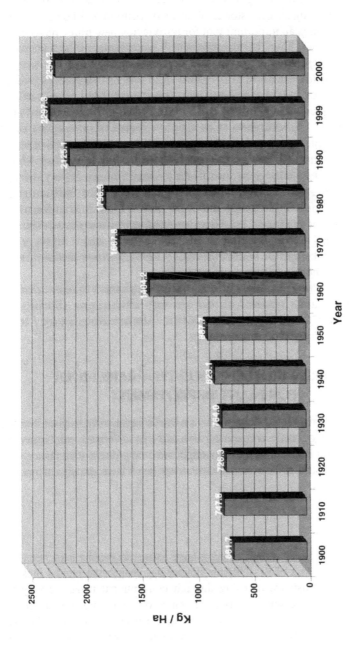

FIGURE 7.1 Yield (Kg/ha) of wheat grain in the United States between 1900 and 2000.

proved very useful in exploiting hybrid vigor (Jauhar, 1981; Jauhar and Hanna, 1998). The speed with which Indian breeders accomplished the development of high-yielding grain hybrids of pearl millet using CMS lines was described as "one of the most outstanding plant breeding success stories of all time" (Burton and Powell, 1968).

Wide hybridization is an important tool for introducing new traits into otherwise desirable crop cultivars. Thus, hybridization with wheatgrasses of the genera *Agropyron* Gaertner, *Thinopyrum* A. Löve, and *Lophopyrum* A. Löve has contributed substantially to genetic enrichment of wheat. The tools of classical cytogenetics greatly facilitated wide hybridization and chromosome-mediated gene transfer from wild species into crop cultivars (Friebe et al., 1996; Fedak, 1999; Jauhar and Chibbar, 1999). Although wide hybridization is an effective means of introducing desirable alien genes into crop plants, it has several limitations, e.g., transmission of unwanted alien chromosomes, and adverse genetic interactions leading to sterility. Moreover, to introduce a single desirable gene into wheat by sexual means is very tedious and time consuming. However, modern biotechnological approaches facilitate the asexual incorporation of desirable genes into crop plants. Genetic transformation by microprojection has, for example, been demonstrated in wheat and most other cereals (Jauhar and Chibbar, 1999; Repellin et al., 2001). Most major crops are being transformed by microprojectile bombardment, by *Agrobacterium*-mediated transformation and other methods of direct gene transfer.

Modern biotechnology certainly offers a useful supplement to conventional plant breeding and the results obtained so far are very encouraging (Borlaug, 1997; Swaminathan, 1999; Cook, 2000; Jauhar, 2001a). Modern biotechnology has great potential for accelerating crop improvement.

MODERN BIOTECHNOLOGY: A RAPID TOOL FOR PLANT IMPROVEMENT

As stated above, modern biotechnology offers a useful supplement to plant breeding. A few examples of how genetic engineering has and can accelerate crop improvement programs are described below. Genetically engineered crop plants with enhanced resistance to pests and diseases and elevated micronutrient levels have already been produced.

RESISTANCE TO PESTS

Bt Corn

Numerous pests attack crop plants, causing huge losses. European corn borer (ECB), for example, causes a loss of $1–2 billion dollars per year in the United States alone (Hyde et al., 1999). Resistance breeding through conventional means is very slow. Moreover, finding a suitable source of pest resistance poses a major limitation. Thus, to breed a corn variety with resistance or even partial resistance to ECB could take 10 to 15 years by conventional breeding — provided a suitable resistance donor is available. Through 12 years of conventional breeding, Syngenta, a Swiss seed and agrochemical company, was able to produce a corn variety with only 10% resistance to ECB (personal comunication). Modern tools of genetic engineering may considerably accelerate this process. *Bacillus thuringiensis* (Bt), a soil-based bacterium isolated by a German scientist from a dead moth in the Thuringia region of Germany, has a gene that confers resistance to ECB. The Bt gene codes for a crystalline protein that, when ingested, kills the larvae of ECB. This gene, when engineered into the corn genome, confers resistance to ECB, thereby largely eliminating damage from this serious pest. Corn engineered with the Bt gene acquired the capacity to produce its own protein that is toxic to ECB. Thus, this gene acts as an efficient biopesticide. It took Syngenta only 5 years to engineer the Bt gene into corn.

Farming benefits of Bt corn. Recently, several seed companies, such as Novartis and Mycogen, have produced ECB-resistant corn inbreds that are being used to produce superior hybrids. Genetic engineering provides an efficient method of eliminating pest damage without adversely affecting grain yields. According to estimates by scientists at the University of Minnesota, farmers average several times higher return on investment by using Bt corn for insect control, compared with the use of a chemical insecticide (Ostlie et al., 1997). Bt corn hybrids had 4–8% greater grain yields than standard hybrids when infested with ECB (Lauer and Wedberg, 1999).

Bt corn and the environment. Even more importantly, the Bt-induced insect resistance in corn is much safer to farmers and other field workers than insecticides. Chemical insecticides not only kill crop pests but beneficial insects as well. However, several studies have shown that the Bt protein is not toxic to humans, domestic animals, fish or other wildlife. Bt-corn is therefore beneficial to the environment. The Environmental Protection Agency (EPA) of the United States registered Bt corn after studying several years of human and animal safety data (Ostlie et al., 1997). After another year-long review of transgenic crops, the EPA announced on 16 October 2001 that it had reauthorized commercial planting of Bt corn varieties (Palevitz, 2001).

Bt Cotton: Its Benefits and Environmental Impact

Genetically engineered Bt cotton has resistance to serious pests like bollworms. Bt cotton could prove beneficial to cotton-growing countries. India has the highest acreage of cotton, followed by the U.S. and China. Almost 70% of all chemical pesticides used in India are used on cotton alone (*India Today*, 23 July 2001). The use of Bt cotton would help reduce the use of chemical pesticides and hence benefit Indian farmers and the environment. Falck-Zepeda et al. (1999) estimated that, with the production of Bt cotton in the United States, the spraying frequency needed for insect control would be reduced from 10 to 12 applications per year to 2 to 3 per year. This would result in economic gains to farmers and would also be beneficial to the environment. Transgenic cotton varieties that were genetically engineered for resistance to insect larvae, herbicides or both accounted for about 78% of the upland cotton planted in the United States in 2001. This is up from 72% of the U.S. cotton acreage in 2000 (USDA, Agricultural Marketing Service Cotton Program, 2001).

Bt Rice

Yellow stem borer (YSB) is a serious pest of rice. Sources of resistance to this insect have not been found in the rice germplasm. Several varieties of rice have been successfully transformed with Bt genes such as *cry 1 A(b)* or *cry 1 A(c)* (Fujimoto et al., 1993; Wun et al., 1996; Datta et al., 1997; Alam et al., 1998). Transgenic rices, highly resistant to the YSB in the laboratory, were evaluated under field conditions in China and showed a high level of resistance to YSB and to natural outbreaks of leaf folders — another lepidopteran insect of rice (Tu et al., 2000). Hybrid rices are widely grown in China, with Shan Yu 63 one of the most popular hybrids. Its restorer parent, Minghui 63, was transformed with Bt gene. Shan Yu 63 with Bt gene showed a high level of resistance in field tests.

RESISTANCE TO DISEASES

Plants are constantly exposed to environmental challenges resulting in biotic and abiotic stresses. Biotic stresses are caused by pathogens including fungi, bacteria, and viruses. These pathogenic organisms have co-evolved with their host plants, resulting in a constant race between the pathogen and its host. This co-evolution has produced much of the earth's biological diversity (see Rausher, 2001). Genetic variability in plants helps them ward off infection by a myriad of pathogens that attack them.

In the last few years, the identification of key regulatory genes involved in plant defense has provided evidence that plants use several different defense pathways against different pathogens (Thomma et al., 2001). Much of the global food security we currently enjoy is constantly at risk because of huge annual losses of food crops (valued at billions of U.S. dollars) due to diseases and insects worldwide (Anderson and Pandya-Lorch, 1999).

As stated earlier, using conventional plant breeding to incorporate disease resistance in crop plants takes a long time. Attempts have been made to engineer durable disease resistance in economically important crop plants (Stuiver and Custers, 2001). Genetic engineering techniques appear promising.

Fungal Diseases

Engineering with antifungal genes could prove to be an important avenue for producing crop plants resistant to fungal pathogens. Among others, pathogenesis-related proteins (PR proteins) have attracted the attention of cereal workers. Genes and cDNA clones for various classes of PR proteins have been isolated from a variety of cereals. Some of these genes and cDNA clones have been used to transform rice (Datta et al., 1997; Datta and Muthukrishnan, 1999), wheat (Bliffeld et al., 1999; Chen et al., 1999; Altpeter et al., 1999; Muthukrishnan et al., 2001) and barley, *Hordeum vulgare* L. (Roulin et al., 1997), and similar work is in progress in sorghum and maize (see Muthukrishnan et al., 2001).

Fusarium head blight (FHB), or scab, caused by *Fusarium graminearum* Schwabe, is a devastating disease of bread wheat, durum wheat (*Triticum turgidum* L.), and barley, causing enormous economic losses for growers of the Northern Plains area of the U.S. since the 1990s (McMullen et al., 1997). The combined direct and secondary economic losses suffered by wheat and barley producers in the scab-affected regions of the U.S. during the 1998–2000 period were estimated at $2.7

billion (Nganje et al., 2001). Current wheat cultivars lack resistance to FHB. Some of the wild relatives of wheat are rich sources of genes for resistance to diseases including FHB. Therefore, hybridization with these wild species offers an important option for transferring FHB resistance into wheat. Most of these wild grasses can be hybridized with wheat, making it possible to introduce genes for resistance to scab and other diseases. Chromosome engineering methodologies, based on manipulation of the pairing control mechanism and induced translocations, have been successfully employed to transfer specific disease and pest resistance genes from alien sources into wheat cultivars (Friebe et al., 1996; Jauhar and Joppa, 1996; Jauhar and Chibbar, 1999).

By hybridizing durum wheat cultivars with wheatgrasses, fertile hybrid derivatives with FHB resistance were produced (Jauhar and Peterson, 2001). However, this process is slow and cumbersome (Jauhar, 2001b). Modern biotechnological tools offer great promise for accelerating this process. The technique of genetic transformation of durum wheat standardized earlier in our laboratory (Bommineni et al., 1997) has paved the way for direct incorporation of antifungal genes into durum wheat. Several antifungal genes have been isolated and some of their products are reported to inhibit Fusarium growth *in vitro* and *in planta*. Attempts to produce transgenic wheat and barley expressing these genes to combat FHB are being made in several laboratories (see Dahleen et al., 2001).

Sheath blight of rice causes serious yield losses. Natural resistance to this disease in rice germplasm is not known. The disease is caused by *Rhizoctonia solani*, which has wide host range. Transgenic tobacco and canola plants with enhanced resistance to *R. solani* have been produced by introducing the bean chitinase gene under the control of the cauliflower mosaic virus (CaMV) 35S promoter (Broglie et al., 1991). Similarly, Logemann et al. (1992) transformed tobacco plants with a barley gene encoding a ribosome inactivating protein under the control of the wound inducible *win-2* promoter from potato.

Chitinases and glucanases degrade the major structural polysaccharides of the fungal cell wall. They attack the growing hyphal tip and are potent inhibitors of fungal growth. About six chitinase genes have been identified in rice (Zhu and Lamb, 1991). Lin et al. (1995) introduced a 1.1-kb rice genomic DNA fragment containing a chimeric chitinase gene through PEG-mediated transformation. Transgenic rice plants showed moderate level of resistance to sheath blight.

Dutch elm disease is reported to have destroyed 20 million trees across Britain in the last 30 years. Scientists in Scotland have produced genetically modified elms that are resistant to the fungus. Professor K. Gartland, head of molecular and life sciences at the University of Abertay in Dundee, states that "this work could help tackle damaged landscapes and ecosystems blighted by tree fungal diseases, such as Dutch elm disease and chestnut blight, throughout the world" (*The Independent*, Scotland, 28 August 2001).

Bacterial Diseases

Bacterial blight of rice occurs throughout the rice-growing areas and causes serious yield losses. At least 23 genes are known for resistance to this disease. Several of them have been incorporated into improved varieties. A gene, *Xa21*, with broad spectrum resistance to bacterial blight, was found in a wild species of rice, *Oryza longistaminata* A. Chev. & Roehr. It was transferred to cultivated rice through backcrossing (Khush et al., 1990). Song et al. (1995) cloned this gene through map-based cloning and Wang et al. (1996) introduced this gene in a rice variety, Taipei 309. Transformed plants also showed a wide spectrum of resistance. Recently, Tu et al. (2000b) introduced this cloned *Xa21* gene into the widely grown rice variety IR72, and transgenic IR72 is now being evaluated for resistance under field conditions.

Viral Diseases

Virus diseases cause enormous losses in crops worldwide. For example, virus diseases of rice in Southeast Asia cause losses of more than $1 billion per year (Herdt, 1991). Conventional breeding techniques to develop resistance are expensive and painfully slow. Transgenic technology offers an excellent option to protect crop plants against viral pathogens. Transformation of plants with portions of viral genomes frequently confers on the plants resistance to the virus from which the viral sequence was derived. The first illustration that a pathogen-derived resistance (PDR) offers a viable means of producing virus-resistant plants was provided by Powell-Abel et al. (1986), who demonstrated that transgenic tobacco plants expressing the tobacco mosaic virus coat protein were resistant to the virus. Such resistance results because gene sequences derived from the pathogen inhibit virus replication and disease development (Sanford and Johnston, 1985). Beachy et al. (1990) suggested that expression of a virus coat protein as a transgene in a plant confers resistance to the virus in direct proportion to the amount of coat protein produced by the transformed plant.

This novel technique opened up new avenues of controlling viral diseases in plants and has major implications in crop improvement (see Wilson, 1993; Mueller et al., 1995; Lomonossoff, 1995; Bendahmane and Beachy, 1999). Thus, coat-protein-mediated resistance has helped to control papaya ring spot in Hawaii (Gonsalves, 1998). This strategy was also employed for developing resistance to rice viruses. Hayakawa et al. (1992) cloned the coat-protein gene of rice stripe virus and introduced it into two japonica rice varieties through electroporation of protoplasts. The transgenic plants expressed the coat protein at high levels (up to 0.5% of the total soluble protein) and exhibited a significant level of resistance to virus infection. Rice yellow mottle virus (RYMV) is another serious disease, particularly in countries in East and West Africa, causing a loss of about 329,000 metric tonnes of rice for the period 1998–2000 (West Africa Rice Development Association [WARDA], 2000). Because of a lack of conventional solution to this problem, a transgenic approach based on PDR was successfully employed to produce an RYMV-resistant variety, Bouaké 189 (Pinto et al., 1999).

GENETIC ENHANCEMENT OF HUMAN NUTRITION

The present world population of 6.1 billion is projected to reach 8 billion by 2030. It is estimated that we will have to produce 50% more food grains to meet the challenge of feeding this population. Breeding high-yielding crops should be an overriding consideration of crop improvement programs. However, improvement of nutritive value of cereal grains should receive equally high priority to alleviate the mineral and vitamin deficiencies that affect the health of poor people who derive most or all of their calories from cereals. Iron deficiency, which generally causes anemia, is recognized as the most common dietary deficiency in the world, especially among the poor nations. It mostly affects children and women of reproductive age; in pregnant women, severe anemia may cause fetal growth retardation and is responsible for a large proportion of maternal deaths (Gillespie, 1998). Iron deficiency also leads to diminished work performance, impaired psychomotor development and intellectual performance, and decreased resistance to infections (Dallman, 1990). Vitamin A deficiency also poses severe health consequences, affecting almost one third of children in developing countries. This deficiency may impair growth, development, vision, and functioning of the immune system and, in extreme cases, vitamin A deficiency leads to blindness and death (WHO, 1995; Sommer, 1990; Sommer and West, 1996; UN ACC/SCN, 1997). About 100 million children under 5 suffer from vitamin A deficiency and hence are prone to eye damage; half a million children become partly or totally blind each year, and many of them subsequently die (Conway and Toenniessen, 1999).

Some examples of the use of biotechnology in enhancing the nutritional status of crops are given below.

Genetic Enrichment of Rice

Genetic engineering techniques have been successfully employed to raise the micronutrient content of rice, the staple food of millions of people mostly in the poor nations. For example, Goto et al. (1999) introduced the soybean *ferritin* gene into the rice variety Kita-ake through *Agrobacterium*-mediated transformation. The promoter for the rice-seed storage protein glutelin *glub-1* was used to localize the expression of soybean gene specifically in the endosperm. The iron content of transgenic seeds was as much as threefold greater than that of untransformed controls.

Another genetic engineering approach for increasing the bioavailability of iron in rice diets is the elimination of phytate. This sugar-like molecule binds a high proportion of dietary iron, preventing its complete absorption by the human body. A Swiss team led by Ingo Potrykus introduced a fungal gene for enzyme phytase that breaks down phytate, thus improving the bioavailability of iron in rice diets (Lucca et al., 2000).

Recently, an exciting breakthrough was witnessed when transgenic technology was employed to genetically upgrade the nutritional status of rice. Rice grains do not normally contain β-carotene, the precursor to vitamin A. However, they do contain geranylgeranyl pyrophosphate, which can be converted to β-carotene by a sequence of four enzymes. The four genes for these enzymes, two from daffodil and two from the bacterium *Erwinia uredovora*, were introduced into the rice variety Taipei 309 through *Agrobacterium*-mediated transformation. One to three transgene copies were found in transformed plants. Ten plants harboring all four introduced genes showed the normal vegetative phenotype, were fully fertile, and had yellow endosperm indicating carotenoid formation. Thus, by engineering rice with genes derived from daffodils and bacteria, Portrykus and his colleagues "instructed" rice to produce vitamin A. By also incorporating the iron-synthesizing capacity in it, they produced rice grains rich in both vitamin A and iron (Ye et al., 2000). This transgenic rice, called "golden" rice, has the potential of saving millions of lives and averting blindness among millions of children. Golden rice is therefore also referred to as the "grains of hope." The rice variety Taipei 309 was used to introduce the β-carotene biosynthetic pathway, as it is easy to transform. However, it is no longer cultivated. The International Rice Research Institute has started the project aimed at introducing the genes into widely grown improved cultivars through transformation. It is estimated that elite rice cultivars containing β-carotene will become available for on-farm production in 3–4 years. Using conventional breeding alone, it would be unimaginable to produce rice cultivars rich in both vitamin A and iron. However, it is possible to employ traditional breeding techniques to transfer the desirable traits of golden rice into many otherwise superior rice cultivars. Thus, a combination of both modern and conventional tools could bring about phenomenal genetic enrichment of crop cultivars, hitherto impossible to achieve.

Genetic Upgrading of Other Crop Plants

Tools of modern biotechnology have been successfully employed to genetically enhance the nutritional status of several crop plants. Two examples are given below.

1. **Potato**. The potato, *Solanum tuberosum* L., is the most important noncereal food crop used for human consumption. The need to improve its nutritional quality cannot, therefore, be overemphasized. Earlier, Raina and Datta (1992) cloned a gene that encodes a seed-specific protein, amaranth seed albumin (*AmAl*) from *Amaranthus hypochondriacus* L. *AmAl*, encoded by a single gene, is a well-balanced protein in terms of amino acid composition and is also nonallergenic, making it an ideal candidate as a donor protein for use in genetic engineering. Chakraborty et al. (2000) reported the tuber-specific as well as constitutive expression of *AmAl* in potato by using granule-bound starch synthase (GBSS) and CaMV 35S promoters, respectively. The authors reported that the expression of *AmAl* in transgenic tubers resulted not only in a significant increase in most essential amino acids but also in higher protein content in tubers compared with control plants. These findings open up avenues for engineering *AmAl* or similar genes to improve the nutritional status of other crops.

2. **Lupin**. Lupin (*Lupinus angustifolius* L.) is an important grain legume used as animal feed. The lupin seed protein is a source of the ten amino acids that are essential in the diet of nonruminant animals but is deficient in the sulfur-containing amino acids methionine and cysteine (Waddell, 1958). Growth of pigs increased significantly when diets containing lupin as the major protein source were supplemented with methionine

(Liebholz, 1984). Molvig et al. (1997) increased the nutritive value of lupin seeds by increasing methionine levels by expressing a seed albumin gene introduced from sunflower (*Helianthus annuus* L.). The authors found that methionine supplementation had no effect on true protein digestibility, but increased biological value and net protein utilization for all cultivars by between 4.6% and 19.7%.

PERCEIVED DANGERS OF MODERN BIOTECHNOLOGY: OVERCOMING PUBLIC MISTRUST

Genetic engineering has great potential for accelerating crop improvement and has already produced encouraging results (Borlaug, 1997; Cook, 2000; Ye et al., 2000; Jauhar, 2001a; Repellin et al., 2001). Unfortunately, this technology is facing resistance from certain sections of the public. Campaigns have been waged to create fear about the potential adverse impact of genetically modified (GM) plants or foods on human health and the environment (Borlaug, 2000; Marchant, 2001; Marris, 2001; Jauhar, 2001a). Although the concerns and perhaps misconceptions of certain groups may be overblown, some issues raised may be of legitimate concern. These issues must be addressed.

Although several beneficial crop plants have been produced, certain groups think that genetically engineered plants may pose a danger to the environment and human health. A perceived, or perhaps even real concern, is the potential for a transgene to move from a genetically engineered herbicide-resistant crop plant to its wild relatives, thereby creating a possible "super weed" that may be hard to control by the use of available herbicides. The possibility of such a genetic pollution through a misplaced transgene is there, but in most cases it is highly unlikely to happen because of the difficulty of hybridization between a transgenic crop plant and its wild relatives. In the case of self-pollinated crops, such as most cereals, the risks of transgene escape are negligible. Even in the case of most cross-pollinated crops, embryo culture is needed to produce wide crosses.

A novel technique to reduce transgene escape from a crop species to its wild relatives is to engineer, for example, herbicide resistance into the crop's chloroplast genome, as has been done in tobacco (Daniell et al., 1998). This challenging technology now works with tobacco and may be applicable in rice also (Khan and Maliga, 1999). Crossing barrier genes could also be used as shown by work on maize. Hybrids between maize and its annual wild relative teosinte can be readily made by application of teosinte pollen to maize silks. A teosinte gene or gene cluster, *Teosinte crossing barrier 1 (Tcb1)* restricts its crossability with maize (Evans and Kermicle, 2001). When introduced into maize, *Tcb1* may be useful for creating reproductive isolation barrier between maize and its wild relatives or for isolating one variety from another.

Another concern is the possibility of health risks posed by GM foods, e.g., allergenicity. Therefore, the GM foods must be carefully tested before release to the public. Recently, a genetically altered corn variety called StarLink somehow contaminated portions of the food supply in the United States. Traces of the Cry9C protein of *B. thuringiensis* were found in taco shells (EPA, 2000a). Because StarLink has not been approved for human consumption, it had to be withdrawn from the market. Such incidents underscore the need for extra precautions on the part of plant and food scientists and the agricultural sector. Crossing barrier genes like *Tcb1* could help obviate such problems. Concerns have been raised about possible adverse environmental impacts of Bt-corn, incorporating resistance to European corn borer, on the environment. A group of researchers claimed that milkweed leaves (a favorite food of butterfly larvae) dusted with heavy concentrations of Bt-corn pollen proved toxic to monarch butterfly larvae (Losey et al., 1999). However, this controversial report was later discounted by several researchers (see Shelton and Sears, 2001). Niiler (1999), for example, showed that the monarch butterfly faces little threat from Bt-corn.

GENETICALLY MODIFIED CROPS FOR ALLEVIATING WORLD HUNGER

In 1950, the world population stood at 2.5 billion; in 1999, it was around 6 billion; and will rise to about 8 billion by 2025 (Dyson, 1999). Almost 800 million people, most in the developing world, are chronically malnourished and 24,000 die of hunger every day. About 40 years ago, there were about a billion hungry people and population projections show that there may still be 600 million hungry people by 2025 (Chrispeels, 2000). Although the 20th century witnessed a phenomenal increase in yields of crop plants, especially cereals, largely by conventional breeding (see, for example, Figure 5.1), food production has not kept pace with the burgeoning population growth. Even now, 100 million preschool children suffer from vitamin A deficiency and some 400 million women between the ages of 15 and 49 have an iron deficiency leading to anemia (Conway and Toenniessen, 1999).

IMPORTANT ACCOMPLISHMENTS OF MODERN BIOTECHNOLOGY

Clearly, the available agricultural technologies that led to the Green Revolution cannot keep pace alone with the present and projected population increases. Modern biotechnological tools, alone or in combination with traditional techniques, hold great promise for augmenting agricultural productivity in quantity as well as quality. Gene transfer technologies developed in the last two decades could revolutionize agricultural production in several ways. The efficacy of transgenic crop varieties in increasing production and lowering production costs has already been demonstrated (Borlaug, 2000; Herrera-Estrella, 2000; Chrispeels, 2000; Prakash, 2001). In 1996 and 1997, the cultivation of virus-, insect-, and herbicide-resistant plants accounted for 5% to 10% increase in yield and also resulted in savings on herbicides of up to 40% and on insecticides of between $145 and $290 per ha (James, 1998). Between 1996 and 1999, the area planted to transgenic crops increased from 1.7 to 39.9 million ha (James, 1999).

Despite efforts in preventing pre- and post-harvest crop losses, pests destroy over half of all world crop production, and post-harvest loss due to insects, most of it in the developing world, is estimated to be 15% of the world production. Modern biotechnology could help alleviate many of these problems. Insect-resistant cultivars, using the δ-endotoxin of *B. thuringiensis*, have been produced in several important crop species including maize, cotton, rice, tobacco, tomato, potato, sugarcane, and walnut. Of these, maize, cotton, and potato are already under commercial cultivation in the United States and several other countries. Genetically engineered virus resistance could greatly benefit farmers in affluent as well as poor countries. Resistance to 30 different viral diseases has been engineered into 20 plant species, using variations of the pathogen-derived resistance strategy (Herrera-Estrella, 2000). Thus, transgenic approaches have been successfully applied to produce virus-resistant papaya (Gonsalves, 1998) and RYMV-resistant rice (Pinto et al., 1999) that should help a large proportion of the poor populace of Africa and Asia.

Another hope of modern biotechnology is possible improvement in fertilizer-use efficiency of crop plants. Transgenic wheat engineered with high levels of Glu dehydrogenase, for example, resulted in 29% more yield with the same amount of fertilizer than did the normal crop (Smil, 1999).

Even more importantly, the new technology has immense power to improve the nutritional quality of crop species for feeding the ever-growing human population. Transgenic technology has been successfully employed to enhance the nutritional status of several crop plants (Kishore and Shewmaker, 1999) including rice (Ye et al., 2000; Portrykus, 2001) and potato (Chakraborty et al., 2000). The transgenic rice, or golden rice, has the potential of saving millions of human lives and preventing blindness among millions of children. It would of course be unimaginable to produce golden rice by conventional breeding.

UNIVERSALITY OF MODERN TECHNOLOGY

It is remarkable that plant biotechnology often generates crop improvement strategies that can be applied to several different crops. Tropical fruit crops suffer severe losses in developing countries because the fruits ripen rapidly and rot because of lack of appropriate storage conditions and efficient means of transporting them to the final consumer. The hormone ethylene regulates a number of developmental processes including fruit ripening. In transgenic tomato plants, antisense inhibition of ethylene biosynthetic genes results in delayed ripening and extended shelf life of perishable fruits and vegetables (see Hackett et al., 2000). Genetically engineered-delayed ripening, as has been tested on a commercial scale for tomato, has great potential application for tropical fruit crops like the mango in India and other tropical countries.

TRANSFER OF MODERN TECHNOLOGY TO DEVELOPING COUNTRIES

An enormous advantage in the application of modern biotechnology in the developing countries is that, in principle, it does not require major changes in the agricultural practices of small farmers. Most of the available technology for producing improved transgenic crop cultivars could be effectively used to improve productivity in these countries. Possible benefits of GM crops to developing countries include:

(1) Improved nutrition and health benefits; (2) improved quantity and quality of food crops and animal products; (3) reduced dependence on costly pesticides and herbicides resulting in valuable savings and of course cleaner environment for underprivileged farmers; and (4) clean, safe and cheap production of edible vaccines.

Herrera-Estrella (1999, 2000) lists several effective ways of transferring the technology to developing countries. First,to train scientists from these countries in universities, research institutes and other suitable laboratories in the developed countries; second, to assist developing countries in establishing their own facilities for biotechnological research; and third, to transfer technology in terms of gene constructs or transgenic plants from universities or private companies to the existing research centers in developing countries. Because many of these technologies have been patented by private industry, a major challenge would be to find mechanisms or resources to transfer this technology to developing nations.

There is urgent need for the development and use of biotechnology in African countries, which stand to gain the most from the new technology (Wambugu, 1999; Machuka, 2001). That modern biotechnology offers considerable promise to tackle agricultural problems in Africa is shown, for example, by fruitful collaboration between John Innes Centre in England and West Africa Rice Development Association, a part of the Consultative Group of International Agricultural Research Centers. This collaborative work led to the production of the virus-resistant rice variety Bouaké, which showed resistance to rice yellow mottle virus isolates from geographically diverse locations in Africa (Pinto, 2000). This example shows how collaborative efforts between a well-equipped research laboratory in the West and an agricultural research center in Africa can solve a serious production problem in rice, a crop that sustains a large proportion of the human population in Africa.

A recent report prepared under the auspices of the Royal Society of London (based on the recommendations of seven of the world's academies of science) concluded: "... GM technology, coupled with important developments in other areas, should be used to increase the production of main food staples, improve the efficiency of production, reduce the environmental impact of agriculture, and provide access to food for small-scale farmers" (Bowles and Klee, 2001).

Genetically modified crops offer a new hope for meeting food needs in the 21st century (Pinstrup-Andersen and Pandya-Lorch, 2000) especially for developing nations (Herrera-Estrella and Alvarez-Morales, 2001). However, the final decision to adopt modern biotechnology should ultimately rest with the developing countries. A source of easily available information on technology transfer should be available to scientists and policy makers in the developing world. Initial efforts in this direction have been carried out by the International Service for the Acquisition of Agrobiotech Application (ISAAA), a nonprofit organization attempting to play the role of an honest broker in facilitating technology transfer to needy countries (Herrera-Estrella, 1999). Adapting biotechnology (imported or homemade) to local crops acquires special significance, considering that local farmers in those countries are more likely to embrace a known crop with desired genetic modification than an unknown foreign crop.

Maize, for example, is a staple food in many parts of the African continent and is also used for human consumption in Egypt and Mexico. South Africa exports maize to southern African countries. Nofal (2002) reports that South African farmers have started planting genetically modified white maize, with an estimated 1,596,005 ha to be planted in 2001–2002. This is perhaps the first example of a GM crop being commercially planted so widely anywhere in the world. In 2000, worldwide commercial plantings of GM crops reached 44.2 million hectares, an 11% increase from 39.9 million hectares in 1999 (Lema, 2001). A large proportion of these hectares are in Argentina, Canada, China and the USA.

DaSilva (2001) states that in many developing countries there is widespread use of GM crops. In China, 13 gene-altered crops, including rice, wheat, corn and potato, have been released in the agricultural sector. Ten countries in Latin America and the Caribbean (Argentina, Belize, Bolivia, Costa Rica, Chile, Cuba, Dominican Republic, Guatemala, Mexico and Peru) were engaged in field trials of transgenic crops including cotton, maize, potato, and soybean. Several other countries, including Brazil, Colombia, Venezuela, China, India, Indonesia, Malaysia, Thailand, Egypt, and Zimbabwe are also engaged in trials with transgenic cotton, corn, potato, soybean and tomato. In Africa, trials are expected to get under way in Kenya, Nigeria, and Uganda (DaSilva, 2001).

CONCLUSION AND PERSPECTIVES

As stated earlier, the world population is expected to double by the year 2050, compounding the already precarious problem of feeding humankind. Food production will need to be doubled or even tripled to meet the ever-growing need of the mushrooming population. Genetic enhancement of nutritive value of crops will help abate malnutrition. Genetic improvement of food crops affected by conventional plant breeding brought about a phenomenal improvement in crop yields. The methods, although slow, will no doubt continue to play a major role in crop improvement programs. More recently, the advent of genetic engineering has provided novel techniques of rapid gene transfer into crop plants. The successful use of transgenic approaches to combat pests and diseases and malnutrition among the poor masses constitutes an exciting breakthrough in genetic amelioration of our crop plants. Thus, the development of golden rice, genetically enriched with vitamin A and iron, may become one of the most important success stories of all time. Adopting conventional breeding methods, these superior traits of golden rice may be easily transferred to other rice varieties consumed by and well adapted to environments in developing countries. Such a nutritional enhancement of a cereal crop would be unthinkable through traditional breeding alone.

Besides nutritional enrichment of food crops, transgenic technology has the potential of producing edible vaccines. Vegetable and fruit crops with appropriate genetic engineering could provide immunization against deadly diseases like hepatitis or tuberculosis. A day may come when, instead of taking an injection, one may need only to eat a banana or perhaps a tomato. Scientists all over the world, including those at the Indian Institute of Science, Bangalore, are attempting to insert appropriate genes into plants like tomato, banana or melon, whose fruit could be eaten uncooked and thus provide an oral dose of a vaccine. Edible vaccines orally administered through GM foods could become available at a fraction of the current costs, estimated at $.02 instead of the usual US$15 for an injectable dose (DaSilva, 2001).

Like any new technology, acceptance of genetic engineering is encountering resistance from certain sections of the public. The potential benefits and risks of the new technology must be carefully weighed because no technology is completely risk free. Floating around are several misconceptions about genetically engineered food plants and their adverse effects on human safety and the environment. Such fears are largely unfounded. Genetically modified foods are safe to human health. In their meeting on 22 March 2001, the Medical Research Council of England stated, "There is no evidence to suggest that GM foods are harmful to human health" (http://www.biotech-info.net/GM_research_med.html). In fact, we have been consuming and continue to consume genetically modified foods on a daily basis. And there is no report of any injury to human health.

Countering the prevailing antibiotech sentiment in Europe, a biosafety report from the European Union, summarizing 81 research projects on GM crops financed by the EU over the last 15 years at a cost of $64 million, suggested that GM crops may even be safer than regular crops. The European Commission (the EU's executive branch) stated that the research has not found "any new risks to human health or the environment beyond the usual uncertainties of conventional plant breeding." The commission further reported: "Indeed, the use of more precise technology and the greater regulatory scrutiny probably make them even safer than conventional plants and foods," (The Associated Press, Arizona Daily Star, 9 October 2001).

Nor do genetically engineered organisms or foods pose apparent danger to the environment. In fact, as stated earlier, the Bt-crops can be beneficial to the environment. The U.S. EPA estimates that the use of Bt-crops in the U.S.A. results in an annual reduction of >7.7 million acre treatments of synthetic insecticides (U.S. EPA, 2000b), which include mostly broad-spectrum insecticides that can affect non-target organisms and potentially cause environmental and human health risks (Shelton and Sears, 2001).

The underlying thrust of opponents of the new technology is that it is unnatural and hence unsafe. Transgenic technology of introducing new genes into plants is considered by some as tinkering with nature. Even high-profile celebrities such as Prince Charles of England think that "tinkering with nature could have disastrous long-term consequences," and that genetic modification of crop plants should be left to God alone (Daily Telegraph, London, 10 June 1998 and 19 May 2000). However, we must remind ourselves that man, in his efforts to improve food production, has been tinkering with nature for centuries. Even traditional plant breeding is, in essence, man-made evolution that has produced crop cultivars that sustain humankind. We should perhaps not suddenly get nervous about genetically altering crop plants now when basically we have been doing pretty much the same thing for centuries.

Any breeding activity is accompanied by genetic modifications, which ultimately involve changes at the DNA level. The newer biotechnological tools of gene transfer are, in fact, a refinement of earlier ones, and gene transfer by these techniques poses no greater danger or risk to the consumer. Many of the current crop cultivars we consume every day do, after all, contain genes of alien origin (Jauhar, 1993; Friebe et al., 1996; Jauhar and Chibbar, 1999). It is heartening to note that GM crops are being increasingly accepted by farmers worldwide. Clive James, ISAAA chairman, recently stated that modern biotechnology is delivering significant benefits to farmers, increasing the number of farmers planting transgenic crops from 2 million in 1999 to 3.2 million in 2000 (BusinessWorld, 21 Dec. 2001).

The GM technology is an important weapon in our war against poverty and starvation. However, the new technology will complement, not replace, conventional plant breeding. The old and new technologies should go hand in hand to accelerate crop improvement to sustain global food security. We have no doubt that, with proper education and awareness, genetic engineering will be widely accepted worldwide. Africa, which lagged behind in reaping the fruits of the Green Revolution, must join in and benefit from the biotechnology revolution. Hopefully, the enormous potential of this technology will be harnessed to the best advantage of the entire human race.

REFERENCES

Alam, M.F., K. Datta, E. Abrigo, A. Vasquez, D. Senadhira and S.K. Datta. 1998. Production of synthetic *Bacillus thuringiensis cry 1A(b)* gene with enhanced resistance to yellow stem borer. *Plant Sci.* 135: 25-30.

Altpeter, F., I. Diaz, H. McAuslane, K. Gaddour, P. Coabonero and I.K. Vasil. 1999. Increased insect resistance in transgenic wheat stably transformed expressing trypsin inhibitor Cme. *Molec. Breed.* 5: 53-63.

Anderson, P.P. and R. Pandya-Lorch. 1999. Securing and sustaining adequate world food production for the third millennium. World Food Security and Sustainability: The impacts of biotechnology and industrial consolidation, National Agricultural Biotechnology Council Report 11 NABC, Ithaca, NY: 27-48.

Beachy, R.N., S. Loesch-Fries and N.E. Turner. 1990. Coat protein-mediated resistance against virus infection. *Annu. Rev. Phytopathol.* 28: 451-474.

Bendahmane, M. and R.N. Beachy. 1999. Control of tobamovirus infections via pathogen-derived resistance. *Adv. Virus Res.* 53: 369-386.

Bliffeld, M., J. Mundy, I. Potrykus and J. Futterer. 1999. Genetic engineering of wheat for increased resistance to powdery mildew disease. *Theor. Appl. Genet.* 98: 1079-1086.

Bommineni, V. R., P.P. Jauhar and T.S. Peterson. 1997. Transgenic durum wheat by microprojectile bombardment of isolated scutella. *J. Hered.* 88: 475-481.

Borlaug, N.E. 1997. Feeding a world of 10 billion people: The miracle ahead. *Plant Tiss. Cult. Biotech.* 3: 119-127.

Borlaug, N.E. 2000. Ending world hunger. The promise of biotechnology and the threat of antiscience zealotry. *Plant Physiol.* 124: 487-490.

Bowles, D. and Klee, H. 2001. Introduction to the special issues on plant GM technology. *Plant J.* 27: 481-482.

Broglie, K., I. Chet, M. Holliday and R. Cressman. 1991. Transgenic plants with enhanced resistance to fungal pathogen *Rhizoctonia solani*. *Science* 254: 1194-1197.

Burton, G.W. and J.B. Powell. 1968. Pearl millet breeding and cytogenetics. *Adv. Agron.* 20: 49-89.

Chakraborty, S., N. Chakraborty and A. Datta. 2000. Increased nutritive value of transgenic potato by expressing a non-allergenic seed albumin gene from *Amaranthus hypochondriacus*. *Proc. Natl. Acad. Sci. USA* 97: 3724-3729.

Chen, W.P., P.D. Chen, D.J. Liu, R.J. Kynast, B. Friebe, R. Velazhahan, S. Muthukrishnan and B.S. Gill. 1999. Development of wheat scab symptoms in delayed transgenic wheat plants that constitutively express a rice thaumatin-like protein gene. *Theor. Appl. Genet.* 99: 755-760.

Chrispeels, M.J. 2000. Biotechnology and the poor. *Plant Physiol.* 124: 3-6.

Conway, G. and G. Toenniessen. 1999. Feeding the world in the twenty-first century. *Nature Suppl.* 402: C55-C58.

Cook, R.J. 2000. Science based risk assessment for the approval and use of plants in agricultural and other environments. In: G.J. Persley and M.M. Lantin (Eds.) *Agricultural Biotechnology and the Poor: Proceedings of an International Conference*, Washington, DC: pp. 123-130.

Dahleen, L.S., P.A. Okubara and A.E. Blechl. 2001. Transgenic approaches to combat Fusarium head blight in wheat and barley. *Crop Sci.* 41: 628-637.

Dallman, P.R. 1990. *Iron: Present Knowledge in Nutrition.* M.L. Brown (Ed.) Washington D.C. International Life Sciences Institute, Nutrition Foundation. pp 241-250.

Daniell, H., R. Datta, S. Varma, S. Gray and S.-B. Lee. 1998. Containment of herbicide resistance through genetic engineering of the chloroplast genome. *Nature Biotech.* 16: 345-348.

DaSilva, E.J. 2001. GMOs and development. *Electronic J. Biotech* [online]. 16 Aug 2001, vol. 2 no. 3. www.ejb.org/content/vol4/issue2/issues/01/index.html.

Datta, S.K., L.B. Torrizo, J. Tu, N.P. Oliva and K. Dutta. 1997. Production and Molecular Evaluation of Transgenic Plants. IRRI Discussion paper 21, International Rice Research Institute, Los Banos, Philippines.

Datta, S.K. and S. Muthukrishnan. 1999. *Pathogenesis-Related Proteins in Plants.* CRC Press, Boca Raton, FL. 291 pp.

Dyson, T. 1999. Prospects for feeding the world / commentary: Bread for the world–another view. *Brit. Med. J.* 319: 988-991.

EPA, U.S., 2000a. Assessment of Scientific Information concerning StarLink Corn Cry9CBt Corn Plant-pesticide. Federal Register 65 (October 31, 2000). Arlington: Environmental Protection Agency, pp. 65246-65251.

EPA, U.S., 2000b. Biopesticides Registration Document: Preliminary Risks and Benefits Sections; *Bacillus thuringiensis* plant-pesticides. Washington, D.C., Environmental Protection Agency and Office of Pesticide Programs, Biopesticides and Pollution Prevention Division.

Evans, M.M.S. and J.L. Kermicle. 2001. *Teosinte crossing barrier1*, a locus governing hybridization of teosinte with maize. *Theor. Appl. Genet.* 103: 259-265.

Falck-Zepeda, B.J., G. Traxler and R.G. Nelson. 1999. Rent Creation and Distribution from the First Three Years of Planting Bt Cotton. International Service for the Acquisition of Agri-Biotech Applications No. 14-1999. 17 pp.

Fedak, G. 1999. Molecular aids for integration of alien chromatin through wide crosses. *Genome* 42: 584-591.

Friebe, B., J. Jiang, W.J. Raupp, R.A. McIntosh and B.S. Gill. 1996. Characterization of wheat alien translocations conferring resistance to diseases and pests: Current status. *Euphytica* 91: 59-87.

Fujimoto, H., K. Itoh, M. Yamamoto, J. Kyozuka and K. Shimamoto. 1993. Insect resistant rice generated by introduction of a modified δ-endotoxin gene of *Bacillus thuringiensis*. *Bio/Technol.* 11: 1151-1155.

Gillespie, S. 1998. Major issues in the control of iron deficiency. Micronutrient Initiative and UNICEF, The Micronutrient Initiative, Ottawa, ON.

Gonsalves, D. 1998. Control of ring spot virus in papaya: A case study. *Annu. Rev. Phytopathol.* 36: 412-437.

Goto, M., T. Yoshihara, N. Shigemoto, S. Toke and F. Taikawa. 1999. Iron fortification of rice seed by the soybean *ferritin* gene. *Nature Biotech.* 17: 282-286.

Hackett, R.M., C.-W. Ho, Z. Lin, H.C.C. Foote, R.G. Fray and D. Grierson. 2000. Antisense inhibition of the Nr gene restores normal ripening to the tomato never-ripe mutant, consistent with the ethylene receptor-inhibition model. *Plant Physiol.* 124: 1079-1086.

Hayakawa, T., Y. Zhu, K. Itoh and Y. Kimura. 1992. Genetically engineered rice, resistant to rice stripe virus, an insect transmitted virus. *Proc. Natl. Acad. Sci. USA* 89: 9865-9869.

Herdt, R.W. 1991. Research priorities for rice biotechnology. In: G.S. Khush and G. Toenniessen (Eds.) *Rice Biotechnology*, Wallingford, UK: CAB Int., pp. 19-54.

Herrera-Estrella, L. 1999. Transgenic plants for tropical regions: Some considerations about their development and their transfer to the small farmer. *Proc. Natl. Acad. Sci. USA* 96: 5978-5981.

Herrera-Estrella, L.R. 2000. Genetically modified crops and developing countries. *Plant Physiol.* 124: 923-926.

Herrera-Estrella, L. and A. Alvarez-Morales. 2001. Genetically modified crops: Hope for developing countries. *Embo Reports* 2: 256-258.

Hyde, J., M.A. Martin, P.V. Preckel and C.R. Edwards. 1999. The economics of Bt corn: Valuing protection from the European corn borer. *Rev. Agri. Econ.* 21: 442-454.

James, C. 1998. Update in the development and commercialization of genetically modified crops. *Int. Serv. Acquisition Agrobiotechnol. Appl. Briefs.* 5: 1-20.

James, C. 1999. Global review of commercialized transgenic crops: 1999. International Service for the Acquisition of Agri-biotechnology Applications Briefs No. 12 Preview. *Int. Serv. Acquisition Agrobiotechnol. Appl. Briefs*, Ithaca, NY.

Jauhar, P.P. 1981. *Cytogenetics and Breeding of Pearl Millet and Related Species.* Alan R. Liss, Inc., New York. 310 pp.

Jauhar, P.P. 1993. Alien gene transfer and genetic enrichment of bread wheat. In: A.B. Damania (Ed.) *Biodiversity and Wheat Improvement*, John Wiley and Sons, Chichester, England: 103-119.

Jauhar, P.P. 2001a. Genetic engineering and accelerated plant improvement: Opportunities and challenges. *Plant Cell Tiss. Org. Cult.* 64: 87-91.

Jauhar, P.P. 2001b. Problems encountered in transferring scab resistance from wild relatives into durum wheat. *Proc. 2001 Nat. Fusarium Head Blight Forum*, Cincinnati, December 8-10, 2001. pp. 188-191.

Jauhar, P.P. and R.N. Chibbar. 1999. Chromosome-mediated and direct gene transfers in wheat. *Genome* 42: 570-583.

Jauhar, P. P. and L. R. Joppa. 1996. Chromosome pairing as a tool in genome analysis: Merits and limitations. In: Jauhar, P. P. (Ed.) *Methods of Genome Analysis in Plants*. CRC Press, Boca Raton. pp. 9-37.

Jauhar, P.P. and W.W. Hanna. 1998. Cytogenetics and genetics of pearl millet. *Adv. Agron.* 64: 1-26.

Jauhar, P.P. and T.S. Peterson. 2001. Hybrids between durum wheat and *Thinopyrum junceiforme*: Prospects for breeding for scab resistance. *Euphytica* 118: 127-136.

Khan, M.S. and P. Maliga. 1999. Fluorescent antibiotic resistance marker for tracking plastid transformation in higher plants. *Nature Biotech.* 17: 910-915.

Khush, G.S. 1999. Green Revolution: preparing for the 21st century. Genome 42: 570-583.

Khush, G.S. and P.S. Baenziger. 1996. Crop improvement: Emerging trends in rice and wheat. In: V.L. Chopra, R.B. Singh and A. Varma (Eds.) *Crop Productivity and Sustainability — Shaping the Future*, Proc. 2nd Int. Crop Sci. Congress, Oxford and IBH Publishing Co. Pvt. Ltd., New Delhi: 113-125.

Khush, G.S., E. Bacalangco and T. Ogawa. 1990. A new gene for resistance to bacterial blight from *O. longistaminata. Rice Genet. Newslet.* 7: 121-122.

Kishore, G.M. and C. Shewmaker. 1999. Biotechnology: Enhancing human nutrition in developing and developed worlds. *Proc. Natl. Acad. Sci. USA* 96: 5968-5972.

Lauer, J. and J. Wedberg. 1999. Grain yields of initial Bt corn hybrid introductions to farms in the northern Corn Belt. *J. Prod. Agric.* 12: 373-376.

Lema, K.L. 2001. Farmers turn to biotechnology. *BusinessWorld,* Dec 21, 2001.

Liebholz, J. 1984. A note on methionine supplementation of pig grower diets containing lupin-seed meal. *Anim. Prod.* 38: 515-517.

Lin, W., C.S. Anuratha, K. Datta, I. Potrykus, S. Muthukrishnan and S.K. Datta. 1995. Genetic engineering of rice for resistance to sheath blight. *Bio/Technol.* 13: 686-691.

Logemann, J., G. Jack, H. Tommerup, J. Mundy and J. Scheel. 1992. Expression of a barley ribosome-inactivating protein leads to increased fungal protection in transgenic tobacco plants. *Bio/Technol.* 10: 305-308.

Lomonossoff, G.P. 1995. Pathogen-derived resistance to plant viruses. *Annu. Rev. Phytopathol.* 33: 323-343.

Losey, J.E., L.S. Rayor and M.E. Carter. 1999. Transgenic pollen harms monarch larvae. *Nature* 399: 214.

Lucca, P., R. Hurrell and I. Potrykus. 2000. Development of iron-rich rice and improvement of its absorption in humans by genetic engineering. *J. Plant Nutr.* 23: 11-12.

Machuka, J. 2001. Agricultural biotechnology for Africa. African scientists and farmers must feed their own people. *Plant Physiol.* 126: 16-19.

Marchant, R. 2001. From the test tube. *Embo Reports* 2: 354-357.

Marris, C. 2001. Public views on GMOs: Deconstructing the myths. *Embo Reports* 2: 545-548.

McMullen, M., R. Jones and D. Gallenberg. 1997. Scab of wheat and barley: A reemerging disease of devastating impact. *Plant Dis.* 81: 1340-1348.

Molvig, L., L.M. Tabe, B.O. Eggum, A.E. Moore, S. Craig, D. Spencer and T.J.V. Higgins. 1997. Enhanced methionine levels and increased nutritive value of seeds of transgenic lupins (*Lupinus angustifolius* L.) expressing a sunflower seed albumin gene. *Proc. Natl. Acad. Sci. USA* 94: 8393-8398.

Mueller, E., J. Gilbert, G. Davenport, G. Brigneti and D.C. Baulcombe. 1995. Homology-dependent resistance; Transgenic virus resistance in plants related to homology-dependent gene silencing. *Plant J.* 7: 1001-1013.

Muthukrishnan, S., G.H. Liang, H.N. Trick and B.S. Gill. 2001. Pathogenesis-related proteins and their genes in cereals. *Plant Cell Tiss. Org. Cult.* 64: 93-114.

Nganje, W.E., D.D. Johnson, W.W. Wilson, F.L. Leistritz, D.A. Bangsund and N.M. Tiapo. 2001. Economic Impacts of Fusarium Head Blight in Wheat and Barley: 1998-2000. Agribusiness and Applied Economics Report No. 464. Agricultural Experiment Station, North Dakota State University, Fargo, North Dakota.

Niiler, E. 1999. GM corn poses little threat to monarch. *Nature Biotech.* 17: 1154.

Nofal, J. 2002. SA (South Africa) starts growing genetically modified white maize. *Business Report,* 3 January 2002.

Ostlie, K.R., W.D. Hutchinson and R.L. Hellmich. 1997. Bt corn and European corn borer: Long term success through resistance management. University of Minnesota Extension Service Bulletin 7055-GO: 17.

Palevitz, B.A. 2001. EPA reauthorizes Bt corn. *Scientist* 15: 11.

Pinstrup-Andersen, P. and R. Pandya-Lorch. 2000. Meeting food needs in the 21st century: How many and who will be at risk? Presented at the AAAS Annual Meeting, February 2000, Washington D.C.

Pinto, Y.M., R.A. Kok D.C. Baulcombe. 1999. Resistance to rice yellow mottle virus (RYMV) in cultivated African rice varieties containing RYMV transgenes. *Nature Biotech.* 17: 702-707.

Pinto, Y. 2000. Transgenic resistance to rice yellow mottle virus disease: Highlighting some of the issues facing the potential introduction of biotechnological products into Africa. AgBiotechNet® vol. 2, ABN 039

Potrykus, I. 2001. Golden rice and beyond. *Plant Physiol.* 125: 1157-1161.

Powell-Abel, P., R.S. Nelson, B. De, N. Hoffmann, S.G. Rogers, R.T. Fraley and R.N. Beachy. 1986. Delay of disease development in transgenic plants that express the tobacco mosaic virus coat protein. *Science* 232: 738-743.

Prakash, C.S. 2001. The genetically modified crop debate in the context of agricultural evolution. *Plant Physiol.* 126: 8-15.

Raina, A. and A. Datta. 1992. Molecular cloning of a gene encoding a seed-specific protein with nutritionally balanced amino acid composition from *Amaranthus*. *Proc. Natl. Acad. Sci. USA* 89: 11774-11778.

Rausher, M.D. 2001. Coevolution and plant resistance to natural enemies. *Nature* 411: 857-864.

Repellin, A., M. Bảga, R.N. Chibbar and P.P. Jauhar 2001. Genetic enrichment of cereal crops by alien gene transfers: new challenges. *Plant Cell Tiss. Org. Cult.* 64: 159-183.

Roulin, S., P. Xu, A.H.D. Brown and G.B. Fincher. 1997. Expression of specific (1,3)-β-glucanase genes of near-isogenic resistant and susceptible barley lines infected with the leaf scald fungus (*Rhynchosporium secalis*). *Physiol. Molec. Plant Pathol.* 50: 245-261.

Royal Society of London, 2000. Transgenic Plants and World Agriculture. http://www.royal-soc.ac.uk/policy/reports.htm.

Sanford J.C. and S.A. Johnston. 1985. The concept of parasite-derived resistance: deriving resistance genes from the parasite's own genome. *J. Theor. Biol.* 113: 395-405.

Shelton, A.M. and M.K. Sears. 2001. The monarch butterfly controversy: Scientific interpretations of a phenomenon. *Plant J.* 27: 483-488.

Smil, V. 1999. Long-range perspectives on inorganic fertilizers in global agriculture. Travis P. Hignett Memorial Lecture, International Fertilizer Development Center, Muscle Shoals, Alabama, U.S.A.

Sommer, A. 1990. Vitamin A status, resistance to infection and childhood mortality. *Ann. NY Acad. Sci.* 587: 17-23.

Sommer, A., and K.P. West. 1996. *Vitamin A Deficiency: Health, Survival, and Vision.* New York and Oxford: Oxford University Press.

Song, W.Y., G.L. Wang, L.L. Chen, H.S. Kim, L.Y. Pi, T. Holston, J. Gardner, B. Wang, W.X. Zhai, L.H. Zhu, C. Fauquet and P. Ronald. 1995. A receptor kinase-like protein encoded by the rice disease resistance gene, *Xa21*. *Science* 270: 1804-1806.

Stuiver, M.H. and J.H.V.V. Custers. 2001. Engineering disease resistance in plants. *Nature* 411: 865-868.

Swaminathan, M.S. 1999. Harness the gene revolution to help feed the world. *International Herald Tribune*, October 23, 1999.

Thomma, B.P., I.A. Penninckx, W.F. Broekaert and B.P. Cammune. 2001. The complexity of disease signaling in *Arabidopsis*. *Curr. Opin. Immunol.* 13: 63-68.

Tu, J., G. Zhang, K. Datta, C. Xu, Y. He, Q. Zhang, G.S. Khush and S.K. Datta. 2000a. Field performance of transgenic elite commercial hybrid rice expressing *Bacillus thuringiensis* δ-endoprotein. *Nature Biotech.* 18: 1101-1104.

Tu, J., K. Datta, G.S. Khush, Q. Zhang and S.K. Datta. 2000b. Field performance of *Xa21* transgenic rice (*Oryza sativa* L.), IR72. *Theor. Appl. Genet.* 101: 15-20.

UN ACC/SCN (United Nations Administrative Committee on Coordination/Subcommittee on Nutrition). 1997. Third report on the world nutrition situation. Geneva: ACC/SCN.

Waddell, J. 1958. Supplementation of plant proteins with amino acids. In: A.M. Altschul (Ed.). *Processed Plant Protein Foodstuffs*. Academic Press, New York, pp. 307-351.

Wambugu, F. 1999. Why Africa needs biotechnology. *Nature* 400: 15-16.

Wang, G.L., W.Y. Song, D.L. Ruan, S. Sideris and P.C. Ronald. 1996. The cloned gene *Xa21* confers resistance to multiple *Xanthomonas oryzae* pv. *Oryzae* isolates in transgenic plants. *Mol. Plant–Microbe Interaction* 9: 850-855.

WHO (World Health Organization) 1995. Global prevalence of vitamin A deficiency. Micronutrient Deficiency Information System Working Paper No. 2. Geneva: WHO.

Wilson, T.M. 1993. Strategies to protect crop plants against viruses: pathogen-derived resistance blossoms. *Proc. Natl. Acad. Sci. USA* 90: 3134-3141.

Wun, J., A. Kloti, P.K. Burkhardt, G.C. Ghosh-Biswas, K. Launis, V.A. Iglesias and I. Potrykus. 1996. Transgenic indica rice breeding line IR58 expressing a synthetic cry 1 A(b) gene from *Bacillus thuringiensis* provides effective insect pest control. *Biol. Tech.* 14: 171-176.

Ye, X., S. Al-Babill, A. Klötl, J. Zhang, P. Lucca, P. Beyer and I. Potrykus. 2000. Engineering the provitamin A (β-carotene) biosynthetic pathway into (carotenoid-free) rice endosperm. *Science* 287: 303-305.

Zhu, Q. and C.J. Lamb. 1991. Isolation and characterization of a rice gene encoding a basic chitinase. *Mol. Gen. Genet.* 226: 289-296.

8 Energy Inputs in Crop Production in Developing and Developed Countries

D. Pimentel, R. Doughty, C. Carothers, S. Lamberson, N. Bora and K. Lee

CONTENTS

INTRODUCTION

FOOD AND POPULATION

Adequate supplies of staple food crops, which people rely on for their health and very survival, are threatened as the human population increases and the resources that support crop production diminish. The staple crops include wheat, rice, corn, soybeans, white potato, sweet potato, cassava and others (Pimentel and Pimentel, 1996). Consider that, worldwide, more than 3 billion people are currently malnourished (WHO, 1996). This is the largest number and percentage of malnourished humans ever recorded in history. The United Nations University (1999) projects that Africa will be able to feed only 40% of its population in 2025. Recent reports from the Food and Agriculture Organization of the United Nations and the U.S. Department of Agriculture, as well as from numerous other international organizations, further confirm the serious nature of the global food shortages (Population Summit of the World's Scientific Academies, 1994).

The world human population is currently at more than 6 billion and, based on current rates of increase, it is projected to double to approximately 12 billion in fewer than 50 years (PRB, 2000). Thus, great pressure is being placed on all the resources essential for food production and especially fossil energy, which is a finite resource.

Through continued use, cropland is degraded, water is polluted, fossil energy supplies diminished and biological resources lost, yet all are vital to human survival. These losses further restrict present agricultural production and its expansion to meet additional food needs (Pimentel et al., 1999). Although recent increases in crop yields have been achieved in fossil-fuel-dependent agriculture, intensive use of cropland production is causing widespread soil erosion (Pimentel and Pimentel, 1996).

WORLD ENERGY RESOURCES

Humans rely on various sources of power for food production, housing, clean water and a productive environment. These range from human, animal, wind, tidal and water energy to wood, coal, gas, oil and nuclear sources. Of these, fossil-fuel resources have been most effective in increasing food production, feeding a growing number of humans and helping alleviate malnourishment and numerous other diseases (Pimentel and Pimentel, 1996).

About 445 quads (1 quad = 10^{15} BTU; 445 quads = 111 x 10^{15} kcal or 384 x 10^{18} Joules) from fossil and renewable energy sources are used worldwide each year for all human needs (DOE/EIA, 1996; *British Petroleum Statistical Review of World Energy*, 1999). In addition, about 50% of all the solar energy captured by photosynthesis and incorporated in biomass worldwide is used by humans. Although this amount of biomass energy is very large (approximately 600 quads), it is inadequate to meet the food needs of all humans (Pimentel et al., 1999). To compensate, about 384 quads of fossil energy (oil, gas and coal) are utilized each year worldwide (DOE/EIA, 1996; *British Petroleum Statistical Review of World Energy*, 1999). Of this amount, 91 quads are utilized in the United States (about

17% in the food system; USBC, 1998). Yearly, the U.S. population consumes about 53% more fossil energy than all the solar energy captured by harvested U.S. crops, forest products and all other vegetation.

The current high rate of energy expenditure throughout the world is directly related to many factors, including rapid population growth, urbanization and high resource-consumption rates. Indeed, fossil-energy use has been growing at a rate even faster than the rate of growth of the world population. From 1970 to 1995, energy use has been doubling every 30 years, whereas the world population has been doubled every 40 years (PRB, 2000; DOE/EIA, 1996). Future energy use is projected to double every 32 years, while the population is projected to double in about 50 years (PRB, 2000; DOE/EIA, 1996).

Some developing nations with high population growth rates are increasing fossil-fuel use in their agricultural production to meet increasing demand for food and fiber. For instance, in China between 1955 and 1992, fossil-energy use in agriculture for irrigation and for producing fertilizers and pesticides increased100-fold (Wen and Pimentel, 1992).

The overall projections of the availability of fossil-energy resources for mechanization, fertilizers and pesticides are discouraging because the availability of fossil fuels is finite. The world supply of oil is projected to last 40 to 50 years (Campbell, 1997; Youngquist, 1997; Ivanhoe, 2000; Duncan, 2001). The natural gas supply is adequate for about 50 years and coal for about 100 years (*British Petroleum Statistical Review of World Energy*, 1999; Youngquist, 1997; Bartlett and Ristinen, 1995). These estimates are based on current consumption rates and current population numbers.

Youngquist (1997) reports that current oil- and gas-exploration drilling data has not borne out some of the earlier optimistic estimates of the amount of these resources yet to be found in the United States. Both the production rate and proven reserves continue to decline. Oil and gas are imported in ever-increasing amounts each year (*British Petroleum Statistical Review of World Energy*, 1999; Youngquist, 1997; DOE, 1991), indicating that neither is now sufficient for U.S. domestic needs and supplies. Analyses suggest that, as of 1998, the United States had already consumed about three quarters of its recoverable oil and that the last 25% is now being used (Ivanhoe, 2000).

To help alleviate the diminishing fossil-energy supplies, available renewable energy technologies, such as biomass and wind power, could provide an estimated 200 quads of renewable energy worldwide (Pimentel et al., 1999; Yao, 1998). Note that 200 quads is only about half of the energy currently consumed. However, producing 200 quads of renewable energy would require transferring some agricultural land, like pastures, to energy production.

METHODOLOGY

The energy expenditures and economic costs of major food crop production systems in both developed and developing countries are analyzed, including some systems dependent on hand labor and draft-animal power. For the developed-country data, information on food crop production in the United States was used because abundant data were available and they are similar to intensive crop production systems in other developed nations. For example, in the U.S., the average energy input for wheat production was about 17.8 GJ, in Germany, the average was reported to be 17.5 GJ and in Greece, the input was 21.1 GJ (Tsatsarelis, 1993; Kuesters and Lammel, 1999). Accounting procedures used in the U.S., Germany and Greece differed somewhat because of the availability of data. In addition, a wide range of technology is used in wheat production in all countries, ranging from low-input organic to high-input irrigated production. The data detailed for the U.S. system are presented.

In developed countries, most of the energy inputs are fossil energy inputs for mechanization and fertilizers, whereas in developing countries, the major energy expenditure is for human labor. For instance, in U.S. grain production, the labor input was approximately 10 hrs/ha, while in many developing countries the labor input was approximately 1,000 hrs/ha. Labor is a vital component of crop production that is substituted for mechanization and other farming activities. More than nine different procedures are used for measuring the cost of labor input in terms of energy (Giampietro and Pimentel, 1990; Fluck, 1992). In this study, 2,000 hrs of labor input per year per person is assumed or 8 hours per day for 250 days. This is an average figure for the U.S., but varies throughout the world (USBC, 1998). The energy input for labor was based on the number of hours of labor per hectare and the average consumption of fossil energy (about 8,100 liters of oil equivalents) per person per year in the U.S. (*British Petroleum Statistical Review of World Energy*, 1999). The fossil energy consumption per person in each country varies. In contrast, in India, the average is only 280 liters per person per year (*British Petroleum Statistical Review of World Energy*, 1999). Large labor inputs in crop production are less costly in India than in the United States.

As with labor, assigning an energy value to manure is difficult. Properly applied manure can be substituted for commercial nitrogen, phosphorus and potassium fertilizers produced using high inputs of fossil energy. But, because different types of manure are used, are handled differently and are applied in various ways, the values obtained by investigators are highly variable. For example, the nitrogen content of manure varies from 3–20% (dry weight) depending on the type of livestock manure used and how it was handled.

Energy inputs for farm machinery, ranging from a hoe to a tractor, are difficult to assess. In the U.S., for example, farm machinery assets per crop hectare total about $538 and last about 10 years, with yearly repairs estimated to add about 25% per year (USDA, 2000). Knowing the weight of the farm machinery used per hectare per year, Doering (1980) provided detailed data on the energy input required for U.S. production. In this analysis, values were based on the data in the published literature (Doering, 1980).

In their relative importance in agriculture, fossil fuels differ with liquid fuels used more extensively than natural gas and coal. However, no attempt was made to rate and identify the amount of liquid fuel (oil) used in each cropping system. For nine of the crops in both developed and developing countries, a detailed accounting of the inputs are listed, and for 11 additional crops, a summary is given of the energy and economic costs.

For economic accounting, data from each particular country were used. The economies of all developed and developing countries differ significantly from one another, and these differences should be considered when examining the reported economic data.

ENERGY INPUTS AND ECONOMIC COSTS
FOR MAJOR CROPS

The crop systems selected for this analysis were rice, corn, wheat, soybeans, cassava, potato, sweet potato and cabbage and they provide most of the world's food supply. In addition, apples, oranges and tomatoes were included as examples of crops that provide limited calories but excellent minerals and vitamins.

CORN

Corn is one of the world's major grain crops (FAO, 1997). Under favorable environmental conditions, it is one of the most productive crops per unit area of land. An analysis of energy inputs and yields suggests that the high yields of intensive corn production are in part related to the large inputs of fertilizers, irrigation and pesticides.

Nevertheless, by investing many hours of labor, a farmer in a developing country can produce 1,200 kg/ha of corn (Table 8.1). For example, corn production by hand in Indonesia requires about 634 hours of labor and 5 hours of bullock power per hectare, making an energy expenditure of 17.0 GJ. With a corn yield of 1,200 kg/ha in Indonesia (18.1 GJ), the energy input:output ratio is 1: 1.07 (Table 8.1). Note that the energy input is slightly higher than it might be if the energy for the bullock power were withdrawn. The bullocks mostly consume forage so little or no fossil energy is expended for them.

TABLE 8.1
Energy Inputs and Costs of Corn Production per Hectare in Indonesia

Inputs	Quantity	MJ	Costs
Labor	634 hrs[a]	5,389[g]	$37.00[a]
Bullock (pair)	5 hrs[a]	46[b]	5.00[c]
Machinery	10 kg[c]	714[d]	1.00[c]
Nitrogen	71 kg[f]	5,544[e]	8.70[a]
Phosphorus	36 kg[f]	622[c]	2.00[a]
Manure	580 kg[a]	4,040[b]	5.00[a]
Pesticides	0.4L	168[d]	0.70[a]
Seeds	33.6 kg[f]	508 [d]	4.60[c]
Total		17,031	$ 64.00
Corn yield = 1,200 kg[a]	18,144[d]	kcal input: output = 1:1.07	

[a] (Djauhari et al. 1988)
[b] (Tripathi and Sah, 2001)
[c] Estimated.
[d] (Pimentel, 1980)
[e] (FAO, 1999)
[f] (Doughty, 2000)
[g] Per capita use of fossil energy in Indonesia is about 405 liters of oil equivalents per year (British Petroleum, 1999).

TABLE 8.2
Energy Inputs and Costs of Corn Production per Hectare in the U.S.

Inputs	Quantity	kcal x 1000	Costs
Labor	11.4 hrs[q]	462[f]	$114.00[h]
Machinery	55 kg[a]	1,018[e]	103.21[m]
Diesel	42.2 L[b]	481[e]	8.87[i]
Gasoline	32.4 L[b]	328[e]	9.40[j]
Nitrogen	144.6 kg[c]	2,688[g]	89.65[i]
Phosphorus	62.8 kg[c]	260[g]	34.54[i]
Potassium	54.9 kg[c]	179[g]	17.02[i]
Lime	699 kg[c]	220[e]	139.80[n]
Seeds	21 kg[a]	520[e]	74.81[l]
Irrigation	33.7 cm[s]	320[e]	123.00[t]
Herbicides	3.2 kg[r]	320[e]	64.00[j]
Insecticides	0.92 kg[r]	92[e]	18.40[j]
Electricity	13.2 kWh[b]	34[e]	2.38[k]
Transportation	151 kg[d]	125[e]	45.30[o]
TOTAL		7,047	$844.38
7,965 kg yield[p]		28,674	

kcal input: output = 1:4.07

[a] (Pimentel and Pimentel, 1996)
[b] (USDA, 1991)
[c] (USDA, 1997)
[d] Goods transported include machinery, fuels and seeds that were shipped an estimated 1,000 km.
[e] (Pimentel, 1980)
[f] It is assumed that a person works 2,000 hrs per year and utilizes an average of 8,100 liters of oil equivalents per year.
[g] (FAO, 1999)
[h] It is assumed that farm labor is paid $10 per hour.
[i] (Hinman et al., 1992)
[j] It is assumed that herbicide and insecticide prices are $20 per kg.
[k] Price of electricity is 7¢ per kwh (USBC, 1998)
[l] (USDA, 1998)
[m] (Hoffman et al., 1994)
[n] (USDA, 1999)
[o] Transport was estimated to cost 30¢ per kg.
[p] (USDA, 1998)
[q] (Nat'l Agric. Statistics Service, 1999)
[r] (Nat'l Agric. Statistics Service, 1997)
[s] (McGuckin et al., 1992)

The energetics of intensive U.S. corn production are distinctly different from those of labor-intensive corn production of Indonesia. The total input of human labor is only 11.4 hrs per hectare compared with 634 hrs in the labor-intensive system of Indonesia (Tables 8.1 and 8.2).

The fossil energy inputs in U.S. corn production, primarily oil for machinery and natural gas for nitrogen fertilizers, are high. Nitrogen fertilizer represents the largest single input, about 40% of the total fossil energy inputs, while 25% is expended for labor-reducing mechanization (Table 8.2). The total fossil fuel input is estimated to be 29.6 GJ/ha (Table 8.2). The corn yield is also high, about 8,000 kg/ha, or the equivalent of 120.4 GJ/ha of food energy, resulting in an input:output ratio of 1:4.07.

While corn yields are higher in the intensive system than the labor-intensive system, the economic investment is also high or $844/ha compared with $62.50/ha for the labor-intensive system (Tables 8.1 and 8.2).

WHEAT

Wheat and rice are the two most important cereal crops grown in the world today; more wheat is eaten by humans than any other cereal grain. Wheat is produced employing diverse techniques, with energy sources ranging from human labor and animal power to mechanization. As with corn production, energy inputs and yields vary with each wheat production system.

For example, wheat farmers in Kenya use human and bullock power (Table 8.3). Total energy input in this system is about 7.7 GJ, which provides a harvest of about 25.4 GJ in wheat (Table 8.3), for an energy input:output ratio of about 1:3.29. Similar to corn production using bullocks, this energy ratio would be higher if the energy for the bullocks were removed from the assessment.

TABLE 8.3
Energy Inputs and Costs of Wheat Production per Hectare in Kenya

Inputs	Quantity	MJ	Costs
Labor	684 hrs[b,e]	710[d]	$15.39[e]
Machinery	10 kg[g]	672[c]	56.19[e]
Diesel	35 L[g]	1,617[c]	7.35[e]
Nitrogen	22 kg[f]	1,327[a]	12.51[b]
Phosphorus	58 kg[b]	647[a]	32.99[b]
Seeds	202 kg[b]	2,545[c]	61.08[b]
Transportation	200 kg[b]	214[c]	15.84[b]
TOTAL		7,732	$201.38
Wheat yield = 1,788 kg[e]	25,414	kcal input: output = 1:3.29	

[a] (Surendra et al., 1989)

[b] (Hassan et al., 1993)

[c] (Pimentel, 1980)

[d] Per capita use of fossil energy in Kenya is estimated to be 522 liters of oil equivalents per year based on African data (British Petroleum, 1999)

[e] (Longmire and Lugogo, 1989)

[f] (Arama, 1994)

[g] Estimated

TABLE 8.4
Energy Inputs and Costs of Winter Wheat Production per Hectare in the U.S.

Inputs	Quantity	MJ	Costs
Labor	7.8 hr	1,327[d]	$78.00[a]
Machinery	50 kg[j]	3,360[e]	182.00[b]
Diesel	49.5 L[k]	2,373[e]	10.40[b]
Gasoline	34.8 L[k]	1,478[e]	9.98[b]
Nitrogen	68.4 kg[c]	5,342[f]	41.93[b]
Phosphorus	33.7 kg[c]	588[f]	18.53[b]
Potassium	2.1 kg[c]	29[f]	0.65[b]
Seeds	60 kg[a]	916[e]	16.77[b]
Herbicides	4 kg[a]	1,680[e]	11.83[a]
Insecticides	0.05 kg[c]	21[e]	0.80[g]
Fungicides	0.004 kg[c]	2[e]	0.20[g]
Electricity	14.3 kwh[e]	172[e]	1.00[h]
Transportation	197.9 kg[i]	517[e]	59.37[i]
TOTAL		17,805	$431.46
Winter wheat yield	2,670 kg[l]		37,947[e]

kcal input: output = 1:2.13

[a] (Willet and Gary, 1997)

[b] (Hinman et al., 1992)

[c] (USDA, 1997)

[d] It is assumed that a person works 2,000 hrs per year and utilizes an average of 8,100 liters of oil equivalents per year.

[e] (Pimentel, 1980)

[f] (FAO, 1999)

[g] It is assumed that insecticides and fungicides cost an average of $40 per kg, or similar to herbicides.

[h] Price of electricity is 7¢ per kwh (USBC, 1998)

[i] The goods transported include machinery, fuels and seeds and it is assumed that they were transported an average distance of 1,000 km that cost about 30¢ per kg. For energy inputs see (Pimentel, 1980)

[j] Estimated.

[k] (Pimentel and Pimentel, 1996)

[l] (USDA, 1998)

As shown in Tables 8.3 and 8.4, wheat production in the United States requires 17.8 GJ of fossil energy inputs compared with 7.7 GJ for the low-input Kenyan production system. Large machinery powered by fossil fuels replaces the animal power and dramatically reduces the labor input from 684 hrs for Kenya to only 7.8 hrs for the U.S. system. The heavy use of fertilizers and other inputs increases wheat yields from approximately 1,788 kg/ha to 2,670 kg/ha (Table 8.4). Yet, the input:output ratio is lower for the U.S. system than that of Kenya, or approximately 1:2.13.

RICE

Rice is the staple food for an estimated 3 billion people, most of whom live primarily in developing countries. This heavy consumption makes an analysis of various rice production technologies particularly relevant.

The rice production system practiced by Indian farmers using human labor and bullocks requires 1,703 hrs of human labor and 328 hrs of bullock labor per hectare, which total about 1.5 GJ. The total rice yield, 1,831 kg/ha (34.8 GJ), results in an energy input:output ratio of about 1:0.80 (Table 8.5).

As in the production of other grains, the United States uses large inputs of fossil energy to produce rice. Although most of the energy expended is used for machinery and fuel to replace labor, fertilizers account for about half of the total fossil energy input. The human labor input of only 24 hrs/ha is much lower than in India, but is 6.720 kg/ha (102.4 GJ of food energy). The fossil energy investment is about 49.7 GJ, resulting in an energy input:output ratio of 1:2.06 (Table 8.6).

SOYBEANS

Because of its high protein content (about 34%), the soybean is probably the single most important protein crop in the world. Two-thirds of all soybeans produced are

---

TABLE 8.5
Energy Inputs and Costs of Draft-Animal-Produced Rice per Hectare in the Valley of Garhwal Himalaya, India

Inputs	Quantity	MJ	Costs
Labor	1,703 hrs[a]	9,996[c]	$129.86[a]
Bullocks	328 hrs[a]	1,499[a]	40.00[a]
Machinery	2.5 kg[b]	172[f]	11.00[b]
Nitrogen	12.3 kg[a]	962[d]	1.30[e]
Phosphorus	2.5 kg[a]	42[d]	0.30[e]
Manure	3,056 kg[a]	21,298[a]	14.91[a]
Seeds	44 kg[a]	672[a]	6.44[a]
Pesticides	0.3 kg[a]	126[d]	1.33[a]
TOTAL		34,767	$194.14
Rice yield = 1,831 kg[a]		27,917[b]	
		kcal input: output = 1:0.80	

[a] (Tripathi and Sah, 2001)

[b] Estimated.

[c] Per capita fossil energy use in the India is 280 liters of oil equivalents per year (British Petroleum,1999)

[d] (FAO, 1999)

[e] The total for fertilizers reported in (Tripathi and Sah, 2001) was $1.60, we allocated $1.30 for nitrogen.

[f] (Pimentel, 1980)

TABLE 8.6
Energy Inputs and Costs of Rice Production per Hectare in the U.S.

Inputs	Quantity	MJ	Costs
Labor	24 hrs[a]	4,082[c]	$240.00[f]
Machinery	38 kg[a]	3,116[d]	150.00[g]
Diesel	225 L[a]	10,807[d]	47.25[b]
Gasoline	55 L[a]	2,344[d]	15.95[b]
Nitrogen	150 kg[b]	11,714[e]	93.00[h]
Phosphorus	49 kg[b]	853[d]	26.95[h]
Potassium	56 kg[b]	769[e]	17.36[b]
Sulfur	20 kg[b]	126[p]	1.00[p]
Seeds	180 kg[a]	3,032[d]	90.00[i]
Herbicides	7 kg[b]	2,940[d]	280.00[j]
Insecticides	0.1 kg[b]	42[d]	4.00[k]
Fungicides	0.16 kg[b]	67[d]	6.40[k]
Irrigation	250 cm[a]	8,984[a]	294.00[l]
Electricity	33 kwh[a]	357[a]	2.31[m]
Transportation	451 kg[a]	487[a]	135.30[n]
TOTAL		49,720	$1,403.52
Rice yield = 6,720 kg[o]		102,451	
		kcal input: output = 1:2.06	

[a] (Pimentel and Pimentel, 1996)
[b] (USDA, 1997)
[c] It is assumed that a person works 2,000 hrs per year and utilizes an average of 8,100 liters of oil equivalents per year.
[d] (Pimentel, 1980)
[e] (FAO, 1999)
[f] We assume that a farm laborer is awarded $10 per hour.
[g] Estimated.
[h] (Hinman et al., 1992)
[i] Seeds were estimated to cost 50¢ per kg.
[j] (Hinman and Schiriman, 1997)
[k] Insecticides and fungicides were estimated to cost $40 per kg.
[l] 1 cm of irrigation water applied was estimated to cost $1.18.
[m] Price of electricity is 7¢ per kwh (USBC, 1998)
[n] Transportation was estimated to be 30¢ per kg transported 1,000 km.
[o] (USBC, 1998)
[p] Based on the estimate that sulfur costs 5¢ per kg (Myer, 1977), it was calculated that the fossil energy input to produce a kg was 1,500 kcal.

grown in the United States, China and Brazil. In the United States, relatively little of the soybean crop is used as human food, but is instead processed for its oil, while the seed cake and soybean meal are fed to livestock. Soybeans and soy products head the list of U.S. agricultural exports (USDA, 1998).

TABLE 8.7
Energy Inputs and Costs of Soybean Production per Hectare in Illinois

Inputs	Quantity	MJ	Costs
Labor	7.1 hrs	1,210[d]	$71.00[g]
Machinery	20 kg	1,512[e]	148.00 [l]
Diesel	38.8 L[a]	1,856[e]	8.15[h]
Gasoline	25.7 L[a]	1,092[e]	7.45[h]
LP gas	3.3 L[a]	105[e]	0.66[h]
Nitrogen	3.7 kg[c]	290[f]	2.29[h]
Phosphorus	37.8 kg[c]	655[f]	38.35[h]
Potassium	14.8 kg[c]	202[f]	4.59[h]
Seeds	69.3 kg[a]	2,327[e]	48.58[m]
Herbicides	1.3 kg[c]	546[e]	26.00[i]
Electricity	10 kwh[b]	122[e]	0.70[j]
Transportation	154 kg[k]	168[e]	46.20[n]
TOTAL		10,085	$401.97
Potato yield = 3,000 kg[m]		50,778	

kcal output/kcal input = 5.04

[a] (Ali and McBride, 1999)
[b] (Pimentel and Pimentel, 1996)
[c] (Economic Research Statistics, 1997)
[d] It is assumed that a person works 2,000 hrs per year and utilizes an average of 8,100 liters of oil equivalents per year
[e] (Pimentel, 1980)
[f] (FAO, 1999)
[g] It is assumed that farm labor earns $10 per hour
[h] (Hinman et al. (1992)
[l] It is assumed that the price of herbicides is $20 per kg
[j] Price of electricity is 7¢ per kwh(USBC (1998)
[k] The goods transported include machinery, fuels and seeds
[l] (College of Agric., Consumer & Environ. Sciences (1997)
[m] (United Soybean Board, 1999)
[n] Transport of goods was assumed to cost 30¢ per kg

In Illinois, typical of soybean cultivation, soybeans yield an average 3,000 kg/ha and provide about 50.8 GJ (Table 8.7). Production inputs, mainly for machinery, total 10.1 GJ/ha, an input:output ratio of 1:5.04.

Like other legumes, soybeans need less nitrogen than other crops because, under most conditions, they biologically fix their own nitrogen. The biological nitrogen fixation process carried out by soil microbes uses about 5% of the sunlight energy captured by the soybean plants, but saves the energy that otherwise would be required for nitrogen fertilizer production.

POTATO

The white potato is one of the 15 most heavily consumed plant foods in the world today. Even in the United States, where a wide variety of vegetables is available, more potatoes are eaten than any other vegetable, about 22 kg per person per year (USDA, 1998). Potatoes contain some protein (1.5 to 2.5%), are high in vitamin C and potassium and are a substantial source of carbohydrates.

In an intensive potato production system, production per hectare is several times greater than that of other carbohydrate producing crops. More importantly, protein production per hectare is two to three times greater than most other crops.

Based on U.S. data, the largest energy inputs are for machinery and fuel; the second largest input is for fertilizers (Table 8.8). The total energy inputs are about 71.8 GJ/ha with a yield of about 38,820 kg/ha (93.4 GJ of food energy) (Table 8.9). The resulting input:output ratio is 1:1.30. Note that the high water content of potatoes (80%) makes transport relatively energy-costly compared with grain crops.

CASSAVA

Cassava is a major food crop worldwide, especially in Africa, Asia and Latin America, and can be grown in soils of low fertility. One of the highest-producing crops in terms of carbohydrate per hectare, it is one of the lowest in terms of protein per hectare.

The data for cassava production are from Thailand, Colombia, Nigeria and Vietnam. The labor input in the cassava system is relatively high or 1,632 hrs/ha and the average yield is 12,360 kg/ha, (196.5 GJ/ha). With energy input of 54.6 GJ /ha, the resulting input:output ratio is 1:3.60 (Table 8.9).

SWEET POTATO

Along with the white potato and cassava, the sweet potato is another major food crop, especially in the tropics. In addition to carbohydrate, the sweet potato is high in vitamin A, iron and abundant carbohydrate.

The production of sweet potato in the Red River Delta, Vietnam, requires 1,678 hrs/ha of labor, plus relatively large inputs of fertilizers. The average yield is 11,867 kg/ha, providing 49.8 GJ/ha of food energy. The energy input in this system is 24.8 GJ/ha, resulting in an input:output ratio of 1:2.01 (Table 8.10).

COLE CROPS

Cole crops, such as cabbage, are grown worldwide and are excellent sources of nutrients, including vitamin A, vitamin C and iron. Typical of U.S. vegetable production, the major energy inputs are for machinery and fuel, with fertilizers being the second-largest input. The average yield is 38,416 kg/ha, providing 81.3 GJ/ha. The total energy input is 46.2 GJ/ha and the resulting input:output ratio is 1:1.76 (Table 8.10).

In contrast, cabbage production in the Garhwal Himalaya region of India requires 1,831 hrs/ha of labor and 294 hrs/ha of bullock power (Tripathi and Sah, 2001). The total energy input is 45.9 GJ/ha, similar to that for U.S. cabbage production. With a total yield of cabbage in India of 11,423 kg/ha (24.2 GJ), the resulting input:output is 1:0.53 (Table 8.10).

TABLE 8.8
Energy Inputs and Costs of Potato Production per Hectare in the U.S.

Inputs	Quantity	MJ	Costs
Labor	35 hrs[a]	6,720[d]	$350.00[g]
Machinery	31 kg[a]	2,411[e]	300.00[h]
Diesel	152 L[a]	7,287[e]	31.92[h]
Gasoline	272 L[a]	11,550[e]	78.88[h]
Nitrogen	231 kg[b]	18,035[f]	142.60[h]
Phosphorus	220 kg[b]	3,826[f]	121.00[h]
Potassium	111 kg[b]	1,520[f]	34.41[h]
Seeds	2,408 kg[e]	6,208[e]	687.00[h]
Sulfuric acid	64.8 kg[a]	0[i]	73.00[i]
Herbicides	1.5 kg[k]	630[e]	13.50[h]
Insecticides	3.6 kg[k]	1,512[e]	14.40[h]
Fungicides	4.5 kg[k]	1,890[e]	180.00[h]
Electricity	47 kwh[a]	567[e]	3.29[j]
Transportation	2,779 kg[c]	9,689[e]	833.70[l]
TOTAL		71,845	$2,810.90
Potato yield = 38,820 kg k		93,425	

kcal input: output = 1:1.30

[a] (Pimentel and Pimentel, 1996)
[b] (USDA, 1997)
[c] A sum of the quantity values for machinery, fuels and seeds (all converted to mass units)
[d] It is assumed that a person works 2,000 hrs per year and utilizes an average of 8,100 liters of oil equivalents per year.
[e] (Pimentel, 1980)
[f] (FAO, 1999)
[g] Farm labor costs were estimated to be $10 per hour
[h] (Hinman et al., 1992)
[i] Sulfuric acid production is an exothermic process. The cost of sulfuric acid was $73.00/ha. (cking@micron.net)
[j] Price of electricity is 7¢ per kwh (USBC, 1998)
[k] (Pimentel et al., 1993)
[l] 30¢/kg of goods transported (USDA, 1998)

TOMATO

Tomatoes are valued for their vitamin C (23 mg per 100 g of fresh tomato), vitamin A and iron. In the U.S., labor input for tomato production is relatively high, or about 363 hrs/ha. The fossil energy inputs are 136.0 GJ, primarily expended for machinery, fuel and fertilizers. The tomato yield of 55,000 kg/ha provides 46.3 GJ of food energy, with the resulting input:output ratio of 1:0.34 (Table 8.10).

Based on data from Pakistan, the major input for tomato production is labor (2,337 hrs/ha) (Haq et al., 1997). The tomato yield is about 14,767 kg/ha, providing nearly 12.4 GJ of food energy and a resulting input:output ratio was 1:0.94, that is more than double that in the U.S. (Table 8.10).

TABLE 8.9
Energy Inputs and Costs of Cassava Production per Hectare in Thailand, Colombia, Vietnam and Nigeria

Inputs	Quantity	MJ	Costs
Labor	1,632 hrs[a]	22,621[c]	$93.42[a]
Draft animal (buffalo)	200 hrs[b]	2,079[j]	9.64[e]
Machinery	5 kg[b]	391[d]	3.83[a]
Nitrogen	46 kg[a]	3,591[f]	28.52[g]
Phosphorus	33 kg[a]	567[f]	18.15[g]
Potassium	43 kg[a]	588[f]	13.33[g]
Manure, organic	3,400 kg[a]	23,684[j]	10.00[b]
Cassava sticks	6,000 sticks (120 bundles)[k]	1,126[i]	40.00[h]
TOTAL		54,647	$216.89
Yield 12,360 kg a		196,510	
		kcal input:output =1:3.60	

[a] (CIAT, 1996)
[b] Estimated
[c] It is estimated that each person uses about 600 liters of oil equivalents per year. This is based on the average per capita use of fossil energy in Central and South America (British Petroleum, 1999)
[d] (Pimentel, 1980)
[e] (CIAT, 1996)
[f] (FAO, 1999)
[g] (Hinman et al., 1992)
[h] (Ezeh, 1988)
[I] Estimates are that it takes about 8 days to collect cassava sticks for planting. Energy input was calculated based on information in (CIAT, 1996)
[s] (Tripathi and Sah, 2001)

ORANGES

Oranges are a valuable fruit in U.S. agriculture, costing about $3,000 per hectare for production (Table 8.10). Although, per hectare, oranges and other citrus fruits provide more than double the vitamin C content of white potatoes, U.S. citizens obtain half of their vitamin C from white potatoes and half from citrus (USDA, 2000). The production of oranges requires the expenditure of 96.3 GJ/ha of fossil energy. Based on the orange yield of 46,065 kg/ha the food energy is 98.8 GJ, resulting in an input:output ratio of 1.03.

APPLES

Apples are another economically valuable crop in the U.S., costing about $7,725 per hectare to produce. The energy input used in orchards is primarily for machinery, while pesticides contribute nearly 20% of the total energy input.

TABLE 8.10
Energy and Economic Costs of Various Crops Produced in Several Developing and Developed Countries (per Hectare)(Pimentel et al., 2001)

Crop	Country	Yield kg	Energy Harvest MJ	Labor hrs	Labor Input MJ	Energy Input MJ	Economic Costs	kcal Input: Output
Soybean	Philippines	988	16,724	744	5,498	11,315	$310.58	1:1.47
Potato	Philippines	5,500	13,238	1,400	10,349	31,844	$655.60	1:0.42
Sweet Potato	Vietnam	11,867	49,841	1,678	12,403	24,776	$908.73	1:2.01
Cabbage	U.S.	38,416	81,320	60	11,227	46,230	$1,341.08	1:1.76
Cabbage	India	11,423	24,184	1,834	10,781	45,913	$206.95	1:0.53
Tomato	U.S.	55,000	46,301	363	61,236	136,034	$7,337.42	1:0.34
Tomato	Pakistan	14,767	12,403	2,337	8,585	13,184	$1,746.73	1:0.94
Orange	U.S.	46,056	98,780	210	39,287	96,269	$3,048.55	1:1.03
Apple	U.S.	54,743	128,755	385	72,030	210,817	$7,724.53	1:0.61
Apple	India	6,000	14,112	610	3,944	9,110	$81.29	1:1.55
Corn, irrig.	U.S.	7,965	120,431	10	1,869	112,736	$1,674.88	1:1.07

Also, the labor input of 385 hrs/ha in apple production, especially during harvest, is high compared with most other food crops grown in the U.S.. The total labor input is about 72.0 GJ/ha of the total of 210.8 GJ of energy expended. Based on the total apple yield of 54,743 kg/ha, this provides 128.8 GJ of food energy, with an input:output ratio of 1:0.61.

Apple production in the high-hills of the Garhwal Himalaya region of India requires 610 hours of labor, nearly twice that of the U.S. (Tripathi and Sah, 2001). Although the apple yield in India is only 6,000 kg/ha (14.1 GJ/ha), this is a much more favorable input:output ratio or 1:1.57 (Table 8.10). The reason is fewer fossil energy inputs.

IRRIGATED CROPS

Producing food crops employing irrigation requires enormous amounts of water plus the expenditure of fossil energy to pump and apply the water (Postel, 1999). For example, a corn crop grown in an arid region requires about 1,000 mm of irrigated water per hectare (Falkenmark and Lindh, 1993). To pump the water from a depth of only 30.5 m (100 feet) and apply it requires about 112.8 GJ of fossil energy per hectare (Table 8.10).

The total energy inputs for irrigated corn, which is planted on half of U.S. irrigated land, is 29.6 GJ for rain-fed corn compared with 112.8 GJ for irrigated corn, or three times the energy needed for rain-fed corn.

In addition to increased energy for irrigation, overall economic costs of production also rise in an irrigated production system because of the high costs of pumping water (Tables 8.2 and 8.10).

ECONOMICS OF FOOD CROP PRODUCTION

The price value at the farm gate of the 10 crops in developing countries and nine crops assessed in developed countries averages about $.12 per kg. Oranges are not included in the developing-country calculation and sweet potato and cassava are not included in the developed-country calculation.

Corn is produced more cheaply in Indonesia ($.05/kg) than in the U.S. ($.11/kg) and rice is produced more cheaply in India ($.11/kg) than in the U.S. ($.21) (Pimentel et al., 2001). Wheat production costs are $.11/kg in India and $.16/kg in the U.S. (Tables 8.3 and 8.4).

Soybeans and potatoes cost more to produce in the Philippines than in the U.S. (Pimentel et al., 2001). Also, tomatoes are more costly to produce in Pakistan than in the U.S. However, apple production is far more expensive in the U.S. than in India (Pimentel 2001) because of large inputs of labor and other inputs in the U.S. apple system.

Compared with developed nations, farm wages are extremely low in developing countries, ranging from $.06 to $.50 per hour. Yet labor is the primary cost for food production in developing countries because of the great number of hours invested, ranging from 600 to 1,800 hours per hectare in production. The primary costs in U.S. food crop production are for mechanization, fertilizers and pesticides. The cost of irrigation is two to three times the cost of all the other inputs in U.S. food crop production (Pimentel and Pimentel, 1996).

No data were presented concerning the relative incomes and purchasing power of people in each nation and this significantly changes the perspectives in each.

CHANGES IN WORLD FOOD CROP PRODUCTION

FOSSIL ENERGY USE AND CROP YIELDS

Since about 1950, when fossil energy became readily available, especially in developed nations, it supported the 20- to 50-fold increase in the use of fertilizers, pesticides and irrigation. From 1950 to 1980, U.S. grain production per hectare increased three to four times (USDA, 1980). For example, where fertilizer use on corn increased from about 5 kg/ha in 1945 to about 150 kg/ha (30 times), corn yields increased by about four times (Pimentel and Pimentel, 1996). The rate of yield increases during the 30-year period from 1950 to 1980 was about 3% per year. However, since 1980, U.S. grain crop yield increases have declined to only about 1% per year (USDA, 1980). This is because crops have limits to the amounts of fertilizers and pesticides that they can tolerate and use. In fact, nitrogen fertilizer application rates of approximately 500 kg/ha or more are toxic and cause crop yields to decline (Martinez and Guiraud, 1990).

The significant achievement of using fossil energy to increase crop yields, the cereal grains in particular, started with the advent of the Green Revolution (Conway, 1997). During the 1950s, plant breeders developed wheat, rice, corn and other cereal crops to have short statures so that large quantities of fertilizers, especially nitrogen, could be applied in production. The short stature was essential to prevent the plants from growing and then falling over (lodging), which formerly resulted in crop loss.

The availability and use of fossil fuels were instrumental in the success of the Green Revolution. As a result, crop yields per hectare were significantly increased for the newly developed grains. Yet, in 75 countries, less grain was produced by 1990 than at the beginning of the decade (Dasgupta, 1998).

At best, world grain yields per hectare are slowly increasing, at the most about 1% per year, while human population numbers and their food needs are increasing at a greater rate than food production can supply (Pimentel et al., 1999). As the world population increases, it outstrips increases in food production. Thus, it is becoming more apparent that the food supply cannot keep up with the needs of a rapidly growing human population.

On a per capita basis, world grain production has declined since 1984. Grains make up about 80% of the world food. Shortages of the basic resources for a productive crop system now currently exist. These worldwide losses in fertile cropland, loss of freshwater and diminishing fossil energy supplies used in mechanization, fertilizers, pesticides and irrigation are having negative impacts on crop production.

Per capita use of fertilizers worldwide during the past decade declined 17% (Worldwatch, 2001), while available cropland resources per capita decreased more than 20% (Pimentel et al., 1999). A total of 560 million ha of the 1,500 million ha of cropland worldwide has been seriously degraded because of soil erosion (Greenland et al., 1998).

Irrigated land area in developing countries declined about 10% over the past decade (Postel, 1999). A total of 20% of the irrigated croplands worldwide suffer from salinization, a result of poor irrigation and drainage practices (Greenland et al., 1998).

FOSSIL ENERGY USE IN CROP PRODUCTION

Of the total fossil energy consumed in the world, or about 384 quads, approximately 270 quads are used in developed countries and 114 quads in developing countries (*British Petroleum Statistical Review of World Energy*, 1999). The population in developed countries is less than 2 billion, while more than 4 billion live in developing countries (PRB, 2000).

Developed countries use approximately 40 quads of fossil energy, but only about 16 quads of this are used directly for both crop and livestock production (Pimentel and Pimentel, 1996). The remaining 24 quads are used for food processing, packaging, distribution and preparation.

In contrast, in developing countries, approximately 28 quads are consumed in agricultural production. Little fossil energy is used in cooking because biomass energy (fuel wood, crop residues and dung) is the prime fuel (Pimentel and Pimentel, 1996). From 2 to 3 kcal of biomass energy are used to prepare 1 kcal of food in developing countries (Pimentel and Pimentel, 1996; Tripathi and Sah, 20001). Therefore, total energy in the food system in developed and developing countries is about 68 quads per year.

Crop production in both developed and developing countries requires from 7.7 to 210.8 GJ/ha. In developed countries, the fossil energy inputs for machinery to reduce the labor input are high, whereas, in the developing countries, the fossil energy inputs for labor are high. Fossil energy inputs for labor are listed in terms of per capita fossil-fuel consumption. Most of the fossil energy used in world food production is oil for farm machinery and pesticides, while natural gas is vital for the production of nitrogen fertilizers.

The total energy expended in the food system of developed countries is approximately 5 J to supply 1 J of food, while, in developing countries, the ratio is approximately 4 J invested to supply 1 J of food. In developed countries, people consume an average of 3,400 kcal of food per person per day, whereas people in developing countries consume 2,400 kcal of food per day per person (FAO, 1999). This 1,000 fewer kcal consumed per person per day in developing countries reflects, in part, the lower total fossil energy inputs in their food system.

RENEWABLE ENERGY

The U.S. is currently consuming about 91 quads (24%) of world's 384 quads expenditure of fossil energy (*British Petroleum Statistical Review of World Energy*, 1999; USBC, 1998). Best estimates are that using a mix of renewable energy technologies about half (45 quads) of the current fossil energy consumption in the

U.S. could be produced employing an array of renewable energy technologies (Pimentel et al., 1994).

Liquid fuel needs for tractors and other farm machinery might be met using hydrogen or pyrolytic oil produced from wood (Pimentel et al., 2001; Pimentel et al., 1994). Nitrogen can be produced using electrical discharge to convert atmospheric nitrogen to nitrate. However, about 200,928 J of energy are required to produce a kilogram of nitrogen by this method, compared with 78,078 J required using fossil-energy-dependent technologies (Treharne and Jakeway, 1980; FAO, 1999). Based on current renewable energy technologies, a quantity of energy produced using renewable technologies costs from five to 10 times more than an equivalent amount obtained from fossil energy sources.

FUTURE TECHNOLOGIES

In the past decades, advances in science and technology have been instrumental in increasing industrial and agricultural production, improving transportation and communications, advancing human health care and, in general, improving many aspects of human life. However, much of this success is based on the availability of resources in the natural ecosystems of the earth.

Technology cannot produce an unlimited flow of the vital natural resources that are the raw material for sustained agricultural production. Genetic engineering holds promise, provided that its genetic transfer ability is wisely used. For example, the genetic modification of some crops, such as rice, to have high levels of iron and beta carotene, would improve the nutrition of millions of people in the future, particularly those in developing countries where rice is the prime grain consumed (Friedlander, 2000). In addition, the possibility exists for biological nitrogen fixation to be incorporated into crops such as corn and wheat. Hopefully, improved technologies, including the more effective management and use of resources, will help increase food production.

Yet there are limitations to what technology can accomplish. In no area is this more evident than in agricultural production. No known or future technology could, for example, double the quantity of the world's fertile cropland available for production. Granted, synthetically produced fertilizers are effective in enhancing the fertility of eroded croplands, but their production relies on sustained supplies of finite fossil fuels. Thus, in countries like the U.S. and China, farmers can be expected to experience rapidly diminishing returns with the further application of fertilizers.

To date, biotechnology that started more than 20 years ago has not stemmed the decline in per capita food production. Currently, more than 40% of the genetic engineering research effort is devoted to the development of herbicide resistance in crops (Paoletti and Pimentel, 1996). This herbicide-tolerant technology has not increased crop yields, but instead generally increased the use of chemical herbicides and polluted the environment. Indeed, this technology could eventually result in increasing labor and decreasing crop yields as weed species acquire additional herbicide resistance (Paoletti and Pimentel, 1996).

SUMMARY

Based on the information presented, if current trends in human population growth and fossil fuel consumption continue into the future, projections for the adequacy of tomorrow's world food supply are not encouraging. When the world population expands to nearly 8 billion as projected in about 15 years, food yields will have to increase by 33% (Greenland et al., 1998). The factors that govern our success in achieving this are dependent on our dedication to conservation and judicious use of our natural resources, increasing political and economic stability and, most vital, reduction in the world population (Pimentel et al., 1999). The basic equation of people versus food and energy intensifies the imbalances between the human food supply and the natural resource needs of a rapidly growing world population.

ACKNOWLEDGMENTS

We thank the following people for reading an earlier draft of this chapter and for their many helpful suggestions: Mario Giampietro, Instituto Nazionale di Ricerce per gli Alimente e la Nutrizione (INRAN), Rome, Italy; Rena Perez, Cuba; Marcia Pimentel, Cornell University, Ithaca, NY; Dazhong Wen, Academia Sinica, Shenyang, China; Walter Youngquist, Petroleum Geologist, Eugene, Oregon; and two anonymous reviewers of the manuscript.

REFERENCES

Ali, M. and W.D. McBride. 1999. *Soybeans: Cost of Production, 1990.* http://usda.mannlib.cornell.edu/data-sets/crops/p4009/

Arama, P.F. 1993. Breeding and selection of bread wheat to *Septoria tritici* blotch, *Current Plant Science and Biotechnology in Agriculture; Durability of Disease Resistance,* Kluwer Academic, Dordrecht: 18:191.

Bartlett, A.A. and R.A. Ristinen. 1995. Natural gas and transportation, *Physics and Society* 24: 9.

British Petroleum 1999. *British Petroleum Statistical Review of World Energy,* British Petroleum Corporate Communications Services, London.

Campbell, C.J. 1997. *The Coming Oil Crisis,* Multi-Science Publishing Company & Petroconsultants S.A, New York.

CIAT 1996. *Cassava Production, Processing and Marketing in Vietnam.* Proceedings of Workshop, Hanoi, Vietnam, October 29-31, 1992, Howler, R. H. (Ed.), Bangkok, Thailand. cking@micron.net, December 2, 1999.

College of Agricultural, Consumer and Environmental Sciences 1997. Machinery Cost Estimates: Summary of Operations, University of Illinois at Urbana-Champaign, http://web.aces.uinc.edu/fbfm/farmmagmt.htm.

Conway, G. 1997. *The Doubly Green Revolution, Food for All in the Twenty-First Century,* Penguin, London.

Dasgupta, P. 1998. The economics of food. In: *Feeding a World Population of More than Eight Billion People,* Waterlow, D.G., L.F. Armstrong and R. Riley (Eds.), Oxford University Press, New York: 19.

Djauhari, A., A. Djulin and I. Soejono. 1988. *Maize Production in Java: Prospects for Improved Farm-Level Production Technology*, CGPRT Centre, Indonesia.

DOE 1991. Annual Energy Outlook with Projections to 2010, U.S. Department of Energy, Washington, D.C., Energy Information Administration.

Doering, O.C. 1980. Accounting for energy in farm machinery and buildings. In: *Handbook of Energy Utilization in Agriculture*, Pimentel, D. (Ed.), CRC Press, Boca Raton, FL: 9.

Doughty, R.S. 2000. Unpublished thesis. Cornell University, Ithaca, NY.

Duncan, R.C. 2001. World energy production, population growth and the road to the Olduvai Gorge, *Population and Environment*, 22: 503.

Economic Research Statistics 1997. Soybeans: Fertilizer Use by State, 1996. http://usda.mannlib.cornell.edu/data-sets/inputs/9X17l/9717/agch0997.txt.

Ezeh, N. 1988. Comparative economic analysis of NAFPP and traditional cassava/maize production technologies in Rivers State of Nigeria, *Agricultural Systems*, 27: 225.

Falkenmark, M. and G. Lindh.1993. Water and economic development. In: *Water in Crisis: A Guide to the World's Fresh Water Resources*, Gleick, P. (Ed.), Oxford University Press, Oxford: 80.

FAO 1997. Quarterly Bulletin of Statistics, Food and Agriculture Organization, United Nations, Rome, Italy: 10.

FAO 1999. Food Balance Sheets, Food and Agriculture Organization. United Nations, Rome, Italy.

FAO 1999. Agricultural Statistics, http://apps.fao.org/cgi-bin/nph-db.pl?subset-agriculture Food and Agriculture Organization, United Nations.

Fluck, R.C. 1992. Energy of human labor. In: *Energy in World Agriculture*, Fluck, R. C. (Ed.), Elsevier, Amsterdam, Holland: 31.

Friedlander, B.P. 2000. Genetically engineered food could be lifeline for developing world. American Association for the Advancement of Science. http://www, news.cornell.edu/Chronicle/00/2.24.00/AAAS.McCouch.html.

Giampietro, M. and Pimentel, D. Assessment of the energetics of human labor, *Agriculture, Ecosystems & Environment*, 32: 257.

Greenland, D.J., P.J. Gregory and P.H. Nye. 1998. Land resources and constraints to crop production. In: *Feeding a World Population of More than Eight Billion People*, Waterlow, D.G. Armstrong, L.F. and Riley, R. (Eds.), Oxford University Press, New York: 39.

Haq, Z.U., S.H. Saddozal, Jahkanzeb and Z. Ullah. 1997. Economics of intercropping: a case study of tomato production in District Nowshera, *Sarhad J. Agric.*, 13, 1997.

Hassan, R.M., W. Mwangi and D. Karanja. 1993. Wheat Supply in Kenya: Production Technologies, Sources of Inefficiency and Potential for Productivity Growth, CIM-MYT Economics Working Papers, 93-02.

Hinman, H., G. Pelter, E. Kulp, E. Sorensen and W. Ford. 1992. Enterprise Budgets for Fall Potatoes, Winter Wheat, Dry Beans and Seed Peas under Rill Irrigation, Farm Business Management Reports, Columbia County, Washington State University, Pullman, WA.

Hinman, H. and R. Schiriman. 1997. Enterprise Budgets, Summer Fallow–Winter Wheat—Spring Barley Rotation, Columbia County, Washington State University, Pullman, WA.

Hoffman, T.R., W.D. Warnock and H.R. Hinman. 1994. Crop Enterprise Budgets, Timothy-Legume and Alfalfa Hay, Sudan Grass, Sweet Corn and Spring Wheat under Rill Irrigation, Kittitas County, Washington, Farm Business Reports EB 1173, Washington State University, Pullman, WA.

International Energy Annual, 1996, DOE/EIA-0220[96]. U.S. Department of Energy, Washington, D.C.

Ivanhoe, L.F. 2000. World Oil Supply – Production, Reserves and EOR, Hubbert Center Newsletter # 2000/1-1.

Kuesters, J. and J. Lammel. 1999. Investigations of the energy efficiency of the production of winter wheat and sugar beet in Europe, *Euro. J. Agronomy*, 11: 35.

Longmire, J. and J. Lugogo. 1989. The economics of small-scale wheat production technologies for Kenya, CIMMYT Working Papers, 89/01.

Martinez, J. and G. Guiraud. 1990. A lysimeter study of the effects of a ryegrass catch crop during a winter wheat — maize rotation on nitrogen leaching and on the following crop, *J. Soil Sci.*, 41: 5.

McGuckin, J.T., N. Gollehon and S. Ghosh. 1992. Water conservation in irrigated agriculture: a stochastic production frontier model, *Water Res. Res.*, 28: 305.

Myer, B. 1997. *Sulfur, Energy and Environment*, Elsevier Scientific, New York.

National Agricultural Statistics Service 1997. Agricultural Chemical Usage 1996 Field Crops Summary, U.S. Department of Agriculture, Economic Research Service, Washington, D.C.

National Agricultural Statistics Service 1999. Farm Labor, National Agricultural Statistics Service, Internet: http//usda.mannlib.cornell.edu., December 1999.

Paoletti, M. G. and D. Pimentel. 1996. Genetic engineering in agriculture and the environment, *BioScience*, 46: 665.

Pimentel, D. 1980. *Handbook of Energy Utilization in Agriculture*, CRC Press, Boca Raton, FL.

Pimentel, D., L. McLaughlin, A. Zepp, B. Kakitan, T. Kraus, P. Kleinman, F. Vancini, W.J. Roach, E. Graap, W.S. Keeton and G. Selig. 1993. Environmental and economic effects of reducing pesticide use in agriculture, *Agric. Ecosys. Environ*, 46: 273.

Pimentel, D., T. Rodrigues, R. Wang, K. Abrams, H. Goldberg, E. Staecker, L. Ma, L. Brueckner, C. Trovalo, U. Chow, U. Govindarajulu and S. Boerke. 1994. Renewable energy: economic and environmental issues, *BioScience*, 44: 536.

Pimentel, D. and M. Pimentel. 1996. *Food, Energy and Society*, Colorado University Press, Boulder, CO.

Pimentel, D., O. Bailey, P. Kim, E. Mullaney, J. Calabrese, F. Walman, F. Nelson, X. Yao. 1999. Will the limits of the Earth's resources control human populations? *Environment, Development and Sustainability*, 1: 19.

Pimentel, D., R. Doughty, C. Carothers, S. Lamberson, N. Bora and K. Lee. 2001. Energy inputs in crop production in developing and developed countries, Environmental Biology, Report 01-2.

Population Summit of the World's Scientific Academies, 1994. National Academy of Sciences Press, Washington, D.C.

Postel, S. 1999. *Pillar of Sand: Can the Irrigation Miracle last?* W.W. Norton, New York.

PRB 2000. World Population Data Sheet, Population Reference Bureau, Washington, D.C.

Surendra, S., P.S. Madhu, R.S. Rana, V.K. Mittal and B. Rupa. 1989. Energy and cost regulation for cultivation of rice (*Oryza sativa*) – wheat (*Tritichum aestivum*)- maize (*mays*) – wheat rotations, *Indian J. Agric. Sci.*, 59: 558.

Treharne, R.W. and L. Jakeway. 1980. Research and development of fertilizer production using renewable energy sources, *Am. Soc. Agric. Eng.*, PR 80-41: 3.

Tripathi, R.S. and V.K. Sah. 2001. Material and energy flows in high-hill, mid-hill and valley farming systems of Garhwal Himalaya, *Agriculture, Ecosystems and Environment*, 86: 75.

Tsatsarelis, C.A. 1993. Energy inputs and outputs for soft winter wheat production in Greece, *Agriculture, Ecosystems & Environment*, 43: 109.

United Soybean Board 1999. *Soybean Yield by State 1998.* http://www.unitedsoybean.org/99soystats/page_18.htm.

UNU/INRA 1999. World Food Day, 16 October 1999, The United Nations University, Institute of Natural Resources in Africa.

USBC 1998. Statistical Abstract of the U.S. 1996, 200th ed., U.S. Bureau of the Census, U.S. Government Printing Office, Washington, D.C.

USDA 1980. Agricultural Statistics 1981. U.S. Department of Agriculture, U.S. Government Printing Office, Washington, D.C.

USDA 1990. Agricultural Statistics, U.S. Department of Agriculture, Washington, D.C.

USDA 1991. Corn-State. Costs of Production, U.S. Department of Agriculture, Economic Research Service, Economics and Statistics System, Washington, D.C., Stock #94018.

USDA 1997. Agricultural Resources and Environmental Indicators, *Agricultural Handbook*, U.S. Department of Agriculture, Economic Research Service, Natural Resources and Environmental Division, Washington, D.C.: 712.

USDA 1997. National Agricultural Statistics Service, U.S. Department of Agriculture, Economic Research Service, Washington, D.C.

USDA 1998. Farm Business Briefing Room 1998, U.S. Department of Agriculture, Washington, D.C.

USDA 1998. Agricultural Statistics, U.S. Department of Agriculture, Washington, D.C.

USDA 1999. Agricultural Statistics, U.S. Department of Agriculture, Washington, D.C.

USDA 1999. Agricultural Prices, 1998 Summary, U.S. Department of Agriculture, National Agricultural Statistics Service, Washington, D.C.

USDA 2000. Agricultural Statistics, U.S. Department of Agriculture, Washington, D.C.

Wen, D. and D. Pimentel. 1992. Ecological resource management to achieve a productive, sustainable agricultural system in northeast China, *Agriculture, Ecosystems and Environment*, 41: 215.

WHO 1996. Micronutrient Malnutrition — Half of the World's Population, World Health Organization 78: 1.

Willet, G.S. and W.G. Gary. 1997. Enterprise Budgets, Summer Fallow — Winter Wheat — Spring Barley Rotation, Columbia County, Washington State, Farm Business Reports, Washington State University, Pullman, WA.

Worldwatch 2001. *Vital Signs 2001*, W.W. Norton & Company, New York.

Yao, X. 1998. Personal communication.

Youngquist, W. 1997. *Geodestinies: The Inevitable Control of Earth Resources Over Nations and Individuals*, National Book Company, Portland, OR.

Part Two

Environment Quality

9 Environmental Conflict and Agricultural Intensification in India

Gurneeta Vasudeva

CONTENTS

ENVIRONMENTAL SCARCITY AND CONFLICT

Over the past decade, in an effort to define a multidisciplinary approach to global, regional and local environmental problems that threaten the social and economic well-being of people, considerable research has been conducted on the links among environment, impoverishment and conflict. The thesis, broadly stated, is that environmental degradation often undercuts economic potential and human well-being, which, in turn, helps fuel violence, civil strife and political tensions (Figure 9.1). Various studies have analyzed causal links between environmental change and conflict with a focus on developing countries, which are most likely to exhibit environmental conflict in the future as a result of the growing pressure on the already scarce natural resources (see de Soysa, I. and Gleditsch, N.P., 1999; Vest, G.D. and Leitzmann, K.M., 1999; Homer-Dixon, T.F., Boutwell, J.H. and Rathjens, G.W., 1993).

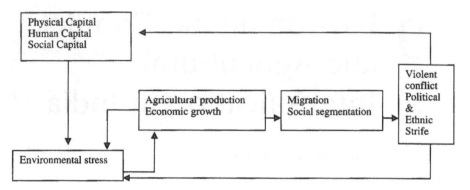

FIGURE 9.1 Causal Links between environmental change and conflict.

The obstacles to developing a conceptual clarity regarding conflict induced by environmental degradation and resource scarcity are quite formidable. Among the elusive elements in this process is an acceptable definition of conflict itself. Ashok Swain has defined conflict as a pervasive social process that occurs at all levels — between states, between groups and between the state and a group (Swain, A., 1996). While most definitions include a component of struggle, strife or collision, Wallensteen has defined conflict "as a social situation in which a minimum of two parties strive at the same time to acquire the same set of scarce resources"(Wallensteen, P., 1988).

Agricultural activities make up as much as 29% of the GDP in India, and as much as 60% of the population depends on the agricultural sector for livelihood. This chapter examines the factors that could create pressure on natural resources and hence, an adverse impact on agricultural productivity and access to food, thereby accentuating the large social and economic inequities and deprivation that already exist in society and have a potential for triggering violent conflict.

Currently, there is concern that activities related to agriculture may be affecting the environment and, conversely, inefficient utilization and management of natural resources could have an adverse impact on agricultural productivity. In intensive production systems — which have become increasingly important in developing countries such as India — the primary environmental concerns arise from land degradation, deforestation, contamination of groundwater due to excessive use of chemical fertilizers and pesticides, and loss in genetic diversity as a result of monoculture.

Similarly, unsustainable agricultural practices resulting in reduced production from agricultural land have, in several cases, led to displacement of small and marginal farmers, forcing them to migrate in search of alternative means for survival. In cases where survival is constrained by environmentally degraded areas and burgeoning pressures on urban areas within the country, migration has transcended national boundaries and led to political tensions, as has been observed in the case of the large-scale migration from Bangladesh to Assam and to the other northeastern states in India.

In recent years, the phenomenon of "environmental refugees," a label that describes human migration as a result of natural resource scarcities, has assumed great significance globally, largely due to the several instances of social, political and economic conflicts as a result of displaced populations. Essam El-Hinnawi, who virtually coined the term in his 1985 UNEP report defines environmental refugees as "... those people who have been forced to leave their traditional habitat temporarily or permanently because of a marked environmental disruption (natural and/or anthropogenic) that jeopardized their existence and /or seriously affected the quality of their life."

Wherever the environmental migrants settle, they are likely to create competition for resources and employment with the native population and communities. The northeastern states in India, in particular, have attracted large-scale migration from Bangladesh, largely due to the formers' low population densities and fertile agricultural land, even though the economic conditions in these states may not be ideal. These factors have contributed to providing cheap unskilled labor and agricultural land as a means of livelihood for the migrants. In many instances, the migrants have benefited at the cost of the development of the original inhabitants, thereby leading to clashes between the natives and immigrants, with consequent adverse impacts on the economic and political stability of the states in question.

Pressure on natural resources is also likely to spur conflict between competing stakeholders and groups. For example, where multiple states within the country are dependent on the same river systems, there have been problems in reconciling their interests, paving the way for interstate disputes over sharing river water. In some instances, these disputes have led to direct violence that necessitated judicial intervention.

It must be noted however, that resource and environmental problems are quite different for the array of agro-ecological conditions that exist in India, creating pressures on the land, water and forest resources in varying degrees. The diversity of the conditions also implies that there cannot be a fixed model that can be imposed to address unsustainable agricultural practices and resolution of conflicts that arise. Instead, the process of innovation and the capacity to adapt in adverse conditions must be made sustainable through an enabling policy environment. Reform measures designed to reap economic benefits, for instance, are also likely to have direct or indirect positive impacts on the environment, but many distortions in the policy framework persist, due to political economy constraints whereby perhaps small but important groups of people derive benefits from the prevailing conditions. The outcome of policy interventions also depends on institutional arrangements, ownership and control of natural resources, which are discussed in the concluding section of this chapter.

POPULATION GROWTH AND FOOD SUPPLY

The rate of growth in agricultural production in India is expected to exceed its population growth rate by as much as three times during the Ninth Five Year Plan (1997–2002), and this trend is likely to continue in the future as well. Still, 200 million Indians are reported to be undernourished, despite the fact that India ranks near the top agricultural exporters, with agriculture composing almost 18% of the country's total exports. Exports of about 5 mt or $1.4 billion worth of cereals and pulses, the staple foods of the Indian diet, were reported in 1998 (FAI, 1999).

On reviewing the relationship between food deprivation and population growth, it is observed that, while most undernourished people live in countries with the highest population growth rates, there is no support for the proposition that high population growth or density are associated with slower rates of per capita food production growth (Figure 9.2) (Dyson, T., 1996). It has been observed, on the other hand, that food deprivation is caused, not as a result of inadequate food production, but because people's claim to food is disrupted as a result of lack of assets or resources to grow or retain enough of their harvests to meet their needs. In the state of Kerala, for instance, which has a population density of 747 persons/sq km, compared with the national average of 267 persons/sq km, there have been significant improvements in indicators of poverty and hunger, compared with the north Indian states of Punjab and Haryana, which have far lower populations densities (401 persons/sq km and 369 persons/sq km respectively) and significantly higher agricultural productivity as a result of the Green Revolution technologies.

Serious questions have been raised about the impact of the Green Revolution in reducing poverty and hunger. While the onset of the Green Revolution since the 1970s has led to significant increases in crop yields, there have been both persuasive supporters and strong critics of the effectiveness of this development strategy as a tool to alleviate hunger and poverty. Since the early years of the Green Revolution, it has been observed that technologies that required purchased inputs such as improved seeds, fertilizers and pesticides inherently favored the rich farmers, and the landless and marginal farmers lacked the resources to

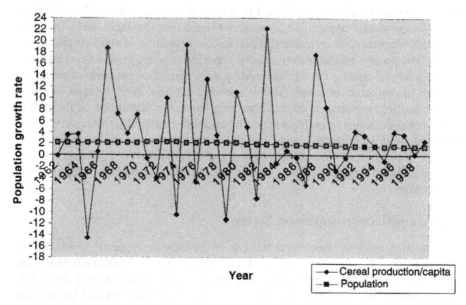

FIGURE 9.2 Population and per capita cereal production trends in India (FAO, 2000).

benefit from this capital-intensive technology. Moreover, the Green Revolution has focused on improving productivity of just two or three crops, thereby leading to a loss in genetic diversity, as well as ignoring the productivity of crops such as pulses and legumes grown by small farmers. The new technologies, in any case, are designed to work on good-quality farmland with irrigation and are inappropriate for marginal lands. The increase in productivity of the larger and richer farmers and the consequent reduction in prices has, in fact, contributed to the economic hardships for the smaller and poorer farmers. Although, in recent years, many poor farmers have adopted modern varieties of crops and technologies that have increased productivity and yields, the delay has been attributed largely to the inefficiencies in institutional mechanisms for financial and technical assistance. It is also commonly believed that the benefits from a technological transformation can be realized only if it is driven by the demands of the local farmers themselves.

Therefore, it may be said that food deprivation is not a direct consequence of population growth but, like population growth, is a consequence of social and economic conditions. Hence, addressing the inequities in terms of access to and control over assets such as natural resources, social capital, human knowledge, physical infrastructure and financial resources is critical to achieving a balance between population growth and food security.

The Rural–Urban Divide

It is indeed paradoxical that, even though the overall food grain production (which is the mainstay of the rural economy in India) has doubled from 108.5 mt in 1970 to 212 mt in 1998, the rural–urban gap has not declined. The rural–urban poverty headcount ratio has increased from 1.09 in 1987 to 1.23 in 1997 (IFAD, 2001). The rural population also continues to be more vulnerable to the consequences of environmental and economic downturns, with consequent spillover effects in the urban areas. This trend is in evidence globally. According to the Rural Poverty Report 2001 of the International Fund for Agricultural Development, 75% of the world's 1.2 billion poor are rural, will remain so for several decades, and the Indian subcontinent accounts for 44% of this population.

It is observed that, even though rural welfare indicators have improved, the rural–urban gap in terms of access to safe drinking water, adequate sanitation and health services remains inequitable and inefficient. Where resources have to be divided between urban and rural spending, the outlay per capita is normally less in rural areas, even though the initial levels of development and well-being are much lower in rural than in urban areas. Therefore, while urban-oriented policies have made urban living more attractive, they have also led to higher congestion costs and attracted migration from rural areas. Investments in rural infrastructure and technologies for reduction in the cost of cultivating staple crops in rural areas, for instance, could benefit both the farmers and urban food buyers, who spend most of their income on food staples. Studies have revealed no corresponding urban output, which, if expanded or made cheaper, benefits the rural poor on a comparable scale (IFAD, 2000).

Development of rural areas is therefore critical to the challenge of food security and prevention of conflict arising from pressures on natural resources. In this regard, some of the key challenges that need to be addressed are (1) equitable and efficient allocation of natural resources such as water and land and higher shares, access and control of these assets by the rural people, (2) widening market access for rural farm and nonfarm products by enhancing skills, technological innovation, improved infrastructure and institutions, and (3) participatory and decentralized management approach and innovative financing mechanisms.

ENVIRONMENTAL DEGRADATION

To analyze the social and economic impacts of agricultural activities, it is essential to examine the extent of environmental impacts of agricultural intensification that could lead to a decline in crop yields and reduction in overall productivity due to higher level of inputs to maintain yields. The adverse environmental impacts of agricultural intensification are amply borne out by the widespread instances of severe land degradation and loss in soil nutrients, which have resulted in instances of decline in rice and wheat yields in certain areas since the 1990s — a contrast to the dramatic increases in crop productivity in the early stages of the Green Revolution. Adverse environmental impacts have also led to the conversion of agricultural land to lower-value uses and sometimes temporary or permanent abandonment of plots, thereby exacerbating the social and economic conditions of the small and marginal farmers.

In India, the main types of land degradation can be categorized as soil erosion from wind and water; chemical degradation in the form of loss of nutrients, soil salinization, sodicity and acidification; and physical degradation in the form of waterlogging, compaction and flooding. As much as 63% of the total land resource is affected by degradation in varying degrees, however, not all of the land degradation results from agricultural practices and may also be determined by factors such as geological formation, rainfall, susceptibility to erosion and vegetation.

In irrigated areas, the major environmental problems are associated with intensive use of water coupled with poor drainage, thereby leading to waterlogged soils and a rise in the water table. In India, as much as 21.7 mha or 7.1 % of the land area is affected by salinity and waterlogging, with the resultant loss in crop productivity estimated at 9.7 mt annually. Studies carried out by the International Rice Research Institute have revealed that perennial flooding of rice paddies and continuous rice culture have led to build-up of micronutrient deficiency, soil toxicity and reduction in nitrogen-carrying capacity of the soil, thereby necessitating increased fertilizer consumption to increase yields from existing paddy fields. Excessive and inappropriate use of pesticides has also led to deterioration in the quality of water in several areas, posing a health hazard for the population. An increasing reliance on a few carefully bred crop varieties contributes to a loss in genetic diversity and to a common vulnerability to the same pest and to susceptibility to weather-related risks. In some cases where large areas have been planted with the same wheat or rice varieties, widespread losses have occurred because of the outbreak of a single pest or disease. The loss in traditional varieties could also lead to a reduction in the genetic pool available for plant breeding (Hazell, P. and Lutz, E., 1998).

In rain-fed areas (which constitute as much as 67% of the total agricultural area), land degradation has been attributed largely to high population densities and widespread incidence of poverty and hence pressures on natural resources. Until recently, natural resources were abundant in these areas, and, once used, farmers could allow these resources to recover through rotation and shifting cultivation. Environmental problems associated with rain-fed agriculture also include conversion of primary forest to agricultural area, thereby resulting in loss of biodiversity and exposure of fragile lands; expansion into steep hillsides, causing soil erosion and lowland flooding; degradation of watershed areas with downstream siltation of dams and irrigation systems; increased flooding and shortened fallows resulting in loss of soil nutrients and organic matter; and increasing pressure on common property resources such as woodlands and grazing areas.

PRESSURES ON LAND AND WATER RESOURCES

Composing 15% of the world's population but only 2.4% of the earth's land area, India has undertaken a path of agricultural intensification that is highly dependent on its land and water resources. The following paragraphs examine the constraints on land and water availability for agricultural purposes and instances of conflict as a result of competition for water resources.

India already has a high proportion of its land under cultivation. In 1998, 180.6 mha or 61% of the total land area in India was reported to be under cultivation. Furthermore, the land area per capita has declined from 0.48 ha in 1951 to 0.15 ha in 2000 (FAI, 1999). Factors such as excessively unsuitable terrain, poor soil quality, and unreliable rainfall have precluded cultivation in areas that are not already under cultivation. While increasing levels in population and the concomitant demand for food production may create the need for expanding the natural resource base, this would be neither possible on a significant scale nor desirable due to environmental considerations. Any further expansion would occur only at the cost of despoiling environmentally fragile areas and without sustainable levels of yields.

Juxtaposed against these limits to the expansion of cropland is the specter of inroads made on agricultural land by nonagricultural uses. While, historically, more potential cropland has been converted to agricultural land than urbanization has taken away, it is likely that the current unprecedented increases in levels of urbanization may constitute a potential threat to the loss of agricultural production as a result of loss in agricultural land.

In 1970, only 20% of the population or 110 million people lived in urban areas. In 2000, this number had grown to 288 million, accounting for 28% of the population, and this is expected to increase at an annual rate of about 15% to 499 million or almost 46% of the total population by 2020. While data on urban absorption of agricultural land is scarce, factors such as type of land converted to urban uses and the final per capita urban land area would influence the actual extent of cropland losses as a result of urbanization. It is estimated that, based on current densities of urban areas, approximately 0.62 mha will be converted to urban use by 2020.

Data for cereal production for the period 1980–1990 and 1990–2000 reveals a decline in the growth rate from 3.3% to 2.1% respectively. Similarly, cereal yields have declined from 3.4% in the period 1980–90 to 2.3% in the period 1990–2000 (FAI, 1999) Therefore withdrawal of land from agriculture for urban uses may contribute to further reductions in productivity in the future, with limited potential to compensate for these losses by expanding into other arable areas. This may also result in spillover effects in the form of further reduction in the size of landholdings and, in some cases, even landlessness for small farmers and hence displacement and migration of populations to environmentally fragile areas as well as to urban areas in search of alternative means of livelihood.

In addition to the concern relating to the availability of sufficient cropland to meet agricultural demand, the accessibility of water would perhaps pose the most serious threat to the future of agricultural productivity. While technological progress would continue to make it possible to increase agricultural production with relatively modest expansion of land in agricultural use, this, however, has not been the experience to date with water consumption and major improvements in water efficiency are unlikely in the medium term.

With agriculture contributing roughly 29% of India's GDP and production from irrigated land composing 56% of total agricultural production, a large percentage of India's GDP can be viewed as closely linked to the availability of water. Groundwater has been increasingly observed to be the preferred choice of farmers for irrigating their land due to a higher degree of control, adequacy and reliability. In 1996/97, groundwater accounted for 62% of the net irrigated area (FAI, 1999). The overuse of groundwater has emerged as a growing concern because aquifers are being continuously depleted, with pumping rates exceeding the rate of natural recharge. As against a critical level of 80%, the level of exploitation is over 98% in the state of Punjab and in other states such as Haryana, Tamil Nadu and Rajasthan. The problem is becoming increasing serious. In the southern India state of Tamil Nadu, for example, excessive pumping is estimated to have reduced water levels by as much as 25–30 meters in one decade. Implications of diminishing availability of groundwater for sustainable agriculture assumes significance when it is observed that the states currently facing the highest levels of groundwater exploitation are also India's agriculturally most important. Overexploitation of groundwater not only lowers its quality by rendering it saline, but also puts fresh water beyond the reach of farmers who depend on traditional technologies for drawing water and cannot make their wells any deeper.

Even though the Himalayan rivers carry a substantial amount of water annually, these rivers have been unable to meet the water demand arising from the agricultural practices of the Green Revolution in the northern states of India. The average amount of fresh water available per capita has declined throughout India from 5277 cubic meters (m^3) in 1955 to 2464 m^3 in 1990 and is estimated to further decline to 1496 m^3 in 2025 (Swain, A., 1998). The country also suffers from uneven distribution of water resources among the various regions. As a result of the seasonal monsoon rainfall, 80% of the rivers' annual runoff occurs in the 4 months from June to September. In addition, the amount of rainfall varies considerably, as a result of which, parts of the country such as Rajasthan in the west may receive as little as 0.2 m of annual rainfall, and Meghalaya in the east may receive as much as 11m. Floods and droughts are recurrances as a result of variation in the rainfall, thereby exacerbating the adverse impacts on agricultural production. The rivers in peninsular India are largely rain-fed and dry up during the summer. Most parts of the Deccan plateau, which receives marginal rainfall, are increasingly dependent on river storage or tanks for irrigation. With the exception of the water-abundant eastern region and the coastal strip along the Western Ghat Mountains, most parts of the country face increasing shortages of water.

Irrigation development continues to dominate the strategy for economic planning and agricultural growth, with more than $4.6 billion earmarked for irrigation schemes. Irrigation has brought significant benefits by allowing crops to be grown year round, thus enabling crop diversification and yields. It has also been the essential prerequisite for expansion of the use of chemical fertilizers and high yielding varieties (HYVs) of wheat and rice. However, with the total irrigation potential estimated at 113.5 mha, and 73.2 mha already under irrigation, the development of irrigation schemes is fast approaching its limits. Moreover, with the total water demand estimated to be almost equal to water availability by 2025 and the demand for water in the industrial and domestic sectors rising at the expense of the agriculture sector, increasing the irrigated output per unit of land and water consumption would be essential to meet the food demand.

RIVER-WATER SHARING DISPUTES

River-water sharing disputes create the potential for many new social and political conflicts, as has been observed in both the northern and southern states in India. In Punjab for instance, with a cropping intensity of about 189.5% in 1996/97, the irrigation requirements are estimated at 43.55 maf. With growing pressure on agricultural production, it has become increasingly difficult for Punjab to accept water transfer to the states of Haryana and Rajasthan from the Indus basin, which meets the irrigation needs in Punjab. The issue has remained largely unresolved and has even been ethnicized for political gains. Similarly, even though the states of Uttar Pradesh, Haryana and Delhi contain 21.5%, 6.1% and 0.4% of the catchment area of the Yamuna River respectively, they are the major users of its waters and have been involved in disputes with other north Indian states such as Himachal Pradesh, Madhya Pradesh and Rajasthan regarding the sharing of the Yamuna River's water.

In the south, the sharing of the Cauvery River has been a contentious issue between the two water-starved states Karnataka and Tamil Nadu. Even though 75% of the catchment area of the Cauvery River lies within Karnataka, traditionally its utilization has been small in Karnataka, and the farmers in Tamil Nadu have used as much as 75% of the river water. However, in the past couple of decades, Karnataka has undertaken several irrigation projects along the tributaries to meet its growing agricultural needs, thereby reducing the amount of water available to Tamil Nadu. The escalation of the dispute between Tamil Nadu and Karnataka regarding the sharing of the river water led to a supreme court decision to set up a Cauvery Waters Disputes Tribunal in 1990, providing interim relief to Tamil Nadu by instructing Karnataka to release water on a weekly basis in the summer months. This decision was subsequently countered by an ordinance issued by the government of Karnataka, despite the supreme court's continued support for the jurisdiction of the tribunal. The ensuing gridlock resulted in the eruption of violence and arson in Karnataka and its eviction of many Tamils. The violence subsequently spread to Tamil Nadu, where many Kannadiga landowners and farmers were driven out. The water-sharing negotiation of the Cauvery has been further complicated by the emergence of a new actor, Kerala, an upper riparian state that has recently demanded an increase in its share to 99.8 thousand million cubic feet (tmc-ft), claiming that it contributes 147 tmc-ft to the river. A politically recalcitrant approach has eluded the resolution of the dispute that shows all signs of aggravating into a violent confrontation, as well as leading to further alienation of the center section from the southern Indian states (Swain, A., 1998).

Therefore, as is evident from the above discussion, with large populations depending upon agriculture for their livelihood, water-sharing issues have increasingly been used as a means for achieving political gains and have often caused an upsurge in local communities and farmers to defend their interests. Water-sharing issues have been further complicated by the disputes arising from displacement and environmental damage caused by water development projects. Increasing water scarcity could therefore further exacerbate the problems of national integration in India, where strong ethnic identities already pose a great threat to political stability. A major institutional challenge for the water resource planning and management of rivers is the establishment of river basin authorities, which need to be viewed as a mechanism not only for addressing the institutional challenges but also for the resolution of interstate conflicts with regard to water sharing.

CONCLUSIONS AND RECOMMENDATIONS

HUMAN RESOURCE DEVELOPMENT

As large segments of the population continue to be economically active in the agriculture sector, there is increasing evidence that development of human capital is vital to increasing agricultural productivity and natural resource management. Moreover, the diffusion of technologies for effective and efficient natural resource management, pest control, irrigation, and biotechnology applications is imperative for modern and intensive agricultural systems. The human development effort in terms of basic health and literacy must also be emphasized, not only for the population engaged in the agricultural sector, but for the entire rural population.

Institutional Mechanisms

Conserving or improving the environment often requires collective action by the various user groups, thereby providing the basis for a participatory and decentralized approach for natural resource management. As a consequence of the separation of ownership and use of natural resources, indigenous institutions and mechanisms at the grassroots level that have been successful in the management of water resources have been gradually wiped out. These include a variety of local-level traditional water-harvesting mechanisms adapted to the varying ecological conditions across the country. The share of tanks in the net irrigated area, for example, has declined steadily after peaking in 1958/59. While this decline can be attributed in part to the higher efficiency of well and canal irrigation, what is of concern here is that the decline has been accentuated by institutional factors (TERI, 1999).

It has been commonly observed that local organizations can often be effective in securing compliance with rules of common property use pertaining to water, common grazing ground, and forests. Also, involvement of local stakeholders in development and management practices and selection of technologies often promotes innovation and effective adoption of appropriate technologies. Moreover, creating conditions where local organizations become more efficient through effective collaboration with the public- and private-sector organizations may reduce the costs of environmental conservation.

While it is recognized that participatory approaches may not be easy to implement on a large scale, especially in the case of watershed management for example, which involves multiple users and stakeholders with competing needs, it is important to note that local organizations could play a key role in building the social capital and creating a consensus about the use of the water resources for diverse, multiple and often conflicting purposes (Lutz, E., 1998).

Public–Private Partnerships

In recent years, there has been a growing realization that the development process's being increasingly market driven necessitates partnerships with the international private sector, thereby opening access to markets and information. Private-sector entities have also shown a growing interest in commercially viable partnerships that seek to improve the quality of life for rural dwellers by providing support for agricultural research, infrastructure development and market access. Partnerships with nongovernment organizations (NGOs), cooperatives and governments can also assist in developing the bargaining power of the farmers through trade and marketing associations.

Equitable Access to Land and Water Resources

In an effort to optimize equitable distribution of land and water resources and increase participation of local stakeholders in the management, development and maintenance of rural development projects, it is essential to increase access and control of local stakeholders over resources such as land and water through a market-driven distribution policy, thereby weakening elite dominance. The present policy framework for the development of groundwater for instance, has often been characterized as largely inequitable, favoring rich farmers who have the financial resources to invest in more powerful pumps. Moreover, to earn a decent return on investments in water extraction mechanisms, a farmer must have a captive irrigable command area of a certain minimum size; large land holdings again have an advantage here. Although the development of groundwater markets are believed to promote equity and efficiency, it can be argued that, in the absence of well defined rights that set limits to water withdrawal, the development of groundwater markets could lead to the faster depletion of aquifers, creating at the same time a powerful monopoly of "waterlords." Although aquifers have been depleted in some states, in others such as eastern Uttar Pradesh and Bihar, groundwater sources have remained underdeveloped due to constraints on availability of electricity and financing (TERI, 1999).

Similarly, the case for land redistribution from large landowners to the landless or small owners rests on three main considerations of equity and efficiency: (1) inequality in land distribution not only creates unequal distribution of income, thereby curtailing access to credit, but also makes the poor vulnerable to social stratification and political power of the rich, (2) total employment and production per hectare increases as farm size decreases and (3) equitable land distribution strengthens the nonagricultural activities and therefore helps in alleviating poverty through increased employment in the nonfarm sector (Alexandratos, N., 1995).

Technological Interventions

To move toward environmentally and socially sustainable agriculture it is important to create the appropriate conditions for technological innovation pertaining to recycling of agricultural inputs, lowering of fertilizer and pesticide consumption, raising crop yields, improving irrigation techniques, limiting soil degradation and promoting energy-efficient, renewable energy sources. The efforts to support agricultural development have so far been based largely on transferring technologies from the developed countries for a narrow range of crops in favorable agroclimatic conditions. Traditional farming techniques have been commonly ignored, and plant breeding has focused primarily on cash crops with the objectives of maximizing yields rather than stabilizing yields. Moreover, soil nutrient replacement has been dominated by the use of mineral fertilizers rather than integrated plant nutrition systems, and soil conservation techniques have been designed using engineering techniques rather than biological approaches and moisture management techniques for soil stabilization. While this strategy has had several positive impacts and boosted food security and agricultural export earnings, there is reason to believe that these benefits cannot be maintained in the long term unless agricultural production shifts to a more sustainable path.

In the next stage of agricultural intensification, biotechnology applications are expected to play an important role for the introduction of higher plant resistance to pests and diseases, development of tolerance to adverse environmental conditions, improvement in nutritional value, and, ultimately, an increase in the genetic yield potential of plants. While conventional breeding can have similar objectives, genetic engineering can create transgenic crops that would include genetic material that would otherwise belong to a certain species only in extremely rare cases.

Like many revolutionary developments, however, biotechnology also brings new risks and problems. Currently, research in biotechnology is dominated by a few private-sector companies and, the International Agricultural Research Centers, after a relatively slow start, have been increasing their research in biotechnology for agricultural applications. Multinational chemical and pharmaceutical companies that are involved in the development of biotechnology products and processes have acquired a large number of patents and control a large market share in transgenic seeds. This may pave the way for a growing dependence on agricultural imports and vulnerability to prices that are controlled by a handful of corporations. Moreover, hardly any biotechnology research is being undertaken on the basic food crops or on the problems of the small and marginal farmers.

Considerable environmental risks are also associated with transgenic crops. Among these is the possible escape of herbicide-tolerant genes to wild relatives of the plant, creating super weeds that would be resistant to control. Moreover, the patenting of crop genes might imply that farmers in the future would be obliged to pay royalties to foreign companies on indigenous varieties. Even though biotechnology applications are expected to have a significant impact on agricultural productivity, concern is growing about the research dependency as a result of widespread patenting of biotechnology products and processes that make it prohibitively expensive for developing country markets to adapt these technological developments to meet their agricultural needs. The high costs could further preclude the poor and marginal farmers' access to the benefits of bio-technology applications. Even though countries such as India and China have made some progress in introducing institutional arrangements and increasing the budget for biotechnology research in recent years, the share of developing countries in biotechnology research continues to be very small, and the research emphasis is often placed on export-oriented crops. The private sector is unlikely to change its focus because of the perceived inability of poor farmers to purchase improved seeds or inputs such as herbicides. It is therefore important to build national and regional capacity to undertake research to ensure that small and disadvantaged farmers and resource-poor areas are not left further behind by the biotechnology revolution. The issues pertaining to intellectual property rights and patents would also need to be resolved in a manner that balances the interests of the private-sector companies as well as ensuring control over indigenous genetic materials.

An Integrated Approach

While fundamentally different approaches to development may be required to address the problems related to poverty, environment and agriculture, it is recognized increasingly that failure or success of these strategies is highly interdependent. Any development strategy must therefore simultaneously address questions of long-term sustainability and small-scale adaptation to local ecological conditions. Moreover, instead of pursuing a single objective to increase food production, for example, a variety of strategies must be devised to disrupt the vicious cycles of poverty and environmental degradation. In some regions, this may be possible by investing in infrastructure and technology to increase productivity and sustainability of agriculture, though not necessarily in food production. In other regions, the focus will need to be on the creation of income-enhancing activities through on-farm or nonfarm enterprises and public works programs. Finally, it can be said that, as we step into the 21st century, the challenges of poverty reduction, environmental protection and agricultural development still remain a daunting reality, which, if unchanged could deny the future generations a peaceful and livable planet.

REFERENCES

Alexandratos, N. (Ed.). 1995. *World Agriculture: Towards 2010*. FAO. Rome.

Dyson, T. 1996. *Population and Food: Global Trends and Future Prospects*. London.

El-Hinnawi, E. 1985. *Environmental Refugees*. Nairobi. UNEP.

FAI. 1999. Fertilizer Statistics. Fertilizer Association of India, New Delhi.

FAO.2000. Food and Agriculture Organization (online database: www.fao.org).

Gaulin T. 2000. To Cultivate A New Model: Where de Soysa and Gleditsch Fall Short. Environmental Change and Security Project Report. Summer 2000. The Woodrow Wilson Center. Washington D.C.

Hazell P. and L. Ernst (Eds.). Integrating Environmental and Sustainability Concerns into Rural Development Policies. Agriculture and the Environment: Perspectives on Sustainable Rural Development. The World Bank. Washington D.C.

Homer-Dixon, T.F., H.J. Boutwell and G.W. Rathjens. February 1993. Environmental Change and Violent Conflict. *Scientific American*.

IFAD. 2001. Rural Poverty Report 2001: The Challenge of Ending Rural Poverty. International Fund for Agricultural Development. New York.

Lappé F.M., J. Collins and P. Rosset. 1998. World Hunger: Twelve Myths. New York.

Leonard, J.H. et al. 1989. Environment and the Poor: Development Strategies for a Common Agenda. US-Third World Policy Perspectives, No. 11. Overseas Development Council. Washington D.C.

Lietzmann, K.M, G.D. Vest. 1999. Environment and Security in an International Context: Executive Summary Report. NATO Committee on the Challenges of Modern Society Pilot Study. Environmental Change and Security Project Report. Summer 1999. Woodrow Wilson Center. Washington D.C.

Rosegrant, M.W, P. Hazell. 2000. Transforming the Rural Asian Economy: The Unfinished Revolution. Asian Development Bank. Hong Kong.

Swain, A. 1996. Environmental Migration and Conflict Dynamics: Focus on Developing Regions. *Third World Quarterly*. 17(5): pp 959-973.

Swain, A. 1998. Fight for the Last Drop: Inter-state river disputes in India. *Contemporary South Asia*.

TERI.1999. GREEN India 2047: Looking Back to Think Ahead. Tata Energy Research Institute. New Delhi.

Wallensteen, P. (Ed.). 1988. *Peace Research: Achievements and Challenges*, Boulder, CO. Westview. p.120.

10 Water Quality and Agricultural Chemicals

Ramesh S. Kanwar

CONTENTS

1-5667-0594-0/02/$0.00+$1.50

INTRODUCTION

Food and water are two essential needs of social security. One of the most important questions facing the global society is how to produce enough food to feed the increasing human population in the world. Another parallel question is how much water will be needed to produce enough food. Answers to these questions are not easy. Increased population rates have added more than 4.4 billion people on earth between 1900 and 2000 and average food production has kept pace with the increases in population. Also, between 1900 and 2000, irrigated area has increased from about 50 million hectares to 250 million hectares (Mha) (Gleick, 2000). India and China together have more than 36% of the world population to feed, with more than 21% of the world population living in South Asia. Although world food-grain production has increased significantly, much improvement in feeding people has occurred in Asia (particularly India) as a result of the Green Revolution and increased water use for irrigation. In spite of these gains, 830 million people remain undernourished – 45% in India and China alone. These data clearly indicate that food production alone cannot solve the local and regional food security needs.

In the year 2000, more than 1 billion ha of the world area was cultivated, of which 26% was irrigated, producing more than 40% of all food grown in the world (Gleick, 2000). Also, irrigation accounts for nearly 85% of all water consumed worldwide, which makes less water available for other uses. Table 10.1 gives a summary of major water resources on earth. This table shows that only 2.5% of the total volume of water available on earth is fresh water. About 70% of this is in the form of glaciers or permanent ice locked up in Greenland and Antarctica, and in deep groundwater aquifers (Shiklomanov, 1993). The main sources of water available for human consumption and agricultural use are rivers, lakes and shallow groundwater, which is less than 1% of all fresh water on earth and only 0.01% of all water present on the planet (Gleick, 2000). This makes the job of water resource planners even more difficult, as much of the fresh water is located away from concentrations of human population. Table 10.2 gives water withdrawal and consumptive uses for the year 2000. This table shows that total water use has increased from 579 km³/yr in 1900 to 3,927 km³/yr in 2000, and that the largest water withdrawal has occurred in Asia. Also, future withdrawal rates are expected to grow 2 to 3% annually until 2025 (Gleick, 2000). Table 10.3 gives global water withdrawal and consumptive use for three major categories (i.e., agricultural, industrial and municipal use), showing

TABLE 10.1
Major Water Reservoir Sources on Earth

Water sources	Volume (1000 km³)	% of total water	% of total fresh water
Salt water sources			
Oceans	1,338,000	96.54	—
Saline groundwater	12,870	0.93	—
Saltwater lake	85	0.006	—
Total		97.48	
Fresh water sources			
Glaciers/ground ice	24,364	1.76	69.56
Groundwater	10,530	0.76	30.06
Lakes	91	0.007	0.26
Rivers	2.12	0.0002	0.006
Marshes/wetlands	11.5	0.001	0.03
Soil moisture	16.5	0.001	0.05
Water vapor	12.9	0.001	0.04
Total		2.52	

Source: Gleick, 2000

clearly that agricultural water use continues to make up 85% of all consumptive use on a global basis.

Table 10.4 gives annual available renewable water resources for countries in South Asia, China and the United States. This table shows that agriculture continues to be the major user of renewable water withdrawals. For some countries in South Asia (especially Nepal, Pakistan, and Sri Lanka), agricultural water use is more than 95% of the total withdrawal. This brings up more questions on the efficiency of water use for agricultural purposes. Increased efficiency in water use in agriculture can save water for other uses. Also, improved water-use efficiency in irrigation can result in more food production without increasing additional demands on fresh water. Maintaining a good standard of living will require renewable water resources capacity of 1000 m³ per person per year in countries with thriving economies (Bouwer, 1993). China is developing future management plans on renewable water supplies of 500 m³ per person per year to sustain its economy, whereas India's planners are using 250 m³ per person per year. Many others will have fewer renewable water resources for their economic growth (Bouwer, 1993).

Table 10.5 gives data on domestic water use for countries in South Asia (Gleick, 2000). A minimum of 50 liters per capita/per person per day (lpcd) is recommended for domestic water use by the World Health Organization and the World Bank (5 lpcd for drinking, 20 lpcd for sanitation and hygiene, 15 lpcd for bathing, and 10 lpcd for cooking). Table 10.5 shows that, except for Pakistan, all other countries in South Asia are using less water for domestic use. Billions of people on the earth lack access to the basic requirement of 50 lpcd. More than 60

TABLE 10.2
Water Withdrawal and Consumption by Continent (1900–2025)

Continent	Historic and Forecast Water Use, km³/yr					
	1900	1940	1960	1980	2000	2025
Europe						
Withdrawal	37.5	185	445	491	534	619
Consumption	17.6	54	158	183	191	217
North America						
Withdrawal	70.0	221	410	677	705	786
Consumption	29.2	84	138	221	243	269
Africa						
Withdrawal	41.0	49	86	168	230	331
Consumption	34.0	39	66	129	169	216
South America						
Withdrawal	15.2	28	69	111	180	257
Consumption	11.3	21	44	71	104	122
Australia and Oceania						
Withdrawal	1.6	6.8	17.4	29	33	40
Consumption	0.6	3.4	9.0	15	19	23
Asia						
Withdrawal	414.0	689	1,222	1,784	2,245	3,104
Consumption	322.0	528	952	1,324	1,603	1,971
Total						
Withdrawal	579	1,065	1,989	3,214	3,927	5,137
Consumption	415	704	1.243	1,918	2,329	2,818

Source: Gleick, 2000

countries in the world with the total population of 2.2 billion report average domestic water use of less than 50 lpcd. The purpose of this chapter is to summarize the presently available information on the effects of intensification of irrigated agriculture on land and water resource degradation in South Asia, with examples from India and Pakistan.

NATURAL RESOURCES OF INDIA

SOIL RESOURCES

India's variety of soils range from very productive to very unproductive. They vary between red sandy soils in south India and productive black soils in Maharastra (see also Chapter 2 in this volume). Velayutham and Bhattacharyya (2000) reported that India's total land area of 328 million hectares (Mha) is predominantly covered with

Table 10.3
Global Water Withdrawal and Use for Selected Categories (1900–2025)

Category	1990	1950	1980	2000	2025
Population (million)	—	2,542	4,410	6,181	7,877
Irrigated area (m. ha)	47.3	101	198	264	329
Agricultural use (km³/yr)					
Withdrawal	525	1,122	2,179	2,560	3,097
Use	406	849	1,688	1,970	2,331
Industrial use (km³/yr)					
Withdrawal	37.8	181	699	768	1,121
Use	3.4	14.4	59	85	133
Municipal use (km³/yr)					
Withdrawal	16	53	207	389	649
Use	4.2	14	42	64	84

Source: Gleick, 2000

TABLE 10.4
Annual Renewable Water Resources and Withdrawal Rates for South Asia and Selected Countries for the Year 2000

Country	Renewable water resource (km³/yr)	Renewable withdrawal (km³/yr)	Agricultural use (km³/yr)
Bhutan	95	0.02	0.01
Bangladesh	1210	14.6	12.6
India	1908	500	460
Nepal	210	29	28.6
Pakistan	429	156	151
Sri Lanka	50	9.8	9.4
China	2830	526	405
USA	2478	469	197

Source: Gleick, 2000

red soils (105.5 Mha), black soils (73.5 Mha), alluvial soils (58.4 Mha), laetrite soils (11.7 Mha), desert soils (30 Mha) and hills and tarai soils (26.8 Mha). Red soils occur in the peninsular region of India and support plantation and horticultural crops. Black soils, which are very productive, occur mostly in central, western and southern India and support cotton, sugarcane, vegetables and other cereal crops. The laetrite soils are traditionally poor soils that are prone to soil erosion and nutrient depletion. Desert soils, located in the western part of India, are poor in soil quality, and are prone to wind erosion. The hills and tarai soils are mostly in the northern and northeastern parts of the country and are characterized by high rainfall and high

TABLE 10.5
Population and Per Capita Domestic Water Use for Countries in South Asia (2000)

Country	Population (million)	Estimated domestic water use (liters per capita per day, lpcd)
Bhutan	2.03	10
Nepal	24.35	12
Bangladesh	128.35	14
Sri Lanka	18.85	18
India	1000.77	31
Pakistan	156.01	55

Source: Gleick, 2000

carbon content. The soils in the Indo-Gangetic Plains, which support intensive agriculture for more than 300 million people, have been brought under irrigation by various canals on the Indus, Gagger, and Jamuna rivers. Long-term irrigation of these soils has degraded a certain percentage of the area by salinity, alkalinity and waterlogging (Velayutham and Bhattacharyya, 2000).

WATER RESOURCES

Rainfall

The distribution of water resources in India is highly variable. The main source of water is rainfall, which ranges from 311 mm in the Rajasthan to more than 13,000 mm of annual rainfall in West Bengal. The average rainfall over the Indian subcontinent has been estimated at 1200 million hectare meters (Mham) and average annual rainfall availability in India is 400 Mham (Gupta et al., 2000).

Surface Water (River Basins)

India has more than 20 major river basins from which total water potential has been estimated at 188 Mham. The largest amount of surface water is available for Ganga/Brahmaputra/Barak giving a total of 117 Mham. One of the major problems India is facing is lack of capacity to harness these vast surface-water resources. Much of the surface water flows into the sea and outside India's borders. India is harvesting about 20% of total surface water through reservoirs but capacity must be increased to have better economic growth (Gupta et al., 2000).

Groundwater

India has a significant number of groundwater resources. Out of 400 Mham of annual rainfall, 215 Mham of rainwater eventually becomes part of shallow and groundwater aquifers. In addition, India's streams, rivers and irrigation networks add another 11 Mham to groundwater. Therefore, the total annual groundwater

TABLE 10.7
Available Water Resources in India

Water resource	Average annual availability (M ha m)
Rainfall (M/ha/m)	400.0
Surface water	187.9
Groundwater	43.1
Total (surface and groundwater)	231.0
Irrigation potential for India (estimated)	139.5

Source: Gupta et al., 2000

resource available for exploitation in India is estimated at 43.1 Mham, out of which potential groundwater available for agriculture and irrigation is estimated at 36 Mham (Table 10.7). Currently, India is pumping 16.5 Mham of groundwater for irrigation and the balance of 24.5 Mham is yet to be developed (Gupta et al., 2000).

Utilization of Surface and Groundwater Resources for Irrigation

India is one of the few countries in the world that is extremely rich in water resources. Surface and groundwater resources totaling 231 Mham are plenty to meet India's growing irrigation and industrial development needs for the year 2050. The ultimate irrigation potential of the country has been estimated at 139.5 Mha. So far, India has achieved a total irrigation potential of 89.5 Mha, including double-cropped areas. The remaining potential needs to be developed if adequate water supplies are to be available to meet India's irrigation needs for the burgeoning population (Gupta et al., 2000).

FOOD SECURITY: LAND, WATER
AND ENVIRONMENT QUALITY

The very first basic questions for the world community are: How much water will be needed for a world population of about 10 billion in 2050 (Bouwer, 1993) and where will it come from? Part of the answer we know pretty well. The total availability of freshwater resources for human use is finite (less than 1% of the total water on the planet) and we do not know how much water will be needed for future food production. In the year 2000, 85% of all fresh water consumed worldwide was used for irrigation to produce food. Without irrigation, natural rain-fed agricultural areas in the world would not be able to feed the world's current population. Currently, more than 500 million people live in countries with insufficient water to produce their own food and will depend on having to import from other countries to meet their food needs. An average American diet needs about 1800 m³ of water per year per person from both natural rainfall and irrigation. In South Asia, however, an average diet needs 770 m³ per person per year (Gleick, 2000).

Another important question is: How much crop land would be needed to feed the growing population and what is the potential to further expand land area for food grain production? Currently about 1,510 Mha area is under cultivation globally and another 3,000 Mha are categorized as pasture and rangeland (Scherr, 1999, UNFAO, 1999). More than 2,600 Mha of land worldwide are available on which grain crops may achieve reasonable yields. Out of a total of 1,510 Mha, 276 Mha were irrigated in 1997 (Gleick, 2000), which nearly doubled from 138 Mha in 1960. The irrigated area expanded better than 2% per year in the 1970s but now has fallen to less than 1.4% annually. Expansion of irrigated areas is becoming more difficult because of lack of available land, limited water resources, cost of irrigation systems, and cost of bringing marginal lands under irrigation. The availability of cropland for growing food is becoming another question for many of the world's fastest growing economies. Loss of prime agricultural land to urban and industrial development is the major concern in China, Indonesia and the United States. Total cropland area per capita in the world has decreased from 0.31 ha per person in 1983 to 0.25 ha per person in 2000 (Gleick, 2000).

Because total area under cropland per person is decreasing, agricultural production systems are becoming more intensive to grow much more food on the same per unit area of land. The intensification of agriculture, especially under irrigated conditions, has brought new environmental-quality problems that include soil erosion, land degradation and water pollution.

To provide food security to a growing population, the final question would be: What are the impacts of intensive agriculture and irrigation systems on the degradation of land and water resources? Ecology and economy are twin elements of global stability. About 25–30 years ago, it was a popular belief that goals of economic development and environmental quality were mutually exclusive. Today this view has largely given way to the belief that we need a better understanding of the relationship between development and the environment. The first and foremost component of a comprehensive environmental assessment policy is that development must be environmentally sound and sustainable. Although population rates have been declining (especially for more densely populated countries like China and India), by 2050, the planet could very well have doubled its present population. A frightening look at the future indicates that earth's population will increase to 10 billion by the year 2050 (Bouwer, 1993). The impact of this increased population will be severe on the environmental quality of land and water resources. While as much as 95% of the world's population growth is expected in the developing countries, this is where, by the year 2050, 87% of the world's population is expected to live. Industrial and agricultural use will add enormous stress on the available land and water resources, while also attempting to maintain environmental quality.

An increasing population will require more water in many areas of the world, especially South Asia, largely through more irrigation. At the beginning of the 20th century, 90% of all water used in the world was for irrigation, and in the year 2000, it was expected to be 60% (Bouwer, 1993). This indicates that we must grow more food with less water through more intensive agricultural production systems using pesticides and inorganic fertilizers. Intensive agricultural production systems were introduced in the 1960s with advances in improved crop varieties, mechanization and increased availability of pesticides and fertilizers. More recent experiences in the developed countries, especially Europe and the U.S., have shown that modern intensive agricultural production systems have increased land degradation and water contamination. Intensive row production systems have increased soil erosion and groundwater contamination (Baker and Johnson, 1983). The greater use of agricultural chemicals increased the level of pesticides and nitrates in surface and groundwater sources in agricultural watersheds. (Kanwar and Baker, 1993; Kanwar et al., 1988, 1997; CAST, 1985; Hallberg, 1989). Higher concentration of nitrates and nitrogen in well water was first recognized as a health problem in 1945 when two cases of infant methemoglobinemia (blue baby syndrome) were reported in Iowa (Comly, 1945) and in South Dakota 22 years later (Johnson et al., 1987). Some evidence exists that high nitrate ingestion is involved in the etiology of human cancer (Fraser et al., 1980; Foreman et al., 1985).

The negative impacts of the use of pesticides and fertilizers to human health and the environment have been a source of concern. The use of agrochemicals in South Asia is widespread and intensive in areas where cropping density is high. A better understanding of land- and water-resource degradation from intensive agriculture is needed to assure food security to the fastest growing population in the region.

IMPACT OF INTENSIVE AGRICULTURE AND IRRIGATION MANAGEMENT PRACTICES ON THE ENVIRONMENTAL QUALITY OF INDIA'S SOIL AND WATER RESOURCES

India and the rest of South Asia are blessed with land and water as the two most important natural resources for their agriculture and economic development. The demand for these resources will continue to escalate to provide food security to its growing population. In the global context, India is feeding 16% of the world population with only 2.4% of the world's geographical area. The per capita availability of land in India has decreased from 0.9 ha in 1951 to 0.25 ha in 2000 (Yadav et al., 2000). It is quite possible to increase the intensity of Indian agriculture by another 300% as India has good quality soil, abundant water resources, plenty of sunshine hours annually, skilled labor, and an excellent network of research and extension institutions in agriculture. India has a land area of 328 Mha; 49% of this area is cultivated and about 17% is irrigated. Agriculture contributes 35% gross domestic product and employs about 65% of the total adult population. Growth in agriculture has a significant impact on the employment and income of the rural population. Since its independence in 1947, India has made some significant gains in food production, with grain production increasing from 50 million tons in 1947 to more than 210 million tons in 2000–2001 (Gleick, 2000). This increase in grain production has been higher than the population growth rate in the 20th century and India is a successful model in the world community for providing food security to its massive population. This increase in agricultural productivity has also helped India increase its per capita income at a rate of 2% per year to reach about $300 per year for the entire country.

It is a well-accepted belief in the broader global community that long-term sustainable agricultural production systems are essential to the overall economic development. India has a growing population of more than 1 billion to feed and over two thirds of its work force depend directly or indirectly on agriculture. India needs to develop its economy by establishing environmentally sound agricultural production systems. Several studies indicate that India's population will grow to 1.5 billion people by 2050, needing more than 300 million tons of food grain. This will require several strategies to increase crop production in India. One thing is very clear: to increase crop yields on the current cultivated lands, more efficient use of water, land, chemicals, and germplasm will have to be made.

India's grain production increased significantly during the 1970s to 1990s. Some of the factors that contributed to increased production included improved crop varieties, expansion of irrigated areas, mechanization of agriculture, increased use of chemicals and improved research and extension services. Irrigation and fertilizer use were the key factors to this increase in grain production. India's 30% cropland area is currently irrigated but producing 56% of the country's grain. Rain-fed agriculture occupies 53% of cultivated area and produces only 44% of food grains. The rest of the cropland area (about 17%) is used for other than raising grain crops. For some of the key grain-producing states in India, percentage of irrigated area is much higher than the national average. For example, 95.2% of Punjab's, 78.2% of Haryana's and 65.8% of Uttar Pradesh's cropland areas are irrigated. Total land area under irrigation has increased from 25 Mha in 1960–61 to more than 57 Mha in 1997 (Table 10.8). Fertilizer use increased from 0.3 million tons in 1960–61 to more than 10 million tons in 1997. Improved technologies and expanded irrigation systems have prompted India and other countries in South Asia to intensify their production systems. The farmers in 65% of India's irrigated areas are growing two to three crops a year. This intensification in agriculture has resulted in India's self-sufficiency in grain production. Although food production in India and the rest of South Asia has increased significantly, India has seen sharp degradation of its natural resources (soil, water, and air). The following paragraphs describe the impact on the environment of the intensification in agriculture (Velayutham and Bhattacharyya, 2000; Gupta et al., 2000; Yadav and Singh, 2000).

DEGRADATION OF LAND AND SOIL

Loss of Forests

Grasslands and forests are very important for the sustainability of ecosystems. India has 15% of the world's population but only 2% of the world's forested land. More recent data have shown that, between 1972 and 1982, India has lost forests at a rate of 1.5 Mha per year to agriculture. At this rate, India's forested land may be reduced to about 10% of its total geographical area. Deforestation and overexploitation of grasslands will increase soil erosion and flooding of lowland areas and bring marginal lands into cultivation. Unless more areas are reforested and better management of

TABLE 10.8
Irrigated Areas in Countries in Southeast
Asia (1977)

Country	Irrigated area (,000 hectares)
Bangladesh	3,693
Butan	40
India	57,000
Nepal	1135
Pakistan	17,580
Sri Lanka	600

Source: Gleick, 2000

grasslands is undertaken, further environmental degradation of land and forest resources is inevitable.

Soil Erosion

The major factor in India's soil degradation is erosion. Two types of factors cause soil degradation: natural and human. Natural factors causing soil degradation include climate, hydrology, soil genesis and natural vegetation. There is little that can be done to correct natural factors. It has been reported that nearly one third of the cultivated area of 161.5 Mha in India suffers from water and wind erosion. Intensive agriculture on steeper soils, forest cutting, hill grazing, grass burning, not using conservation tillage methods and heavy rainfall are the main reasons for severe soil erosion. On many lands, soil erosion rates vary from 20 to 100 tons/ha/yr, averaging around 16 tons/ha/yr. Soil erosion is a serious problem in the rain-fed agricultural areas and needs immediate attention to improve soil's productive capacity (Velayutham and Bhattacharyya, 2000).

Maximum efforts should be made to correct human actions or factors to halt further degradation of soils. Increased population pressure is causing deforestation and bringing marginal lands into cultivation. Innovative methods of agricultural production must be developed for intensification of agriculture to bring better yields per unit area without degradation of land and water quality. Experience has shown that no-tillage and other conservation tillage systems should be practiced on highly eroded soils. These tillage systems will help to increase organic matter (Kanwar et al., 1997) and reduce soil erosion. Other practices include contouring, terracing, grass waterways, strip cropping and innovative crop rotations. In addition, planting trees and shrubs on hill slopes and promoting rotational grazing along stream banks and forest areas will significantly reduce erosion. All these practices must be implemented within watersheds that are prone to erosion due to agricultural activities. Another important program would require significant public investment to construct a series of reservoirs to minimize the effect of floods during heavy monsoon rains.

These floods are the major cause of large-scale soil erosion. With proper watershed management practices, the degree of flooding could be reduced.

Loss of Soil Fertility

Chemical deterioration of soils can occur due to loss of organic matter and soil nutrients caused by long-term agriculture. Intensive cultivation of soils to grow two to three crops a year, especially in irrigated and rain-fed agriculture, is allowing continuous depletion of the soil's natural fertility. Although fertilizer use in India has increased from 0.7 million tons in 1950–51 to more than 20 million tons in 2000 (Yadav and Singh, 2000), fertilizer use is highly skewed and still below the minimum required for raising crop yields. For example, the average nitrogen fertilizer use in Punjab is close to 160 kg/ha compared with only 5 kg/ha in Assam. Also, fertilizer use is heavy on some crops and light on others. These differential rates of fertilizer application are causing groundwater degradation in areas like Punjab, and mining natural soil nutrients in other areas. Nutrient-supplying capacity of soils is continuously eroding because nutrient management plans, especially in rain-fed agricultural areas, are not in place.

Salinization

Chemical degradation of soils has occurred in India with increased salinization and alkalization under long-term irrigation practices. The net irrigated area in India increased from 22 Mha in 1957 to about 57 Mha in 2000 (Gupta et al., 2000) Improper management of irrigation (such as low irrigation efficiencies, inadequate drainage for canal irrigation and overexploitation of groundwater for irrigation) has resulted in rising water tables and accumulation of salts near the surface. Abrol and Bhumbla (1971) estimated that about 7 Mha soils are affected by salinity and alkalinity in the Gangetic plain alone and nearly 50% of canal-irrigated soils are degraded by salinization and alkalization due to poor drainage, inefficient irrigation systems and for socio-political reasons. In many coastal regions of Gujrat, overexploitation of groundwater has caused seawater intrusion, bringing salinity problems. Excessive irrigation, especially canal irrigation, is causing waterlogging problems and adding further to environmental degradation. In some areas of the country, salinity problems are increasing at such a fast rate that these areas may become totally unfit for producing any vegetation and will bring social and economic discomfort to the people.

DEGRADATION OF WATER RESOURCES

Rise in the Water Table (Waterlogging and Salinity)

Excessive irrigation on flatlands and poor internal drainage under heavy rainfall conditions are two of the causes of waterlogged conditions. Table 10.6 shows that nearly 11.6 Mha soils are affected with some degree of waterlogging. Poor planning and mismanagement of irrigation systems in India and Pakistan have resulted in rising water tables' causing salinization and waterlogging problems in some of the

TABLE 10.6
Total Area Affected by Different Types of Soil Degradation in India

Degradation type	Area affected (m ha)	% of total land area
Erosion		
Water erosion	148.9	43.5
Wind erosion	13.5	4.1
Chemical degradation		
Salinization	10.1	3.1
Loss nutrients	3.7	1.1
Water logging	11.6	3.5
Total affected area	**187.7**	**57.7**
Soils with no degradation	90.5	27.5
Stable terrain	32.2	9.8
Soils/land not fit for agriculture	18.2	5.5
Total geographical area	**328.7**	**100.0**

Source: Velayutham and Bhattacharyya, 2000

most productive agricultural lands. The problem of rise in the water table is more severe in arid and semi-arid regions of India and certain other areas of central Asia. Uzebekistan and Kazakhstan are two other countries where excessive irrigation has raised water tables by as much as 29 m. In India and Pakistan, the rise in water tables is mainly in areas irrigated by canals and open distributaries. Many of the large irrigation projects in these two countries in South Asia were constructed without proper drainage systems. Lack of subsurface drainage to collect excessive percolation water from surface irrigation methods has resulted in the rise of water tables. One of the best examples in India is where the Indira Gandhi Canal was brought in 1961 to irrigate the driest areas of Rajasthan. Water tables rose at a rate of 1 m per year after the beginning of intensive irrigation practices in Rajasthan (Hooja et al., 1994). In other areas of Rajastan, excessive canal irrigation made areas unfit for cultivation during monsoon season (Rao, 1997). Water table rise of 0.3 m to 0.8 m has been observed in Haryana, Uttar Pradesh and Punjab. Rise in water tables has caused waterlogging in more than 8.5 Mha and has added salinity problems to an additional 3.9 Mha (Singh and Bandyopadhay, 1996). Yadav (1996) has reported that annual increase in waterlogged areas varies from 6,500 ha in Gujrat to 195,000 ha in Uttar Pradesh. Also, Uttar Pradesh has reported an annual addition of 50,000 ha area to salinity buildup due to mismanagement of irrigation practices (Yadav, 1996).

Fall in the Water Table (Water Quality and Seawater Intrusion)

In several areas of India, Pakistan, Nepal, and Sri Lanka, groundwater is the major source of irrigation water. In Punjab, Haryana, and western Uttar Pradesh, shallow groundwater is pumped and used to meet an extensive irrigation-system network. This has caused significant water withdrawal from good-quality aquifers. In Punjab, nearly 70% of the area is irrigated with tubewells (shallow water-table wells) where the number of tubewells increased from 192,000 in 1971–72 to 800,000 in 1993, resulting in an average lowering of water tables by 0.2 m per year (Yadav et al., 2000). This has resulted in more discharge from aquifers than their annual recharge from rainwater. This mismanagement of irrigation is making less water available for currently practiced cropping systems.

Another water quality problem being observed is in the coastal areas. Decline in water tables due to excessive pumping for irrigation has led seawater intrusion into the groundwater aquifer, causing severe damage to the groundwater quality. This has resulted in thousands of wells out of use for irrigation in Gujrat, Orissa, and Andhra Pradesh (Gupta et al., 2000).

Nitrate and Pesticide Pollution of Groundwater

Generally, chemical use in India is low, so nitrate and pesticide pollution of groundwater is not a serious problem for the majority of the groundwater aquifers. However, in certain areas of India (such as Punjab, Haryana, and Uttar Pradesh), intensive grain production systems have used fertilizer and pesticide application rates similar to that of the United States. In these areas, nitrate and pesticide pollution of groundwater sources have been reported. In Punjab, the average NO_3-N concentrations of 2.25 mg/l to 10 mg/l in shallow groundwater have been reported (Gupta et al., 2000). However, Bajwa (2001) mentioned that nitrate concentrations of more than 100mg/l have been observed at times in selected tubewell waters during excessive irrigation in Punjab. Several studies have indicated that 11 to 48% of applied nitrogen in maize–wheat production systems have leached into groundwater systems. Gupta et al. (1999) have found nitrate concentrations of 12 to 16 mg/l in Talkatora Lake near Jaipur, Rajastan, as a result of urban pollution.

Highly soluble chemicals such as nitrate could quickly leach into the soil with rain or irrigation water before they can become part of surface runoff. Subsurface drainage water can then transport these chemicals into surface and groundwater. Chemicals in the strongly adsorbed group include herbicides such as paraquat and triflualin, as well as many of the now banned insecticides (such as DDT, dieldrin, and heptachlor). The majority of herbicides used today fall into the moderately adsorbed group. Several studies have shown that herbicides such as atrazine, alachlor, and cyanazine are mainly lost in surface runoff (Baker and Johnson, 1983, CTIC, 1994) but have also been found in shallow groundwater sources (Kanwar, 1991, Kanwar et al., 1993: Kalita et al., 1997). Kanwar et al. (1997) reported NO_3-N and pesticide losses to shallow groundwater systems under intensive agriculture for conventional and conservation tillage systems. Table 10.9 gives the yearly NO_3-N losses with drain water, which ranged from 4.8 kg/ha in 1992 to 107.2 kg/ha in 1990. The 3-year average (1990–92) NO_3-N losses with drain water were much higher under continuous corn than with corn–soybean rotation for all tillages. Although NO_3-N concentrations were greater under conventional tillage (moldboard plow + disking) than under a no-till system, total NO_3-N losses with subsurface drain flow were higher under the no-till and chisel plow systems because of greater volume of water moving

TABLE 10.9
Average Yearly NO$_3$-N Losses with Subsurface Drainage Water as a Function of Tillage and Crop Rotation (1990-92) NO$_3$-N loss, kg/ha

Year	Crop rotation	Chisel plow	MB plow	Ridge-till	No-till
1990	Continuous corn	100.0	58.1	83.4	107.2
1991	same	76.0	62.7	58.2	61.7
1992	same	17.0	16.6	10.2	14.9
Average (1990–92)	same	64.3a	45.8a	50.6a	61.2a
1990	Corn-soybean	52.4	38.0	30.3	36.5
1991	same	36.3	35.5	29.4	30.3
1992	same	15.3	9.1	11.2	4.8
Average (1990–92)	same	32.1a	27.5a	23.7a	23.9a

Source: (Kanwar, 1994; Kanwar et al., 1997).

through the soil. The data on average monthly NO$_3$-N concentrations in the individual plot piezometers indicate that, under continuous corn, NO$_3$-N concentrations at 1.8 and 2.4 m depths were higher in comparison with the corn–soybean rotation. Such data are not available for principal agro-ecoregions of India or elsewhere in South Asia.

Table 10.10 gives the total yearly losses of herbicides with drain water as a function of tillage and crop rotation for 1990. These data indicate that atrazine losses were greatest in comparison with other herbicides. Also, no-till and ridge-till systems caused greater losses of atrazine, cyanazine and metribuzin because of the preferential movement of these herbicides through macropores. The total yearly average losses for atrazine and alachlor ranged from 2.2 to 7.3 g/ha and 0.06 to 0.62 g/ha, respectively. Similar data are needed for countries of South Asia.

BEST MANAGEMENT PRACTICES TO CONTROL ENVIRONMENTAL DEGRADATION OF SOIL AND WATER RESOURCES

Best management practices (BMPs) are those that control soil erosion, minimize nonpoint source pollution and are economically, socially and environmentally acceptable. The following BMPs could possibly control environmental degradation of soil and water resources:

BMPs TO CONTROL SOIL DEGRADATION

Erosion is the number-one soil degradation problem in South Asia. Soil erosion causes many problems for agriculture and the environment. The loss of fertile soil decreases the production potential of farmland. Most of the eroded soil finds its way

TABLE 10.10
Average Herbicide Loss with Subsurface Drain Water as a Function of Tillage and Crop Rotation (1990-92)

Crop Rotation	Herbicide Type	Herbicide loss with subsurface drain water g/ha.			
		Chisel Plow	MB Plow	Ridge-Till	No-Till
Continuous corn	Atrazine	4.4	2.17	5.9	7.3
	Alachlor	0.36	0.06	0.34	0.31
	Cyanazine	0.10	0.25	0.19	0.16
Corn-soybean	Alachlor	0.05	0.62	0.39	0.16
	Metribuzin	1.7	1.70	3.4	2.5
Soybean-corn	Alachlor	0.79	0.06	0.11	0.16

Source: Kanwar, 1994; Kanwar et. al., 1997

to rivers and streams, causing environmental degradation of water resources for fisheries and other aquatic life. Erosion from intensive farmlands can carry pesticides and fertilizer residues to rivers and streams, adversely affecting the aquatic environments. The extensive soil loss and damage caused by intensive farming methods create agricultural systems that are not likely to have long-term sustainable soils or landscapes.

Vegetative Land Cover

One of the most effective practices to control soil erosion is the maintenance of permanent cover on the land surface. This practice is known as conservation cover. Conservation cover requires establishing and maintaining perennial vegetative cover to protect the soil on land that is not used for agricultural crop production. This will require planting and maintaining of locally suitable grasses, trees, shrubs, vines or legumes in areas on landscapes and around fields and streams that are susceptible to erosion or are in the pathways of field erosion. These control measures may involve restoration of riparian zones that have been cleared by agriculture over the years. Riparian zones are important because they are living filters; they trap and stabilize stream banks, improve water quality of rivers, and establish wildlife habitats (NRCS, 2001). Because large areas of land on sloping grounds are either being farmed for crops or used for animal grazing, other conservation cover methods, such as rotational grazing and conservation tillage practices, are necessary.

Conservation Tillage Systems

The concept of conservation tillage was started in the United States in the 1950s, but was not widely used and accepted until some 30 years later (NRCS, 2001). Any tillage system that leaves at least 30% of the soil surface covered with crop residue after harvesting is defined as a conservation tillage system (CTIC, 1994). This practice has helped replace conventional plowing in many areas of the United States to reduce soil erosion from water runoff. Conservation tillage has not yet been promoted on a large scale in South Asia. Several conservation tillage systems (namely no-till, ridge-till, and chisel plow) are being used to reduce soil erosion and energy input costs, but these systems may require more pesticide use. In recent studies conducted at Iowa State University (Kanwar et al., 1993; 1997; Kanwar and Baker, 1993), it was concluded that conservation tillage systems increase infiltration, organic matter, adsorption of pollutants, microbial activity, and decrease chemical leaching to groundwater. Conservation tillage is an effective BMP for controlling groundwater pollution and reducing soil erosion.

Cropping Systems

Diverse cropping systems are currently used in South Asia. Narrow row width and densely planted crops such as small grains and legumes affect infiltration and runoff volumes. These cropping systems seem to reduce soil erosion and chemical concentrations in the runoff water. Crop rotations also affect the use of chemicals. For example, corn–soybean rotation will not use nitrogen fertilizer in the soybean years, whereas continuous corn practice will use nitrogen year after year. Also, crop rotations offer a greater diversity of pesticide use within a watershed to control nonpoint source pollution. Kanwar et al. (1993) concluded that growing continuous corn increases soil erosion, needs higher N application rates and results in higher NO_3-N losses to the groundwater.

Contouring, Terracing, Filter and Buffer Strips, and Well Buffer Zone

Land topography and soil types confound runoff volumes and soil erosion rates. Terracing and contour farming have been used widely for centuries to create better farming conditions and control and conserve soil and water. Contour farming is a practice that involves farming (planting, cultivating, and harvesting) along contour lines on a sloping land. This method establishes terraces or diversions that are effective in slowing down runoff and reducing loss of sediment (EPA, 2001). Contour farming can reduce soil erosion by as much as 60 to 80% compared with the traditional up-and-down method of farming. Also, vegetative filter and buffer strips and waterways are potential BMPs to mitigate water pollution problems. Vegetative filter strips and grassed waterways have been found to reduce soil erosion and the pesticide loss from 19 to 22% with runoff water (Kanwar, 1994). Also, grassed strips of 5m to 10m around open leaky water wells can filter sediments and chemicals from runoff water and reduce the contamination potential of well water, typically a source of drinking water in rural areas of South Asia.

Integrated Fertility and Nutrient Management

In many areas of South Asia, especially India, growing crops on soils with low natural nutrient supply and restricted input is a challenge. Uptake efficiency of nutrients by different crops depends on several factors, including crop varieties, soil moisture, natural supply of macro- and micronutrients and other hydrogeologic factors. Good nutrient management is essential for sustainable agriculture. For arid regions, with better crop nutrition, less water is required to meet transpiration needs of plants (Velayutham and Bhattacharyya, 2000). In humid regions, nutrient management must address unavoidable nutrient leaching losses to groundwater and runoff losses with excessive irrigation. In acid soils, maintaining a soil pH of 5.5 to 6.0 through nutrient management is needed. Practices of liming with rock phosphate should be practiced, especially for sulfate soils. On saline and sodic soils, nutrient management involves selection of the right crops for planting. Acid-tolerant crops such as coffee, tea, rubber, pineapple, and jackfruit do very well. On some of the waterlogged acid soils of eastern India, indigenous varieties of rice and sorghum have done well. For example, rice, wheat and legume rotation systems have given very high yields on freshly reclaimed sodic soils of Uttar Pradesh and Punjab (Velayutham and Bhattacharyya, 2000). Several researchers have reported that organic matter contents have gone down considerably in most of India's soils. To sustain increased yield potential, long-term efforts are needed to slowly increase organic matter contents (which may take hundred of years) and natural soil fertility.

BMPs to Control Degradation of Water Resources

Irrigation and Drainage Systems

In India and Pakistan, one of the major problems is low irrigation efficiency. In these countries it was a mistaken belief that water from irrigation supplies alone would perform miracles in increasing crop yields. Farmers and policy makers there have now realized that appropriate soil and water management with other inputs is needed to maximize the benefits of irrigation water. The main factors responsible for poor irrigation management systems in South Asia are: government control of irrigation projects, subsidized water pricing policy, undependable water supply, lack of legislation on groundwater development for irrigation, lack of training to farmers on irrigation methods, high seepage, leakage, and percolation losses, low irrigation efficiencies, and adverse environmental impacts of waterlogging and soil salinity. Farmers need to be given adequate training to improve their irrigation methods. Also, introduction of new techniques, such as sprinkler and drip methods, can save water and will significantly improve irrigation efficiencies. New methods of irrigation will increase nutrient efficiency and reduce water contamination caused by agricultural chemicals. Irrigation and drainage practices are typically considered production practices rather than BMPs for water quality enhancement. Irrigation management is important in controlling water quality-related problems. The rate, amount, and timing of irrigation are important considerations. Several irrigation systems that include considerations are surface flow, furrow disking, reuse pits to collect irrigation tail water for reuse, low-energy precision application method, drip and subirrigation methods and use of chemigation. Local hydrologic and geologic factors must be considered before selecting the irrigation management practice. Kalita et al. (1997), and Kalita and Kanwar (1993) found that better water table and subirrigation management practices could be used to reduce the risk of groundwater contamination.

Practices for Minimizing Waterlogging

Much of the waterlogging in India has occurred due to canal irrigation. In addition, waterlogging has occurred in India and Pakistan due to inadequate drainage, leakage and seepage from surface channels, changes in cropping pattern in favor of crops like rice in Punjab, irrigation with poor quality groundwater such as brackish water, and poor farm-water management. The best practice to minimize the increase of waterlogging in canal-irrigated areas is to provide an improved drainage system to lower high water tables in the waterlogged areas. The installation of subsurface drainage systems and pumped drainage could provide the needed relief to bring waterlogged areas under productive agriculture. Once water tables are lowered, deep-rooted crops can help reclaim these areas permanently. Overall, an integrated approach is needed, including preventive and reclamative strategies to be implemented to reclaim waterlogged areas.

BMPs for Minimizing Salinity

One of the best-known BMPs to correct the salinity or alkalinity problem is the reclamation of sodic and saline soils through chemical and biological amelioration. Sodic soils can be easily reclaimed using gypsum, and solubilizing calcium and sodium salts and flushing them out of the active root zone. These methods have been found to be extremely successful in India and Pakistan. Other methods are biological controls such as growing appropriate vegetation and adopting proper management of irrigation and drainage practices. Without adequate drainage systems, accumulated salts cannot be flushed out. In addition, mechanical treatments could include deep tillage to provide better infiltration of water through the soil profile to flush salts. Leaching and drainage are the essential parts of reclamation for chemical and biological methods.

Placement of Chemicals

One of the approaches to reduce the leaching of chemicals to groundwater or surface water under rain-fed and irrigated conditions would be to incorporate the chemical into the soil. If a chemical is broadcast, it will mix with the incoming rainfall or irrigation in a thin mixing zone (about 10 to 20 mm thick) and will either leach to subsoil layers or become part of the runoff water. If a chemical is incorporated, it is less susceptible to runoff losses. With banding practice, the rate of chemical application can be reduced to more than 50% or more. Kanwar and Baker (1993) have shown that banding of herbicides has a significant effect on water quality improvement.

Timing of Chemical Application

Appropriate timing of chemicals under irrigated agriculture can increase the efficient use of chemicals (especially N) and result in decreased leaching losses to groundwater. Kanwar and Baker (1993) observed that split-N applications resulted in lower residual N in the soil profile, and NO_3-N concentrations in the subsurface drainage water were at or below 10 mg/l during the 9-year study period. Hydrologic factors, primarily rainfall patterns, have significant interactions with the timing of chemical applications.

Chemical Rates and Methods of Applications

Reductions in the rate and total quantity of chemical applied could reduce the amount of chemical available for leaching to groundwater or runoff. Kanwar et al. (1988) and Baker and Johnson (1983) have summarized the results of several field studies indicating that higher NO_3-N concentrations in groundwater were related to higher N applications.

Wetlands

In the United States, use of wetlands is becoming a very good practice for erosion control, flood reduction, and water quality improvements (USCE, 2001). This practice has tremendous potential for South Asia to control floods and soil erosion and minimize water quality problems. Wetlands can be established to trap sediments, nutrients, pesticides, and other organic compounds and create cleaner aquatic environments. Water that is treated and discharged from wetlands is considerably improved over its initial quality (Mitsch and Gosselink, 1993). Wetlands can store large quantities of water during flooding and thus could minimize the damage from floods. Also, temporary storage will decrease runoff velocity and reduce flooding peaks. Wetlands have been considered to be the most productive ecosystems in the world (Mitsch and Gosselink, 1993). Wetlands contain large varieties of microbes, plants, insects, reptiles, birds, and fish. Combining conservation techniques with the benefits of wetlands has potential to create farming systems that are sustainable and can meet the food security needs of society.

CONCLUSIONS

Food security is still a major problem for a large portion of the world's growing population. Poor farmers of South Asia have a monumental task to produce more food to meet the increasing needs of the population in view of the declined productivity of soils, decreased land holdings per person, increased costs of inputs, decreased availability of canal and groundwater for irrigation, lowering of water tables through unlimited pumping of water for irrigation, poor groundwater quality for irrigation and degradation of land due to salinity and alkalinity. To maintain a steady supply of food, countries in South Asia must make sure that all efforts are made to develop best management practices and the needed governmental policies to preserve soil and water resources.

South Asia has been blessed with two major natural resources, relatively productive land and a good reservoir of water resources. At the same time, South Asia has 21% of the world's population and one of the highest population densities and population growth rates. Increased population pressure is expected to shrink per capita cultivatable land still further in the years to come. Demands on finite water resources are increasing and, with the increase in population, contamination of water resources is on the rise. Also, increase in the population in South Asia means intensification of agricultural production systems to feed the growing population. This means demand for irrigation water and agricultural chemicals will increase to produce more food, resulting in the pollution of soil, water, air, and other natural environments even further. Intensification of agriculture in India and Pakistan has increased soil erosion due to deforestation, waterlogging due to poorly managed irrigation systems, increased soil salinity and pollution of drinking water supplies. All these factors have placed enormous stress on available land and water resources. Unless best irrigation and cropping management systems are developed in agricultural watersheds to protect degrading land and water resources in South Asia, social and food security is at very much risk.

Intensive agricultural production systems in South Asia have caused significant soil degradation and nonpoint source pollution. Some of the best management practices that are needed to control the degradation of soil and water resources in the area include the use of conservation tillage systems on terraced and contour farming fields to control soil erosion. To control nonpoint source pollution, farmers need to use integrated nutrient and pesticide management practices to reduce the leaching of chemicals to surface and groundwater resources. To control waterlogging and salinity problems, unwise canal irrigation and pumping of groundwater need to change through government legislation and by developing new policies on water subsidies and drilling of groundwater wells. Farmers' training programs on safe application of chemicals, nutrients management, use of wetlands, and improved irrigation application methods must be developed and offered regularly in villages.

REFERENCES

Abrol, I.P. and D.R. Bhumbla. 1971. Saline and Alkali Soils in India: Their Occurrence and Management. Paper presented at FAO/UNDP seminar on soil fertility research. FAO World Soil Research Rep. 41:42-51.

Baker, J.L. and H.L. Johnson. 1983. Evaluating effectiveness of BMPs from field studies. In: Agricultural Management and Water Quality. Iowa Sate University Press, Ames, Iowa.

Bajwa, M.S. 2001. (Oral communication) Nitrate pollution of groundwater in Punjab. Punjab Agricultural University, Ludhiana, India.

Bouwer, H. 1993. Sustainable irrigated agriculture: water resources management in the future. Irrigation J. 43(6):16-23.

Comly, H.H. 1945. Cyanosis in infants caused by nitrate in well water. JAMA 129:112-117.

Conservation Technology Information Center. (1994). Best Management Practices for Water Quality. CTIC, West Lafayette, Indiana., pp. 43.

Council for Agriculture Science and Technology (CAST). 1985. Agriculture and Groundwater Quality. CAST report no. 103. Ames, Iowa.

Environmental Protection Agency (EPA). 2001. Erosion and Sediment Control Management. http://www.epa.gov/OWOW/NPS.

Foreman, D.S., Al- Dabbagh, and R. Roll. 1985. Nitrates, nitrifies, and gastric cancer in Great Britain. Nature 313: 620-625.

Fraser, P., C. Chilvers., V. Beral and M.J. Hill. 1980. Nitrate and human cancer: a review of the evidence. J. Epidemiol. 9:3-9.

Gleick, P.H. 2000. The World's Water 2000-2001, the Biennial Report on Freshwater Resources. Island Press, Washington, D.C.

Gupta, A.B., R.Jain and K. Arora. 1999. Water quality management for the Talkatora Lake, Jaipur – a case study. Water Sci. Tech. 40(2):29-33.

Gupta, S.K., P.S. Minhas, S.K. Sondhi, N.K. Tyagi and J.S.P. Yadav. 2000. Water resources management. In: *Natural Resource Management for Agricultural Productivity in India* (J.S.P Yadav and G.B. Singh, Eds.). Indian Society of Soil Science, New Delhi, India, pp. 137-244.

Hallberg, G.R. 1989. Pesticide pollution of groundwater in humid United States. *Agric. Ecosyst. Environ.* 26:299-367.

Hooja, R.V., S. Niwas, and G. Sharma. 1994. Waterlogging and possible remedial measures in Indira Gandhi Canal command area development project. National Seminar on Reclamation and Management of Waterlogged Soils, Karnal, India.

Johnson, C.J. and B.C. Kross. 1990. Continuing importance of nitrate contamination of groundwater and wells in rural Iowa. *Am. J. Ind. Med.* 18:449-456.

Johnson, C.J., P.A. Bonrud, and T.L. Dosch. 1987. Fatal outcome of methemoglobinemia in an infant. *JAMA* 257:27296-2797.

Kalita, P.K. and R.S. Kanwar (1993). Effect of water table management practices on the transport of Nitrate–N to shallow groundwater. Transactions of the ASAE, 36(2):413-422.

Kalita, P.K., R.S. Kanwar, J.L. Baker and S.W. Melvin. (1997). Groundwater residues of atrazine and alachlor under water table management practices. Transactions of the ASAE 40(3):605-614.

Kanwar, R. S., J.L. Baker and D.G. Baker. (1988). Tillage and split N-fertilization effects on subsurface drainage water quality and corn yield. Transactions of the ASAE 31(2):453-460.

Kanwar, R. S. (1991). Preferential movement of nitrate and herbicides to shallow groundwater as affected by tillage and crop rotation. In: *Proceeding of the National Symposium on Preferential Flow* (T. J. Gish and A. Shirmohammadi, Eds.). *Am. Soc. Ag. Engr.*, pp. 328-337.

Kanwar, R. S. and J.L. Baker. (1993). Tillage and chemical management effects on groundwater quality. In: *Proceedings of National Conference on Agricultural Research to Protect Water Quality*, SCS, Ankeny, IA, pp. 490-493.

Kanwar, R. S., D.E. Stolenberg, R. Pfiffer, D.L. Karlen, T.S. Colvin and W.W. Simpkins. (1993). Transport of nitrate and pesticides to shallow groundwater systems as affected by tillage and crop rotation practices. In: *Proceedings of National Conference on Agricultural Research to Protect Water Quality*, pp. 270-273.

Kanwar, R.S. 1994. Environmental Evaluation of Surface and Groundwater Resources. A Platinum Jubilee Lecture, 81st India Science Congress, Jaipur, India.

Kanwar, R. S., T.S. Colvin, and D.L. Karlen. 1997. Effect of ridge till and three tillage systems and crop rotation on subsurface drain water quality. *J. Prod. Agric.* 10:227-234.

Mitsch, W.J. and J.G. Gosselink. 1993. *Wetlands.* Van Nostrand Reinhold Publishers, New York..

Natural Resource Conservation Service (NRCS). 2001. Sedimentation and Soil Erosion. http://www.nrcs.usda.gov.

Rao, K.V.G.K. 1997. Man's interference with environment in water-use problems of waterlogging and salinity. In: *National Water Policy – Agricultural Scientists' Perception. Proceedings of the Round Table 10 Conference*, August 12-14, 1994. National Academy of Agricultural Sciences, New Delhi, India, pp. 68-80.

Scherr, S. J.1999. Soil degradation: A threat to developing country food security by 2020. International Food Policy Research Institute, 2020 Brief 58, Washington, D.C.

Shiklomanov, I.1993. World freshwater resources. In: *Water in Crisis: A Guide to the World's Fresh Water Resources* (P.H. Gleick, Ed.). Oxford University Press, New York, pp.13-24.

Singh, N.T. and A.K. Bandhopadthyay. 1996. Chemical degradation leading to salt affected soils and their management for agricultural and alternative uses. Soil Management in Relation to Land Degradation and Environment. Bulletin Indian Society of Soil Science, New Delhi, India, pp. 89-101.

United Nations Food and Agriculture Organizations (UNFAO). 1999. FAOSTAT database.htttp://faostat.fao.org.

United States Corps of Engineers (USCE). 2001. Wetland delineation manual.http://www.wetlands.com

Velayutham, M. and T. Bhattacharyya. 2000. Soil resource management. In: *Natural Resource Management for Agricultural Productivity in India* (J.S.P Yadav and G.B. Singh, Eds.). Indian Society of Soil Science, New Delhi, India.

Yadav, J.S.P. 1996. Extent, nature, intensity, and causes of land degradation in India. In: Soil Management in Relation to Land Degradation and Environment. Bulletin Indian Society of Soil Science, New Delhi, India, pp. 1-26.

Yadav, J.S.P. and G.G. Singh. 2000. Natural Resource Management for Agricultural Productivity in India. Indian Society of Soil Science, New Delhi, India.

11 Environmental Quality: Factors Influencing Environmental Degradation and Pollution in India

Clive A. Edwards

CONTENTS

INTRODUCTION

The productivity of Indian agriculture has grown rapidly in recent decades, from severe food shortages in the 1960s to overall national food surpluses in the 1990s (Evenson, et al, 1999). Nevertheless, Indian agriculture still faces many challenges to keep food production in pace with its extremely rapid rates of population increase. For instance, it has been forecast that the demand for cereals in India could exceed domestic production by 24 million metric tons by 2020, by which time India's population has been predicted to grow by a further 3 billion (Edwards and Pimentel, 2001). Although India currently appears to be self-sufficient for food, this may be a fallacy, because as many as 53% of children have been reported to be malnourished, and clearly this problem is liable to increase. India's population has nearly tripled over the last half century, growing from 350 million in 1950 to 1 billion in 2000, with UN predictions of future population growth expected to add another 515 million people by 2050 (Edwards and Pimentel, in press). This will cause major problems in food supply, health and environmental issues. Population growth decreases the availability of land per capita and the amount of land available to each farmer. In India, the number of farms of less than 5 acres (2 ha) increased from 49 million to 82 million between 1970 and 1990, meaning that more than half of India's population has barely enough land to subsist.

As a result of the Green Revolution, rice yields have doubled in India since 1956 from approximately 1 ton per ha to 2 tons per ha (Brown, 1997). Double cropping, e.g., growing wheat and rice in the same season, is commonplace in northern India, and this has also contributed to increased crop yields. However, crop production increasingly suffers from a lack of irrigation water, exacerbated by the fact that India has shorter days during the growing season than some other Asian countries (Abramovitz, 1996). Although India has quadrupled its wheat yield per acre since 1950, mainly through extensive use of fertilizers and pesticides, the progressively increasing prices and availability of these chemicals raises questions on how long such increases can continue. India has also been remarkably successful in increasing milk production by converting crop residues into animal feed. Milk production expanded from 19 million tons in 1966 to 79 million tons in 2000.

A second major problem is the progressive pollution of the Indian environment, soils, waters and air, with agricultural chemicals and liquid, solid and gaseous emissions from automobiles and industrial operations. The rapid increases in agricultural production have been largely based on heavy use of agrochemicals, and the rapid industrialization in India has led to production of effluents and gaseous emissions in very large quantities, before methods of limiting them like those in the U.S. have become fully developed. It may take many years for some ecological compartments of the environment to recover from this pollution.

India has been included in a group of large industrialized and developing countries, termed by the World Watch Institute as E-9 countries (Table 11.1) (World Watch Institute, 2001). This group of countries accounts for 57% of the world's population and 80% of the total economic output. India has the second largest population among this group, but is only sixth in terms of gross national product (GNP), and has critical environmental problems as a result of agricultural and industrial activities. For instance, in India, 19% of the population has no access to clean water and 84% have no toilet facilities available. Adult illiteracy rates are high, and 45% of men and 57% of women can be considered to be illiterate (World Watch Institute, 2001). In terms of air pollution, India ranks fifth of the E-9 countries for sulfur dioxide pollution, and sixth for suspended particulates (Flavin, 2001).

TABLE 11.1
The E-9 States: A Population and Economic Profile

Country or Grouping	Population, 2000 (millions)	Gross National Product, 1998 ($ billions)
China	1,265	924
India	1,002	427
European Union	375	8,312
United States	276	7,903
Indonesia	212	131
Brazil	170	768
Russia	145	332
Japan	127	4,089
South Africa	43	137

Source: World Bank, *World Development Indicators 2000* (Washington, DC: 2000), 10-12; Population Reference Bureau, "2000 World Population Data Sheet," wall chart (Washington, DC: June 2000).

LAND DEGRADATION AND POLLUTANTS

LAND DEGRADATION

All over the world, intensive agricultural crop production has eroded, compacted, waterlogged, or contaminated enormous areas of productive cropland (Gardner, 1996). It has been calculated that, globally, 552 million ha of cultivated land have been damaged to some degree by agricultural mismanagement since 1950 — and this is believed to be a conservative estimate. In Asia, about 38% of agricultural land has become moderately or seriously degraded between 1945 and 1990 (Gardner, 1997). Much of this degraded land remains in production, although it is much less fertile than it was originally. It has been calculated that the loss of productivity of lightly and moderately degraded land is about 10% on average; but, when the land has become very degraded, productivity losses can be as much as 18% (World Watch Institute, 1995).

Soil degradation affects one quarter of India's agricultural land (Gardner, 1996). A total of nearly 70 million ha have been affected by water erosion and nearly 15 million ha by wind erosion. Erosion associated with shifting cultivation has degraded about 27,000 square kilometers of land east of Bihar. About 14 million ha have become subject to serious decreases in soil fertility and about 4 million ha of land have been lost completely through salinization. In India, most of the available cropland is already cultivated and there is relatively little potential for expansion to new areas, as is possible in some countries. Crop production can be increased significantly only by intensification of cropping, which often would have to be on land that is already somewhat degraded.

SOIL PESTICIDE POLLUTION

Insecticides dominate the Indian pesticide market, with a share of about 74% of total pesticides sold. Farmers are using increasing amounts of pyrethoid insecticides and the demand for organophosphate insecticides is decreasing (AGROW, 1997). Herbicides and fungicides account for about 12% each and, in recent years, both have been increasing their market share by about 1% annually (Table 11.2). India is one of the few remaining countries still engaged in the large-scale manufacture, use and export of some of the most toxic persistent organochlorine pesticides, such as DDT (1,1-bis(4-chlorophenyl)-2,2,2-trichloroethane), HCB (hexachlorobenzene) and pentachlorophenol. Approximately 125 companies manufacture more than 60 technical-grade pesticides in India, including BHC (benzene hexachloride), DDT and methyl parathion, with an estimated total production capacity of 126,000 tons. In 1996, companies in India produced 86,000 tons of pesticides and the volume of pesticides sold in India rose by 5% to 83,400 tons, including 18,000 tons of BHC (Agrow, 1996). Insecticides composed 67% of the market, fungicides 22% and herbicides 10%. The following year, 1997, saw a growth of 7.5% in the value of the Indian pesticide market to US$602 million, including 20,000 tons of BHC (Agrow, 1997) (Table 11.2).

Although government regulations have been in place to restrict the use of organochlorine insecticides in India since the 1970s, these pesticides are still used and continue to be serious pollutants of soils and groundwater, either through persistent residues or unrecommended or illegal use (Edwards et al., 1980). For instance, after half a century of spraying these insecticides, in the eastern India states of West Bengal and Bihar, the India Central Pollution Control Board reported DDT in groundwater at levels to be as high as 4,500 µg per liter,

TABLE 11.2
Estimated Indian Production of Selected Insecticides (Tons)
1996 & 1997

Pesticide	1994/95	1996/97
Organochlorines		
BHC	32,000	20,000
DDT	4,300	4,400
Endosulfan	6,700	7,000
Organophosphates		
Malathion	2,800	4,000
Mancozeb	4,100	4,200
Monocrotophos	8,000	10,000
Methyl parathion	2,100	2,400
Phorate	4,100	4,100

Source: AGROW, World Crop Protection News, July 12 and September 27, 1996 and September 12 and October 3, 1997

which is several thousand times higher than the level generally acceptable for human health. This is in spite of a body of evidence that these persistent pesticides break down faster under India's warm, humid climate conditions (Edwards, et al, 1980).

Many pests are now resistant to a range of pesticides, often from different chemical classes. This rapid increase in pesticide resistance produces the paradox that despite enormous increases in pesticide use since the 1950s, the overall share of crops lost to pests has not changed much. The rapid commercialization of genetically modified crops containing *Bacillus thuringiensis* is also increasing the resistance to this microbial pesticide (Wood et al., 2000). One recent global estimate is that as many as 1000 major agricultural pests are now totally immune to pesticides (Brown et al., 1999), including 394 species of insects and mites and 71 weed species, and others have some degree of resistance. This leads to a constant pressure on growers to increase pesticide dosage rates or to adopt usage of newer, and often more expensive, alternative pesticides. These are not pest management strategies that are sustainable in the long-term.

Much of the success of the Green Revolution depended upon the use of pesticides, and pesticide use globally as well as in Asia continues to increase, although it is beginning to show signs of leveling off (Edwards, 1994). It seems unlikely that, in the long-term, e.g., after the end of the current century, pesticides will still be an available answer to pests, disease and weed problems (Edwards and Pimentel, 2000). Currently, most synthetic pesticides are based on fossil fuel compounds, and various predictions indicate that most known sources of fossil fuels will become virtually exhausted or prohibitively expensive over the next 50 years (Edwards and Pimentel, in press). A major problem in crop production systems, based on heavy pesticide use, is that a pesticide dependency develops progressively, due to the gradual elimination of natural enemies of pests that would limit pest increases in the absence of pesticide use. On a visit to Bangalore, India in 1974, this author noted that only three pesticides were used on cotton, but on a subsequent visit in 1978, it appeared that the use of pesticides on cotton had increased dramatically, and many farmers were spraying cotton weekly with insecticides. As a result, whereas in 1974 there were only three serious pests of cotton, in 1978 there were as many as seven causing serious losses, presumably due to the elimination of their natural enemies.

Heavy pesticide use has other negative aspects. It can cause serious human health problems, particularly under tropical conditions, when it may be too hot and humid to wear adequate protective clothing and where many farmers, some of whom may be illiterate, may not be aware of the serious health problems associated with many pesticides (Edwards, 1994). Other serious environmental concerns relate to the toxicity of many pesticides to a range of biological species other than those directly targeted, including insects, soil organisms, aquatic invertebrates and fish, plants, mammals and birds.

India imports relatively small quantities of commercially formulated pesticides. It produces the active ingredients for a number of pesticides and formulates most of the rest (Table 11.2). Only the active ingredients and finished pesticide formulations of the most recently released pesticides developed are imported currently.

SOIL FERTILIZER POLLUTION

Although the amounts of nutrients can gradually decrease in soils if they are not replenished biologically or by inorganic amendments, excessive or mistimed applications of inorganic fertilizer, or even organic fertilizers, can lead to nutrient runoff or leaching and consequent groundwater pollution problems with nitrates (Evenson et al., 1999). Northern India has been ranked as at risk from groundwater nitrate pollution arising from fertilizer runoff, livestock operations, septic systems and human feces, commonly termed "night soil." Nitrates can cause sickness in humans and produce algal blooms and eutrophication in freshwater systems. Globally, fertilizer use has been in a gradual decline since 1989. However, in general, considerably larger rates of fertilizers are used in Asia than in North America and Europe (Table 11.3), even though they are expensive and mostly imported. This is mainly because of the poorer soils in India and other parts of Asia that tend to lose nutrients more rapidly.

Fertilizers, like pesticides, are usually produced from fossil fuels, so are open to the same long-term future constraints as pesticides, because it seems probable that fossil fuels will become seriously depleted or used up in the next 50–100 years, and, long before that, will become increasingly expensive (Edwards and Pimentel, 2001).

WATER SHORTAGES AND POLLUTION

Worldwide, water availability and water pollution are major issues, first in terms of supplies of clean drinking water, but also in terms of water suitable for crop irrigation, as 80–90% of the world's freshwater supplies are used for crop irrigation (Clarke, 1991). Serious increasing regional shortages of water have been predicted for the next century (Postel, 1996).

TABLE 11.3
Distribution of Use of Inorganic Fertilizers in Various Regions

Region	Inorganic fertilizer (kilograms per hectare)			
	N	P_2O_5	K_2O	Total
North America	57.1	21.6	23.1	101.8
Latin America/Caribbean	26.7	18.3	17.1	62.1
Europe	89.7	32.2	36.5	158.4
Former Soviet Union	14.0	4.5	2.3	20.8
West Asia/North Africa	39.7	18.1	3.3	61.1
Sub-Saharan Africa	6.1	3.4	2.1	11.6
East Asia	130.7	51.1	83.2	265.0
South Asia	62.9	19.3	6.6	88.8
Southeast Asia	50.2	16.6	17.0	83.8
Oceania	17.7	25.5	6.8	50.0
World	53.2	21.0	15.5	89.7

Source: FAO Stat 1999

WATER AVAILABILITY

The demands for water in India are already exceeding the available supplies (Rosegrant, 1997). Water tables are falling rapidly in many parts of the country, mainly due to overuse of water for crop irrigation, because pumping costs to supply agricultural crops are low. This lack of water leads to the drilling of ever-deeper wells. For instance, in the Punjab, water tables are falling by 20 cm annually across two thirds of the state. In Gujarat, groundwater levels declined significantly in 90% of wells that were maintained during the 1980s.

Most Indian irrigation systems use water inefficiently because of a lack of incentive for farmers to treat water as a scarce resource. Efficiency can be defined as the ratio of water actually used by crops in relation to the quantity applied. There is an urgent need for much better management of water in India. Harvesting and storing water during the monsoon season could increase the availability of water for irrigation during the dry season. Improved efficiency can be achieved in a number of ways: through government policies and regulation, and technologies such as field leveling, low-energy precision application, drip irrigation and moisture conservation.

Between 30 and 40% of the world's crop production comes from the 17% of the global land that is irrigated (Postel, 1996). For example, in India, the irrigated areas are one third of all cropland, but these account for 60% of overall crop production. Hence, the continued availability of irrigation is a critical issue (Sampat, 2001). India is the leading country in total irrigated area of cropland. The number of shallow tube wells used to pump groundwater increased from 3,000 in 1950 to 6 million in 1990 (Clarke, 1991). Currently, aquifers supply water to more than half of India's irrigated land. About 40% of India's agricultural output comes from areas irrigated with groundwater. This could cause serious future problems, because groundwater depletion is very severe and appears to be increasing.

WATER POLLUTION

Industrial effluents can pollute drinking water. For instance, in India, it was reported that scores of factories across five progressively industrializing state — Gujarat, Haryana, Punjab, Andrah Pradesh and Karnataka were injecting industrial effluents illegally into tube wells that were used for drinking and irrigation water. It has also been concluded that 30 million people in northwestern India are drinking water with high fluoride levels. Pollution of drinking water with arsenic has been reported to be serious in West Bengal, with many patients showing symptoms of chronic arsenic poisoning (Sengupta, 1999). The aquifer sediment in the Ganges delta is rich in arsenic, and when water sources were switched from surface to groundwater, exposure to arsenic poisoning increased significantly.

Salt pollution, when too much water is pumped from coastal aquifers, is another problem often introduced by human activity (Gander, 1996). When this happens, the process may be reversed and the aquifer can become seriously salinized, making the water unusable for drinking; such contamination which has been reported from parts of Gujarat state, and the city of Madras had to abandon some aquifers indefinitely due to salinization. At least 2 million ha of salinized land in India has been abandoned completely (Gardner, 1996).

Pesticides also pollute groundwater and rivers seriously in India through percolation and surface runoff (Eaglesham et al., 2000). Many rivers, such as the Ganges, are still seriously polluted with DDT, although the use of this insecticide is officially restricted. Residues of many other organochlorine insecticides such as dieldrin and heptachlor are present at unacceptable levels in river sediments and may persist there for many years.

Inorganic fertilizers are often used in excessive amounts, not necessarily based on actual needs, particularly in northern India. When heavy irrigation is also used, the movement of nitrates from inorganic fertilizers into drinking water via groundwater or surface runoff is common and can pollute drinking water well above World Health Organization (WHO) acceptable levels.

Microbial pollution of rivers and drinking water is also a very serious problem. The universal practices of sewage disposal, bathing and disposing of ashes of cremated corpses into rivers are common and cause serious microbial pollution. Only 17% of India's cities have partial or full sewage treatment facilities (Clarke, 1991). A 48-km stretch of the Yamuna River, which flows through New Delhi, contains 7,500 coliform organisms per 100 ml of water before entering the capital, but an enormous 24 million coliforms per 100 ml after it leaves, when the safe level of coliform organisms is about 100 per ml of water. Containment of such pollution will be an enormous and long-term problem facing the government.

AIR POLLUTION

The rapid industrialization that is occurring in India brings with it similar problems that Western industrialized nations are currently facing, in terms of emissions of sulfur dioxide, nitrous compounds and carbon dioxide, and their effects on the ozone layer and global climate.

GASEOUS EMISSIONS

Clearly, automobile production and use is increasing dramatically in India; for instance, in 1996, automobile production increased by 26% and these production rates will probably increase even further in the future. In urban areas, this will exacerbate an already high level of air pollution from automobile exhaust emissions, which is not yet under any form of regulation. Vehicles contribute between 50 and 60% of the pollution by key urban air pollutant. For example, Kanpur has particulates in the air that are more than five times the accepted health standards (Brown, et al, 1999). These vehicle emissions also contain those pollutants that contribute to global climate change to a significant degree (Parry, 1990).

Another pollutant produced from vehicles, unless the petroleum is refined, is lead, which can impair kidney, liver and blood-forming organs and cause brain damage to children. Recent studies suggest that 64% of children in Delhi have unhealthy blood levels of lead, probably resulting from automobile emissions (Brown, et al, 1999).

It has been estimated that air pollution in 35 large Indian cities killed 52,000 people in 1995, a 28% increase from the early 1990s. Calcutta has been quoted as one of the five worst cities in the world, in terms of exposure of children to air pollution by sulfur dioxide, particulates and nitrogen oxides. It is quite common to see people in Delhi and other major cities wearing face masks in the city and traffic to avoid breathing the heavy gaseous emissions.

CLIMATIC CHANGES

Emissions of carbon dioxide and effluent gases such as nitrous oxide (N_2O) from industry and automobiles, and methane (CH_4) emissions from various sources as well as carbon dioxide (CO_2) are believed to be having significant global impacts on the ozone layer (Parry, 1990). A United Nations review in October 2000 concluded that, if global emissions are not curtailed drastically, the earth's surface temperature might increase from as little as 2.7°C to as much as 11°C over the next century (Edwards and Pimentel, 2001). Such increases would have major impacts on ocean levels, water availability and crop production. India is one of the eight nations making the highest contributions to global emissions and, it seems likely that, with the rapid rates of increase in industrialization and automobile use, gaseous emissions from the Indian subcontinent will increase significantly in the future. In India itself, this could result in major changes in precipitation that would have serious effects on irrigated crops, although some current scenarios suggest precipitation might increase in India if temperatures rose.

CONCLUSIONS

Although Indian agriculture has been very successful in increasing crop productivity in the past, it faces severe obstacles in maintaining such increases to keep pace with predicted population increases in the long term. There is a finite limit on how much crop productivity per unit area of land can increase. It also seems likely that supplies of fertilizers and pesticides may become progressively attenuated or prohibitively expensive. Indian agriculture and agricultural research should focus on biological and cultural inputs to crop productivity and pest control, including the use of genetically modified crops, if these can be made available on terms acceptable to Indian farmers and economists.

There is considerable scope for much better utilization of a wide range of organic wastes to provide plant nutrients and improve soil structure and fertility. There should be a strong emphasis on the universal development of sewage treatment systems that process biosolids into materials that can be used safely on the land as sources of plant nutrients and for improving soil structure.

Much of food production depends on the availability of energy and renewable energy sources such as ethane and hydrogen-based systems, which could substitute for fossil fuels without the adverse effects of gaseous emissions from fossil fuels, which could produce negative climate changes.

There is an urgent need for a national infrastructure that would provide much better management of critical water supplies to ensure that they are used to maximum benefit. Indeed, the whole of Indian food productivity is linked to the issues of water availability, economy and management.

The increasing gaseous emissions from industry and automobiles must be controlled and, if possible, decreased, by use of alternative, renewable fuels, regulatory activities and government supervision. The only other option is to face productivity issues associated with climate changes produced by these emissions.

If these critical issues are not addressed urgently by research and government action, environmental degradation and pollution will inevitably prevent achieving the much-needed increases in productivity to feed the growing population.

REFERENCES

Abramovitz, J.N. 1996. Imperiled Waters, Impoverished Future: The Decline of Freshwater Ecosystems. World Watch Paper 128. Washington, D.C. 80 pp.

AGROW, 1996 World Crop Protection News. July 12 and September 21, 1996.

AGROW. 1997. World Crop Protection News. 3 October 1997.

Brown, L.R. 1997. The Agricultural Link: How Environmental Deterioration Could Disrupt Economic Progress. World Watch Paper 136. Washington, D.C. 73 pp.

Brown, L.R., Gardner, G., Halwell, B. 1998. Beyond Malthus: Sixteen Dimensions of the Population Problem. World Watch Paper 143. Washington, D.C. 89 pp.

Brown, L.R., Renner, M., Halwell, B. 1999. *Vital Signs 1999*. W.W. Norton Co., New York. 197 pp.

Clarke, R. 1991. *Water: The International Crisis*. Earthworm Publications, London, U.K. 193 pp.

Eaglesham, A., Brown, W.F., Handy, R.W.F. 2000. The Biobased Economy of the 21st Century: Agriculture Expanding into Health, Energy, Chemicals and Materials. NABC Report 12. Ithaca, New York. 159 pp.

Edwards, C.A. 1994. Pesticides as Environmental Pollutants. In *World Directory of Pesticide Control Organizations*, G. Ekstron (Ed.). Royal Society of Chemistry, Cambridge, U.K. 1-24.

Edwards, C.A., Pimentel, D. 2000. Global aspects of agricultural sustainability. In *The Future of the Universe and the Future of Civilization*, V. Burdyhaza (Ed.). UNESCO Publication, Paris, France. 280-290.

Edwards, C.A., Pimentel, D. 2001. The Future of Human Populations: Food, Energy and Water Availability in the 21st Century. In *Ecological Integrity: The Ethics of Maintaining Planetary Life*, P. Miller (Ed.). Rowman and Littlefield, Lanham, MD. (in press)

Edwards, C.A., Veeresh, G.K., Krueger, H.R. 1980. *Pesticide Residues in the Environment in India*. Raja Pareer Press, Bangalore, India. 524 pp.

Evenson, R.E., Pray, C.E., Rosegrant, M.W. 1999. Agricultural Research and Productivity Growth in India. IFPRI Research Report 109. Washington, D.C. 88 pp.

Flavin, C. 2001, Rich Planet, Poor Planet. In *State of the World: 2001*. World Watch Institute, New York. London, 3-20.

Gardner, G. 1996. Shrinking Fields: Cropland Loss in a World of Eight Billion. World Watch Paper 131. World Watch Institute, Washington, D.C. 55 pp.

Gardner, G. 1997. Preserving Global Cropland. In *State of the World 1997*. World Watch Institute, Washington, D.C. 42-59.

Parry, M. 1990. *Climate Change and World Agriculture*. Earthscan Publications, London, U.K. 157 pp.

Postel, S. 1996. Dividing the Waters: Food Security, Ecosystem Health and the New Politics of Scarcity. World Watch Paper 132. Washington, D.C. 76 pp.

Rosegrant, M. 1997. Water Resources in the Twenty First Century: Challenge and Implication for Action. IFFRI Discussion Paper No. 20. Washington, D.C. 24 pp.

Sampat, P. 2001, Uncovering Groundwater Pollution. In *State of the World 2001*. W.W. Norton & Co. New York and London, 19-42.

Sengupta, A.N. 1999. WHO. Arsenic in Drinking Water, Fact sheet 210. WHO, Geneva, Switzerland, 4 pp.

Wood, S., Sabastian, K., Scherr, S.J. 2000. *Pilot Analysis of Global Ecosystems: Agroecosystems*. World Watch Institute, Washington, D.C. 110 pp.

World Bank, 2000, World Development Indicators 2000. World Bank, Washington, D.C. p. 10-12.

World Watch Institute. 2001. *State of the World 2001*. Washington, D.C. 275 pp. World Watch Institute. 1995. In Resources for the Future. Washington, D.C.

12 Agricultural Chemicals and the Environment

David Pimentel

CONTENTS

INTRODUCTION

The rapidly burgeoning human population in India requires increased amounts of food from a shrinking agricultural land base (Ramesh and Dhaliwal, 1996). There are nearly 1 billion people in India on a land base that is about one third that of the United States (Population Reference Bureau, 2000). Malnourishment is a problem in India, but the number of malnourished is not known exactly. The World Health Organization of the United Nations (WHO) recently reported that more than half of the 6 billion people on earth are currently malnourished (WHO, 1996; WHO, 2000). This is the largest number and proportion ever reported in history. This malnourishment is making people more susceptible to other serious diseases such as malaria, schistosomiasis, AIDS, and diarrhea (Pimentel et al., 1999). Most human diseases are on an increase worldwide.

Agriculture is the number one business in India, valued at an estimated $100 billion in net national product per year (AS, 1992). Agricultural production represents about 32% of the net national product and approximately 70% of the population is engaged in agriculture.

This chapter examines the role of chemicals in agriculture and the impacts on the environment.

1-5667-0594-0/02/$0.00+$1.50
© 2002 by CRC Press LLC

AGRICULTURAL FERTILIZERS AND PESTICIDES

Fertilizers and pesticides play an important role in food production in India and other nations (Pimentel et al., 2001). The agricultural inputs for three major crops (rice, cabbage, and apples) in India are listed in Tables 12.1, 12.2 and 12.3. Chemical fertilizers, as well as livestock manure, provide significant amounts of nutrients for these crops. Pesticides also play a role in protecting these and other crops in India against pest attacks and damage. Rice, as a major crop in India, utilizes about 23% of all pesticides used in crop production (Srivstava and Patel, 1990). Production of fruit, such as apples, often requires large inputs of pesticides. Of course, this increases the costs of production because pesticides themselves are expensive.

CROP LOSSES AND PESTICIDES

Worldwide, despite the application of nearly 3 billion kg of pesticides to agriculture, pests destroy more than 40% of all potential crop production (Pimentel et al., 1997). The losses to various groups of pests are estimated to be 15% to insects, 13% to plant pathogens and 12% to weeds.

In India, an estimated 59 million kg of pesticides are applied to agriculture annually (Srivastava and Patel, 1990). Most of the pesticides, or 52 million kg, are insecticides, with 4 million kg as fungicides, 2 million kg as herbicides, and 650 million kg as rodenticides. Note, about 67% of the insecticides consist of DDT and BHC, two chlorinated insecticides (Srivastava and Patel, 1990). Starting in 1990, DDT was to be phased out of use for agricultural purposes, but DDT would still be used for public health purposes in India.

Despite the use of 52 million kg of insecticides, Dhaliwal and Ramesh (Pimentel et al., 1997) report that, on average, insect pests are causing the loss of 25% of potential food and fiber production. It is assumed that plant pathogens and weeds are causing losses nearly as high as pest insects, which would amount to approximately 75% loss. This appears high, but an even more conservative estimate of crop losses would be 10% higher than the world average.

Added to this preharvest loss of 50% is the postharvest loss that is estimated to be 25% for developing countries (Dhaliwal and Ramesh, 1996). Adding the postharvest to the harvest of 50% brings the total loss of all food to pests in India to nearly 63%. I would add at least a 2% loss from rats and other rodents. This estimated 65% of potential food is a terrible loss of food in a country that needs all the food it can have available (Ramesh and Dhaliwal, 1996).

PESTICIDES REACHING TARGET PESTS
AND CONTAMINATING FOOD PRODUCTS

Few appreciate the fact that less than 0.1% of the pesticide applied actually reaches the target pests, which means that more than 99.9% contaminates the environment (Cao et al., 1986). In many cases, the leaf surface areas of the crop plant must be covered with fine pesticide particles to protect the plant from microbe and small insect pests.

TABLE 12.1
Energy Inputs and Costs of Draft-Animal-Produced Rice per Hectare in the Valley of Garhwal Himalaya, India

Inputs	Quantity	kcal x 1000	Costs
Labor	1,703 hrs [a]	2,640 [c]	$129.86 [a]
Bullocks	328 hrs [a]	357 [a]	40.00 [a]
Machinery	2.5 kg [b]	41 [f]	11.00 [b]
Nitrogen	12.3 kg [a]	229 [d]	1.30 [e]
Phosphorus	2.5 kg [a]	10 [d]	0.30 [e]
Manure	3,056 kg [a]	5,071 [a]	14.91 [a]
Seeds	44 kg [a]	95 [a]	6.44 [a]
Pesticides	0.3 kg [a]	30 [d]	1.33 [a]
TOTAL		8,446	$194.14

Rice yield = 1,831 kg [a]   kcal input:output = 1:0.79

[a] Tripathi and Sah (2001).
[b] Estimated.
[c] Per capita fossil energy use in India is 310 liters of oil equivalents per year
[d] (BP, 1992).
[e] FAO (1999).
[f] The total for fertilizers reported was $1.60, we allocated $1.30 for nitrogen.
[g] Pimentel (1980).

Aerial application of pesticides is one of the most ineffective means of applying pesticides to the target crop. For instance, using ultra-low-volume spray equipment gets only about 25% of the pesticide into the target area (Pimentel and Levitan, 1986). This means that 75% drifts off into the nontarget environment. It should be emphasized that this is assuming that the aerial application is made under ideal application conditions, with little or no wind.

In the United States, approximately 35% of all foods in supermarkets have detectable pesticide residues and 1 to 3% have residues above the Food and Drug Administration's acceptable tolerance level (Srivastava and Patel, 1990). In contrast, in India, 97.5% of the foods sold in markets have detectable pesticide residues and 25% of the foods have residues above the acceptable tolerance level (Pimentel et al., 1991).

PUBLIC HEALTH IMPACTS OF PESTICIDES

Since the advent of DDT use for crop protection in 1945, global growth of pesticide use in agriculture has been phenomenal. In 1945, about 50 million kg of pesticides were applied worldwide. Exhibiting an approximate 60-fold increase, global usage is currently at about 3 billion kg per year. In the United States, the use of synthetic pesticides since 1945 has grown 33-fold (Pimentel, 1995) to about 0.5 billion kg (Pimentel et al., 1998). Unfortunately, the increase in hazards is even greater than

TABLE 12.2
Energy Inputs and Costs of Cabbage Production per Hectare Using Bullocks in the High-Hill Region in Garhwal Himalaya, India

Inputs	Quantity	kcal x 1000	Costs
Labor	1,834 hrs [a]	2,934 [c]	$139.85 [a]
Machinery	5 kg [b]	80 [b]	5.00 [b]
Bullocks	294 hrs [a]	310 [a]	35.87 [a]
Nitrogen	27 kg [a]	502 [d]	2.14 [e]
Phosphorus	3.3 kg [a]	14 [d]	0.43 [e]
Potassium	0.2 kg [a]	1 [d]	0.05 [e]
Manure	4,478 kg [a]	7,452 [a]	21.85 [a]
Seeds	1 kg [a]	5 [b]	1.71 [a]
Pesticides	0.01 kg [a]	1 [f]	0.05 [a]
TOTAL		11,299	$206.95
Cabbage Yield = 11,423 kg [a]		5,758	
		kcal input:output = 0.51	

[a] Tripathi and Sah (2001).
[b] Estimated.
[c] It is assumed that a person works 2,000 hrs per year and utilizes an average of 310 liters of oil equivalents per year (BP, 1992).
[d] FAO (1999).
[e] The total cost of fertilizers was $2.62. We allocated this amount to N, P, and K.
[f] Pimentel (1980).

it might appear because the toxicity of modern pesticides has increased by more than 10-fold over those used in the early 1950s (Pimentel, 1995).

In 1945, when synthetic pesticides were first used, there were apparently few pesticide poisonings. Globally, pesticide use had increased to a high of 1.3 billion kg per year by 1973. At this time, the number of human pesticide poisonings reached an estimated 500,000 (with about 6,000 deaths) (Labonte, 1989). Two decades later, WHO reported approximately 26 million human pesticide poisonings each year worldwide (WHO, 1992). Approximately 220,000 cases each year are fatal and about 750,000 result in chronic illnesses.

With about 17% of the world population and having the same ratio of human pesticide poisonings as the rest of the world, India is estimated to have more than 4 million nonfatal pesticide poisonings each year. The number of fatalities is estimated to be 30,000 per year.

Chronic effects of pesticides are varied, with impacts on most systems of the human body. U.S. data indicate that 18% of all pesticides and about 90% of all fungicides are carcinogenic (NAS, 1987). Although both DDT and dieldrin were banned in 1972, their levels in the atmosphere in 1994 created cancer risks of 9 and 12 per million people, respectively (GAO, 1994). The maximum acceptable risk level is 1 per million people.

TABLE 12.3
Energy Inputs and Costs of Apple Production in High Hills of India

Inputs	Quantity	kcal x 1000	Costs
Labor	610 hrs [a]	1,040 [b]	$61.10 [d]
Machinery	20 kg [c]	32 [e]	6.16 [d]
Manure	6 t [g]	10 [f]	12.85 [d]
Nitrogen	20 kg [h]	372 [i]	7.93 [h]
Phosphorus	13 kg [h]	54 [i]	4.27 [h]
Potassium	10 kg [h]	32 [i]	1.83 [h]
Insecticides	6 kg [j]	600 [e]	10.98 [j]
Fungicides	1.3 kg [j]	130 [e]	2.40 [j]
TOTAL		2,270	$81.29
Apple Yield = 6,000 kg k		kcal output/kcal input = 1.57	

[a] Estimated. Based on 10¢ per hour.

[b] A laborer is assumed to work 2000 hours per year and each person in India consumes 310 liters of oil equivalents per year (Tripathi and Sah, 2001).

[c] Estimated.

[d] Swarup and Sikka (1987).

[e] Pimentel (1980).

[f] Estimated based on the fuel required to move 20 t of manure about 2 km.

[g] $21.07 was calculated to purchase 20 t of manure.

[h] Fertilizer inputs were reported in (d) to cost $22.56. This cost was then estimated to provide the amounts of N, P, and K listed in the table.

[i] FAO (1999).

[j] Inputs of insecticides and fungicides were reported in (d) to cost $22.24. This cost was then estimated to provide the amounts of insecticides and fungicides in the table.

[k] The yield was estimated based on the total yield of apples valued at $267.00 per hectare.

Pesticides are also estrogenic, a fact that has linked them to the increased breast cancer rate among U.S. women (McCarthy, 1993). The rate rose from 1 in 20 in 1960 to 1 in 8 in 1995 (McCarthy, 1993). Estrogenic effects have also contributed to the 50% decline in the average sperm count in men over the last 50 years.

Several studies illustrate the effect of pesticides in the respiratory system. For example, among a group of people who applied pesticides commonly, 15% suffered asthma, chronic sinusitis, or chronic bronchitis as compared with 2% for people who only lightly used pesticides (Weiner and Worth, 1972).

Several pesticides, especially the organophosphates and carbamate classes, affect the nervous system by inhibiting cholinesterase (Ecobicho et al., 1990). This is particularly critical among children as a child's brain is more than five times larger in proportion to its body weight than an adult's (Wargo, 1996). In California, 40% of the children working in agricultural fields have blood cholinesterase levels below normal, indicating organophosphate and carbamate pesticide poisoning (Repetto and Baliga, 1996).

Occupational exposure to pesticides and other toxic chemicals has been the best source of information about chemical-related diseases. Problems exist with data on the effects in the general population because of complicating factors that include low concentrations, synergistic effects of multiple contaminants and low-level chronic exposures. This complexity makes it extremely difficult to identify the causative chemicals.

This difficulty is illustrated in India. A medical doctor in the Kerali village observed relatively high rates of disorders in the central nervous systems of children, including cerebral palsy, congenital anomalies and mental retardation (CSE, 2001). Finally, the doctor discovered alarmingly high rates of the insecticide endosulfan in the population. One woman's blood was found to have endosulfan at a level 900 times higher than the acceptable level for drinking water.

ENVIRONMENTAL EFFECTS OF PESTICIDES

The environmental effects of pesticides are extremely complex because there are about 10 million nontarget organisms, water and soil contamination, air pollution and more than 700 pesticide chemicals in use. Because of the complexity of the agricultural and natural ecosystems, little is known concerning the environmental impacts of pesticides. However, sufficient information is available for us to be concerned. A brief discussion of the range of pesticide effects is discussed using data primarily from the United States.

In addition to the pesticide problems that affect humans, thousands of domestic animals are poisoned by pesticides each year in the U.S. and elsewhere in the world. Dogs and cats represent the largest number of domestic animals being poisoned because they often wander freely about the home and farm and therefore have a greater risk of coming into contact with pesticides than other domestic animals (Pimentel et al., 1991).

In both natural and agro-ecosystems, many species, especially predators and parasites, control or help control pest populations. Indeed these natural beneficial species make it possible for natural and agro-ecosystems to remain "green." Without natural enemies and biological control in agriculture, losses of food to pests would increase as much as 50%. This has been confirmed in U.S. agriculture when pesticides have been documented to reduce or totally destroy the natural-enemy populations. This resulted in an explosion in pest-insect populations. In the United States, the destruction of natural enemies in agro-ecosystems is costing the nation more than $500 million each year (Pimentel and Greiner, 1997).

In addition to destroying natural-enemy populations, the extensive use of pesticides has often resulted in the development of pesticide resistance in insect pests, plant pathogens, weeds and rats. Almost 25 years ago, the United Nations Programme reported that pesticide resistance was one of the four most serious environmental problems (UNEP, 1979). Worldwide, more than 500 insect and mite species have been reported to be resistant to pesticides (Green, 1987). The estimated cost of pesticide resistance in pests in the United States is at least $1.4 billion annually (Pimentel and Greiner, 1997).

Wild bees are vital for pollination for about one third of fruits, vegetables, and other crops worldwide (Pimentel and Hart, 2001). Bee pollination in the U.S. has estimated benefits of about $40 billion per year (Pimentel and Hart, 2001). Damage to wild bee populations in the United States alone is estimated to be about $320 million per year (Pimentel et al., 1998).

Basically, pesticides are applied to protect crops from pests in order to increase yields, but sometimes the crops are damaged by pesticide treatments. This occurs when: (1) the recommended dosages suppress crop growth, development and yield; (2) pesticide drift from the targeted crop damages adjacent crops; (3) residual herbicides either prevent chemical-sensitive crops from being planted in rotation or inhibit the growth of crops that are planted; (4) excessive pesticide residues accumulate on crops, necessitating the destruction of the harvest. The estimated crop and tree losses in the U.S. due to recommended pesticide use is nearly $1 billion per year (Pimentel and Greiner, 1997).

Ground- and surface-water contamination from pesticides is also a serious problem worldwide. Just in the United States, if an adequate job were to be carried out to test water resources for pesticide residues, it would cost the nation about $1.3 billion annually. In addition, there are serious fish kills and contaminated fish that cannot be eaten. A conservative estimate on fish losses is about $400 million per year.

In the United States, about 3 kg of pesticides are applied per ha per year (Pimentel and Greiner, 1997). Wild birds and mammals are damaged by these pesticide applications. It is estimated that nearly 70 million birds are killed each year from pesticide applications (Pimentel et al., 1993a). The economic value of these birds is approximately $2 billion each year.

Thus, a conservative estimate of the total effects of damages to the environment and public health is about $9 billion each year.

CONCLUSION

There is a clear need to apply pesticides with greater concern for public health and the environment in India, the United States and generally worldwide. Several countries, such as Indonesia and Sweden, have implemented programs to reduce pesticide use by more than 50% and they have been highly successful (Pimentel et al., 1993b). Pesticide use has been reduced without any reduction in crop yields and, in some cases, with 10% to 15% increase in crop yields.

An estimated 25% of the food material in storage in India is lost to pests such as rats, insects and microbes. Greater effort is needed to protect foods in storage. One problem in India is that wholesalers sometimes add pesticides to stored grains to protect them from insects. The result is that people are being fed grains containing large quantities of pesticides. The Indian government needs to implement policies that will prevent wholesalers from adding pesticides to stored grain.

Knowledge concerning the substitution of nonchemical controls abound (Pimentel et al., 1993b). What needs to be done is to use and implement this knowledge to substitute for pesticides and other chemical use in Indian agriculture.

REFERENCES

AS. 1992. *Agricultural Statistics at a Glance*, Directorate of Economics & Statistics, Department of Agriculture & Cooperation, Ministry of Agriculture, Government of India New Delhi, India, John Wiley, Chichester, U.K.

BP. 1992. British Petroleum Statistical Review of World Energy, British Petroleum Corporate Communication Systems, London.

Cao, D., D. Pimentel and K. Hart. 2001 Postharvest food losses (vertebrates). In *Encyclopedia of Pest Management*, Marcel Dekker, New York, (in press).

CSE 2001. Center for Science and Environment, http://www.cseindia.org/html/extra/inviker-ala.htm.

Dhaliwal, G.S. and A. Ramesh. 1996. An estimate of yield losses due to insect pests in Indian Agriculture, *Indian J. Ecol.*, 23:70.

Ecobichon, D.J., J.E. Davies, J. Doull, M. Ehrlich, R. Joy, D. McMillian, R. MacPhail, L.W. Reiter, W. Slikker and H. Tilson. 1990. Neurotoxic effects of pesticides, In *The Effects of Pesticides on Human Health*, S.R. Baker and C.F. Wilkinson (Eds.), Princeton Scientific Publishing Company. pp. 131.

FAO. 1999. *Agricultural Statistics*, http://apps.fao.org/cgi-bin/nph-db.pl?subset-agriculture, Food and Agricultural Organization, UN.

GAO. 1994. *Reducing Exposure to Residues of Canceled Pesticides*, General Accounting Office, Washington, D.C.

Green, M.B. 1987. Energy in pesticide manufacture, distribution and use. In *Energy in Plant Nutrition and Pest Control, Energy in World Agriculture*, Z.R. Helsel (Ed.), pp165.

Labonte, R.N. 1989. Pesticides and healthy public policy, *Can. J. Publ. Hlth*, 80: 238.

McCarthy, S. 1993. Congress takes a look at estrogenic pesticides and breast cancer, *J. Pestic. Reform*, 13:25.

NAS. 1987. *Regulating Pesticides in Food*, National Academy of Sciences, Washington, D.C.: 272.

Pimentel, D. 1980. *Handbook of Energy Utilization in Agriculture*, CRC Press, Boca Raton, FL, pp455.

Pimentel, D. 1995. *Chemicals in Developing Countries*, Pontifical Academy of Sciences and the Royal Swedish Academy of Sciences, Rome, Italy.

Pimentel, D. and L. Levitan. 1986. Pesticides: amounts applied and amounts reaching pests, *Bio Sci.* 36:86.

Pimentel, D., L. McLaughlin, A. Zepp, B. Lakitan, T. Kraus, P. Kleinman, F. Vancini, W.J. Roach, E. Graap, W.S. Keeton and G. Selig (Eds.). 1991. *Environmental and Economic Impacts of Reducing U.S. Agricultural Pesticide Use*, 2nd ed., CRC Press, Boca Raton, FL, pp. 679.

Pimentel, D., H. Acquay, M. Biltonen, P. Rice, M. Silva, J. Nelson, V. Lipner, S. Giordano, A. Horowitz and M. D'Amore. 1993a. Assessment of Environmental and Economic Costs of Pesticide Use, In *The Pesticide Question: Environment, Economics and Ethics*, D. Pimentel and H. Lehman (Eds.), Chapman and Hall, New York. 47.

Pimentel, D., L. McLaughlin, A. Zepp, B. Kakitan, T. Kraus, P. Kleinman, F. Vancini, W.J. Roach, E. Graap, W.S. Keeton and G. Selig. 1993b. Environmental and economic effects of reducing pesticide use in agriculture, *Agric. Ecosyst. & Environ.* 46:273.

Pimentel, D., C. Wilson, C. McCullum, R. Huang, P. Dwen, J. Flack, Q. Tran, T. Saltman and B. Cliff. 1997. Economic and environmental benefits of biodiversity. *Bio Sci.* 47: 747.

Pimentel, D. and A. Greiner. 1997. Environmental and socio-economic costs of pesticide use. In *Reducing Pesticides: Environmental and Economic Benefits*, D. Pimentel (Ed.), John Wiley & Sons, Chichester, U.K. pp. 51.

Pimentel, D., A. Greiner and T. Bashore. 1998. Economic and environmental costs of pesticide use, In *Environmental Toxicology: Current Developments*, J. Rose (Ed.), Gordon and Breach Science Publishers, Amsterdam, Netherlands. 121 pp.

Pimentel, D., O. Bailey, P. Kim, E. Mullaney, J. Calabrese, F. Walman, F. Nelson and X. Yao. 1999. Will the limits of the Earth's resources control human populations? *Environ., Dev. & Sustainability* 1:19.

Pimentel, D., R. Doughty, C. Carothers, S. Lamberson, N. Bora and K. Lee. 2001. Energy inputs in food crop production in developing and developed countries, *Agric., Ecosyst. & Environ.* (submitted).

Pimentel, D. and K. Hart. 2001. Pesticide use: ethical, environmental, and public health implications. In *New Dimensions in Bioethics: Science, Ethics and the Formulation of Public Policy*, W. Galston and K.M. Shurr (Eds.), Academic Publishers, Boston.79 pp.

PRB 2000. *World Population Data Sheet*, Population Reference Bureau, Washington, D.C.

Ramesh, A. and G.S. Dhaliwal. 1996. Agroecological changes and insect pest problems in Indian Agriculture, *Indian J. Ecol.,* 23:109.

Repetto, R. and S.S. Baliga. 1996. *Pesticides and the Immune System: The Public Health Risks*: World Resources Institute, Washington, D.C. 103 pp.

Srivastava, U.K. and N.T. Patel. 1990. *Pesticides Industry in India: Issues and Constraints in its Growth*. Oxford & IBH Publishing Co. Pvt. Ltd., New Delhi, India. 343 pp.

Swarup,R. and B. Sikka. 1987. *Production and Marketing of Apples*, Mittal Publications, Delhi, India. 157 pp.

Tripathi, R. and V. Sah. 2001. Material and energy flows in high-hill, mid-hill and valley farming systems of Garhwal Himalaya, *Agric. Ecosyst. & Environment* 86. 75 pp.

UNEP 1979. The State of the Environment: Selected Topics-1979. United National Environment Programme, Governing Council, Nairobi.

Wargo, J. 1996. *Our Children's Toxic Legacy: How Science and Law Fails to Protect Us from Pesticides*, Yale University Press, New Haven, Connecticut. 380 pp.

Weiner, B. P. and R.M. Worth. 1972. Insecticides: household use and respiratory impairment, In *Adverse Effects of Common Environmental Pollutants*, MSS Information Corporation, New York. 149 pp.

WHO. 1992. Our Planet, our Health: Report of the WHO Commission on Health and Environment, World Health Organization, Geneva, Switzerland.

WHO. 1996. Micronutrient Malnutrition–Half of the World's Population Affected, World Health Organization, Geneva, Switzerland, 78:1.

WHO. 2000. *Malnutrition Worldwide*, http://www.who.int/nut/malnutrition_worldwide.htm, July 27, 2000.

13 Applying Grades and Standards for Reducing Pesticide Residues to Access Global Markets

K.V. Raman

CONTENTS

INTRODUCTION

Agriculture, the backbone of the Indian economy, has contributed substantially to its growth. Sustained efforts during the last five decades have resulted in the country's not only becoming self sufficient in food grain but also in catering to the global markets. The internalization of the world economy has created a major opportunity for Indian and multinational companies to access new technologies to supply products to the dynamic global market. In no sector is this more true than for agribusiness, where the relaxation of market barriers for food trade has invigorated trade, while the more flexible policy on foreign investment allows for growth from internal and external sources.

India's agroclimatic conditions are suitable for growing all kinds of fruits, vegetables and grains. It also has some well-trained human resources in all areas of agriculture. These characteristics are now attracting several multinational and domestic companies to undertake contract farming of horticultural crops and establish manufacturing facilities to develop value-added products from tropical fruits such as papaya, grapes, pomegranate, banana, pineapple, and mango. Several Indian and

1-5667-0594-0/02/$0.00+$1.50
© 2002 by CRC Press LLC

multinational companies also export spices, legumes, rice, shrimps, fruits and vegetables to the Middle East.

EUROPE AND THE UNITED STATES

India hopes to emerge as a significant exporter of fresh and processed products, and, to make this happen, the government has drawn up a progressive policy framework to facilitate the growth of the food processing industry. Ten major agricultural products identified by India's commerce ministry to boost future exports are: rice, wheat products, coarse grains, spices, cashew, oil meals, sugar, horticultural products, floricultural products, and processed foods. (Federation of the Indian Chambers of Commerce and Industry – FICCI, 2000).

Food exporters in India are aware that, with the changing global scenario, new markets are available and that they will be rewarded when they export products of high quality. Unfortunately, because of lack of appropriate technologies and infrastructure, most Indian exporters are unable to apply appropriate grades and standards (G&S) to assure quality, safety and authenticity of the products. Many factors are considered in establishing G&S, some of the more important ones include: weight, size, shape, density, firmness, insect and disease damage, pesticide residues, cleanliness, contamination, odor, blemishes, etc. With the opening of international markets, exporters are now realizing that, without appropriate G&S, they will not be able to succeed in accessing global markets (Giovannucci and Reardon, 2000). While many factors affect G&S, this chapter addresses the issue of pesticide residues in India's food exports and discusses the kinds of interventions that are needed to meet accepted safety standards required for India to compete for global markets.

THE KEY ISSUES

The use of pesticides has increased several-fold in India and the prospects for future product development will depend on anticipated market size and profitability. The pattern of pesticide usage in India is different from that in the world in general. 76% of the pesticide used is insecticide, compared with 44% globally. The use of herbicides and fungicides is correspondingly less heavy (Mathur, 1999). Approximately 125 companies manufacture more than 126,000 tons of 60 technical-grade pesticides in India, including "dirty dozen" pesticides such as: Benzene hexachoide (BHC), 2,2,bis (p-chlorophenyl)1,1,1, tricholoroethane (DDT) and methyl parathion. In 1997, the Indian pesticide market was valued at US $602 million. India is one of only two countries worldwide (along with the U.S.) to have applied more than 100,000 tons of DDT since its initial formulation (Harris, 2000).

The chemical parameters for tolerable limits of pesticide residues on food products were established by the Prevention of Food Adulteration Act in 1968 and the Agricultural Produce, Grading and Marketing Act, monitored by the Bureau of Indian Standards in 1971. Currently, 144 pesticides are registered under these restrictions. The 1968 act governs the manufacture, transport, and application of pesticides in the interest of safety to human health and of protecting the environment. The enforcement of these acts is difficult, as many pesticides in developing countries, including India, are commonly misused because information on their proper use is not generally available. Their labeling is poor and often not in the native language, extension advisory services are usually unavailable, and proper usage outside of the temperate zone environments may be unknown even by the manufacturers (Bull, 1982).

There are now demonstrated cases of indiscriminate use of pesticides in many developing countries that have led to high residue levels in food. Even small quantities of these residues present in food can lead to high levels in the body when these foodstuffs are consumed over long periods of time. The effects of pesticide consumption are many. They vary from minor health problems to carcinogenicity to endocrine disruption. The long-term effects could also be reduction of lifespan and fertility, increase in cholesterol levels, high infant mortality rates and several metabolic and genetic disorders (Atkin and Leisinger, 2000). Initial animal and human studies link many pesticides to myriad effects, including low sperm counts, infertility, genital deformities, hormonally triggered human cancers such as those of the breast and prostate gland, neurological disorders in children such as hyperactivity and deficits in attention and development, and reproductive problems in wildlife (Colburn, et al. 1996). For additional details on health risks associated with pesticides see article published by Pimentel et al. (2000).

The total intake of pesticides (organochlorines) by Indians is the highest in the world. There are also reports to indicate that dangerous amounts of pesticides that are banned in other countries are ingested by Indians and deposited in the body. According to recent studies on pesticide residues by the Industrial Toxicological Research Institute, high levels of pesticides have been found in just about everything necessary for life, from food to air and water. It is estimated that infants ingesting breast milk in Delhi receive roughly 12 times the daily allowable intake of DDT (Center for Science and Environment 2000). Hundreds of people die from pesticide poisoning each year. A survey of pesticide residues in food samples collected in 12 Indian states found residues in 85% of samples with 43% above the recommended doses. A 7-year study by the Indian Council of Medical Research released in 1993 analyzed 2205 cow and buffalo milk samples from 12 states. Hexachloro cyclohexane (HCH) commonly known as lindane or BHC, was detected in about 85% of the samples with up to 41% of the samples exceeding tolerance limits. DDT residues were detected in 82% of the samples and 37% contained residues above the limit of 0.05 mg/kg, in some cases 44 times higher at 2.2 mg/kg (Harris, 2000). These results, though alarming, don't agree with the studies made by Seth et al. (1998) who indicate that levels of DDT and BHC residues found in agricultural produce such as milk, fats, fodder and meats in India have been mostly below the stipulated maximum residue level (MRL). They also indicate, at times, that higher levels of pesticide residues observed have been the result of improper use or deliberate overuse.

Indian exporters of food to the U.S. and Europe are now increasingly concerned about detention of their products because of pesticide residues and other factors such as contamination by salmonella or aflatoxin. The European Commission and U.S. authorities — the Federal Drug Administration (FDA), the United States Department of Agriculture–Animal Plant Health Inspection Service (USDA-APHIS), and the

Environmental Protection Agency (EPA) have now started applying stricter quality legislation on imported food products, including animal feed, especially pertaining to permissible levels of aflatoxin, hexane, pesticide residues and other harmful foreign materials that enter agricultural produce during postharvest and processing. EPA sets tolerances to ensure food safety and develops practical methods for detecting pesticide residues. The FDA enforces tolerances set by EPA and tests food imported from other countries for compliance with pesticide residue limits. Both USDA and FDA have programs that validate information on pesticide residues for use by EPA. USDA staff notifies the FDA of violations of tolerances in their data collection program. These three agencies assure the general public in the U.S. that imported food is safe from pesticide residues.

The strict regulations followed by these agencies are expected to affect India's exports, particularly groundnut, cashew, walnut, pepper, chilies, oilcake, and marine products. Both aflatoxin and pesticide residues are a very serious problem for many of these agricultural commodities exported from India. During 1998, Germany rejected a shipment exported from India of more than 80 containers of turmeric containing high levels of pesticide residues. The total consignment was valued at over US$1 million (Mistry, 1998). During 2000, the Consumer Education and Research Center in Maharashtra, India, analyzed 13 major brands of wheat flour and three samples of loose flour and found that organochorine pesticides such as DDT, its breakdown products and lindane, banned from use on crops, were present in these samples. DDT was present in five different brands of wheat flour, and lindane in all the brands and in one of the brands of loose flour. Two other organochlorine pesticides, aldrin ($C_{12}H_8Cl_6$) and dieldrin ($C_{12}H_8C_{16}O$) also banned from use, were found in two samples (*Economic Times*, 2000).

All detentions made by the FDA in the U.S. for food products exported from India are posted in FDA's import detention report (IDR), which is updated monthly. During 1999, there were 167 detentions from India for products contaminated with salmonella or pesticides or because products arrived in a filthy state and were unfit for consumption. There were also some problems associated with misbranding (FDA, 2000).

Twelve pesticides, usually referred to as the "dirty dozen," are prohibited and banned for use in agriculture. These include: ethyl parathion, DDT, aldrin, dieldrin, endrin, toxaphene, paraquat, lindane, chlordane, galecron, 2,4-5T, and pentachlorophenol. For additional details on the chemistry and toxicology of these pesticides see the compendium of pesticides published by the U.K. in its website (http://www.hclrss.demon.co.uk/). During 2000, the FDA detained many shipments from India containing illegal pesticide residues. These included shipments of mushrooms, basmati rice, sesame seeds, crushed chilies, moong (legumes), and dried whole peppers (FDA 2000). The use of DDT in India is restricted to public health purposes only and HCH was banned from use in April 1997 because of its long-term persistence in the environment (Seth et al., 1998). The presence of these pesticide residues in wheat flour and other agricultural products indicates improper use of these insecticides under field conditions. A more detailed study of pesticide residues in India's agricultural exports is justified.

INTERNATIONAL G&S BENCHMARKS

Internationally recognized benchmarks such as the Codex Alimentarius Standards and those of the United Nations Economic Commission for Europe (UNECE) are commonly used as a basis for many G&S. The specific issue of "pesticide residues" is considered under the World Trade Agreement on the Application of Sanitary Measures known as the Sanitary and Phytosanitary (SPS) Agreement. Each country that is a member of the World Trade Organization (WTO) has obligations relating to "transparency."

For example, countries are required to publish all sanitary and phytosanitary measures (SPS measures) and notify of any changes to SPS measures. In implementing the agreement, countries are required to identify a single central government authority to be responsible for the notification requirements of the SPS Agreement (the notification authority). Also, countries are required to establish an enquiry point responsible for answering questions from other countries about SPS measures and related issues (the enquiry point).

A monitoring agency referred to as the Codex body is based in Rome and financed jointly by the Food and Agriculture Organization (FAO) and World Health Organization (WHO). The standards, guidelines and recommendations established by the Codex Alimentarius Commission on pesticide residues are updated regularly and provide information for MRL for pesticide residues allowed in all important food products either for domestic consumption or export (WTO, 2000; Codex, 2000). By following Codex established guidelines, countries can set their own guidelines and regulations based on science.

The Codex commission encourages its member countries to use international guidelines and recommendations where they exist. However, members are allowed to use measures that result in higher standards if there is scientific justification. They can also set higher standards based on appropriate assessment of risks so long as the approach is consistent, not arbitrary. The agreement still allows countries to use different standards and different methods of inspecting products.

So how can an exporting country be sure that practices it applies to its products are acceptable in an importing country? If an exporting country can demonstrate that the measures it applies on its exports achieve the same level of health protection as in the importing country, then the importing country is expected to accept the exporting country's standards and methods (Codex, 2000).

Several Indian food exporters are unaware of these regulations. Specific training in pesticide regulatory and residue management programs is needed. This will provide the necessary knowledge to Indian exporters to meet G&S in different markets. It would also facilitate the increasing demand for product trace-back and certification of production methods.

MEETING PESTICIDE RESIDUE STANDARDS IN EXPORT

Food export markets present a set of challenges somewhat different from domestic food safety regulation. Exports of fresh food products such as meat, fish, fruit and vegetables represent a desirable growth opportunity because these products are in high demand and have fewer trade barriers than staple commodity agricultural exports. Fresh food products are also more likely to encounter sanitary and phytosanitary (SPS) barriers to trade. Delivering safe food to distant markets requires process controls throughout the production process and mechanisms to certify to buyers and government regulators that such controls are effective. Developing-country exporters need to know how to meet standards in different markets and how to meet the increasing demand for product trace-back and certification of production methods.

Food safety investments for export markets will be influenced by the growing recognition that a farm-to-table approach is necessary to address food safety. There are several activities needed throughout the food production chain to ensure food safety at the farm production level and during transport, packing, storage, processing and retail. To design effective food safety interventions in developing countries, a summary of these important factors is dealt with in detail by Unnevehr and Hirschhorn (2000). Because many hazards can enter the food chain at different points and it is costly to test for their presence, a preventive approach that controls processes is the preferred method for improving safety. The Hazard Analysis and Critical Control Points (HACCP) system is increasingly used as the basis for food safety regulation and for private certification of food safety (EWG, 1995). In the U.S., HACCP is being used to tackle microbial contamination problems in the seafood industry, and recently, food companies and other industries adopted it. HACCP is a systems approach to food safety, emphasizing quality control from the start of the process and through each critical stage. Under HACCP, responsibility for ensuring safety of the food supply is shared between the government and industry (EWG, 1995).

For exporters in India faced with pesticide-residue problems, one needs to evaluate whether the existing standards have evolved to serve a useful function. They ought to be evaluated to determine how clear, thorough, up-to-date and equitable they are. The results of such evaluations can yield the basic factual and statistical knowledge with which to leverage or influence necessary changes. Such assessments must be conducted by an outside agent to ensure independence. For Indian exporters who are now faced with export rejections due to pesticide residues, the following questions could serve as a guide:

- What are the established pesticide use regulations and inspection requirements?
- What specific procedures are required as part of the application process to the existing standards committee to allow pesticide-residue problems to be recognized on a national basis?
- How difficult and how long is the process for establishing maximum allowable pesticide-residue levels?
- What factors will be considered to determine pesticide-residue levels? Are they appropriate? Are they clear?
- What is the purpose of establishing limits on pesticide residues (wholesale, export, import, retail)?

- Which pesticide residue standards are in use and to what extent are they being applied by countries in the region and by key trading partners, competitors, and relevant associations or participants in the production and supply chain?
- Are current pesticide regulations in place and do they fit well with the requirements of the exporting country?
- What is the role of international organizations in helping develop appropriate pesticide use regulations?
- What is the role of producer, processor, or trade associations in assuring that the food is safe from illegal pesticide residues?
- What is the role of dominant buyers in the market and how will they help in reducing pesticide residues?

BEST PRACTICE AND EXAMPLES

The Environment Working Group in the U.S. has analyzed the records of FDA monitoring data on 42 fruits and vegetables from 11 different countries. The highest percentage of illegal pesticides came from Mexico, Argentina, Columbia and Guatemala (EWG, 1995). Surprisingly, shipments from Chile have had no detentions. Why is it that there are such low pesticide residue violation rates in fruit exports made from Chile? A closer evaluation of the Chilean experience will be useful for Indian exporters to enhance their current G&S system and to meet competition.

Chile produces 1,150 tons each of fresh fruit and processed fruit. The exports are valued at US $1,180 million, and the main markets are the U.S. and Europe. After 1974, Chile made major reforms to improve its fruit sector. An increased emphasis is now placed on applying G&S to promote exports. Medium and large producers and exporters of fruit are linked with the government to promote all aspects of G&S. The issue of pesticide residues is dealt with via on-the-job training to all major fruit producers. Several courses, seminars and technical assistance are provided on how to apply SPS measures to reduce pesticide residues. The Coordinating Committee promotes these activities for fruit and vegetable producers and exporters. This committee, plus the National Agricultural Association, has developed " a code of good practice" for production, processing, and distribution of fruit for export. The committee members work with the Ministry of Agriculture and the national Codex entity to influence Chilean health and safety laws, infrastructure and to influence international Codex discussions.

It is because of such a system that Chile is able to implement effective pesticide-monitoring programs on all its export shipments to the U.S. and Europe (Giovannucci and Reardon, 2000). Adopting the Chilean model in India and elsewhere can save enormous time and resources because many issues are common ones, and adaptations for specific needs can readily be made. For example, studying the grape exports and pesticide-residue-detection systems of Chile could help India's grape exporters, who are now faced with increasing rejections in European markets because of pesticide residues. Common international systems, even with minor local or national variations, can greatly facilitate both import and export functions. Accepted international standards will also eliminate any efforts necessary to ensure WTO compliance.

Another good approach to reducing the use of pesticides is through the use of Integrated Pest Management (IPM). IPM techniques usually focus on the ecology of pests and on the agro-ecosystem as a whole and, where feasible, it tries to incorporate biological and genetic resistance as alternatives to pesticides. European countries (Sweden, Norway, Denmark, the Netherlands) and Canada have all adopted effective IPM programs to reduce pesticide use by 50–75%. In the U.S., the use of IPM is a national priority, and it is estimated that pesticide use can be reduced as much as 50% at an estimated savings of at least $500 million per year without reducing crop yields or substantially reducing the "cosmetic standards" of fresh fruits and vegetables (Kennedy and Sutton, 2000).

In Indonesia, for example, the investment of US$1 million per year in IPM research, in conjunction with extension programs to train farmers to conserve natural enemies, is paying large dividends. Pesticide use for rice in Indonesia has been reduced by 65%, while rice yields have increased by 12%. As a consequence, the Indonesian government has been able to eliminate $20 million in pesticide subsidies (Pimentel et al., 2000).

The past 20 years have seen a substantial increase in knowledge of IPM for a wide range of crops. This includes both new technologies to replace unsustainable use of chemical pesticides, as well as ecological knowledge, which reveals how these technologies can be combined locally and effectively to suppress pest populations. However, much of this IPM knowledge remains at the level of researchers, and not at the farm level where it is of greatest value. The failure to create this channel of information flow between researcher and farmer has also allowed research to become isolated from the real needs of farmers.

The Indian Council of Agricultural Research (ICAR), realizing the benefits of IPM, has now created a national center for IPM to orient work more toward the training of farmers in IPM methods and particularly the conservation and use of natural enemies. Unfortunately, no strong private initiatives within India promote IPM at the farm level.

Another option to limit the use of pesticides is IPM labels. Also referred to by some as Eco-Labels, IPM labels are gaining acceptance by the general public in the U.S. For products to use IPM labels they must be grown using IPM practices (appropriate cultural practices, resistant varieties, biological control and rational use of approved pesticides). All participating farmers in this program comply with agreed-upon IPM elements, and at the time of marketing the product, mass media, brochures and in-store movies developed by the supermarkets enlighten consumers about the term IPM and how it benefits all.

Elements of IPM and the lists of methods to be used by growers who supply IPM-labeled produce, are now available for 13 crops in New York: fresh blueberries, raspberries, strawberries, sweet cherries, sweet corn, and greenhouse-grown tomatoes; processed beets, cabbage, carrots, peas, snap beans, and sweet corn and dry beans. Consumers have shown substantial support for such a program, and the IPM labels are accomplishing a number of goals, the most important of which are: (1) improving the sales of an eco-labeled product, (2) encouraging farmers to account for the environmental impact of their products, (3) making consumers more aware of the environmental issues, and (4) helping protect the environment, which is the ultimate benefit. (Cornell University Web Page: http://www.nysaes.cornell.edu) (NYSIPM, 1997, 2000).

India and other developing countries should consider how an IPM or eco-labeling program would work best in their own situation. Certain broad considerations to include in a program like this are to first select the IPM label that would make the most significant improvement to the environment. For example, if the national program in India has pest-resistant varieties of fruits or vegetables, then these should be promoted by demonstrating the advantages of using fewer or no toxic pesticides when producing this product. This would enable consumers to understand better that by using such approaches their products will meet international standards of pesticide tolerances. The chances of rejections in export will be minimal.

CONCLUSION

The General Agreement on Trade and Tariffs (GATT) provides significant opportunities for India to increase its food exports to Europe, North America and Japan. To be successful, Indian exporters have to understand that the public in these regions is concerned about the chemicals to which it is exposed in food, air, and water. There is a growing awareness on the importance of organic foods, and, on an individual level, some consumers prefer to buy pesticide-free and IPM labeled foods and are willing to pay a premium for these. India, to compete in these markets, will have to follow internationally approved standards to assure food safety. Pesticide residues in shipments made from India to Europe and the U.S. continue to be a major problem. What is needed is a fundamental restructuring that shifts much of the responsibility for assuring food safety and compliance with the food industry. Individuals and corporations who sell or export food treated with pesticides must assure that the levels remaining on the food comply with internationally established Codex limits. Currently, there is little reason to believe that this is the case.

Government agencies in most developing countries are not equipped for pesticide-residue monitoring and enforcement. Some of the exporters in India have requested the government to establish private pesticide testing and monitoring centers. If this is done, the data from such centers should be monitored by the respective in-country regulatory authorities to assure that the data are in compliance with Codex-established standards.

Proper record keeping and public disclosure for data on pesticides applied to crops are also needed. For example, lists should be developed to include information on pesticides used, both those that degrade below levels normally detected by routine analysis and those likely to leave residues. Such lists must be made available to the public upon request. The FDA interception data for pesticide residues is currently posted on a Web site that is generally not known to regulators and exporters in developing countries. This information could be quite useful if translated into local language and made available to interested exporters and regulatory agencies in the developing world.

Strong education programs in the area of pesticide use and residues will promote G&S for exports. New systems must be created to provide growers with up-to-date information about the acceptable use of pesticides on food. For importers, the regulatory authority within a country must provide this information in the local primary language used in that country. The tolerance levels allowed for pesticides should be made available on line, and, in countries where the Internet is still not available or is slow, printed reports should be provided monthly. Questions about pesticide label rates and tolerances in the U.S. are usually provided by the state regulatory agencies. In developing countries, such a system could be promoted through the in-country extension network, radio or TV. Future research on how human health is affected by increasing exposure to pesticides must also be continued. What is clear is that the issue of pesticide residues will be a central feature of the agrifood system for the foreseeable future.

IPM will allow the reduction in pesticides, but, for it to be implemented, strong support is needed from the national governments to reduce pesticide subsidies, strengthen extension and IPM research, and provide adequate infrastructure support. Farmers will need to be sensitized to the reliability of IPM and should be provided with incentives to adopt it. Consumers who are conscious of the hazards of pesticide residues in their food should be willing to support policies and actions aimed at giving farmers options such as credit and higher prices for alternative practices.

Greater opportunities now exist to use biotechnology, biological control, resistant varieties, botanical insecticides and cultural controls to reduce the use of pesticides. In the area of biotechnology, the use of transgenic crops developed by genetic engineering for resistance to insects, weeds and diseases is gaining wide acceptance in the U.S. In 1999, the global area of four principal crops — soybean, canola, cotton and corn — totaled 273 million ha, of which 15%, equivalent to 39.9 million ha, were planted with transgenic varieties (James 2000). Today's insect-resistance technology, developed through the use of *Bacillus thuringiensis* (Bt) toxin gene could substitute for nearly $3 billion worth of the $9 billion market in insecticides, increasing yields by as much as 5–10% as well. In cotton alone, nearly $1.2 billion in insecticides could be substituted today with currently available technology. India has now begun testing genetically engineered insect-resistant cotton in multilocation trials. The cotton crop in India continues to be the number one user of insecticides. Once the newly developed Bt cotton is commercialized in India, insecticide use in cotton will drop significantly. Similarly, herbicide-tolerant soybeans used in the U.S. have led to 33% reduction of overall herbicide use in 1997 on these transgenic soybeans (Krattiger, 1998). Botanical insecticides such as Neem-based azadirachtin are also now gaining importance in India, Europe and the U.S. New formulations of enriched fractions with azadirachtin, plus other active constituents from the neem-seed kernel, have now been patented in India, the U.S. and Europe. These have now received EPA registration in the U.S. These products have faster biodegradability, leave no residues and are fairly easy to process (Agnihotri et al., 1999). Other new and safer pesticides, which are considered natural products obtained by fermentation of the soil fungus species *Streptomyces*, the avermectins and milbemycins, are also now gaining acceptance (Jiang Lin and Ma Cheng Zu, 2000). There are also good prospects for the development of further pesticides from microbiological sources that will have high biological activity in target species and low mammalian toxicity. Activity has now increased in the private sector to remove old chemicals from use in agriculture along with the continued introduction of newer chemicals that require shorter withholding periods and have greater specificities in toxicity (National Research Council, 1987; Australian Science and Technology Council, 1989).

In most instances, the use of resistant crop varieties developed through conventional breeding requires no more time than the development of a new pesticide, and the expenditure of time and resources has been well worth it in many cases. For example, resistant cultivars of cereal crops have been the mainstay of disease protection for many years. Success in crop breeding includes disease resistance of corn to southern corn blight and other blights and wilt, alfalfa to bacterial wilt, pears to fire blight, tobacco to bacterial wilt and sugarcane to mosaic disease. Resistant cultivars have also been the major means of controlling parasitic nematodes, especially some species of root knot, cyst-causing and stem nematodes. Many insect-resistant crop varieties have been released and used in many countries to control leafhoppers, thrips, mites and aphids.

Biological control involving the intentional release or introduction of any biological organism, such as viruses, predators, pathogens and parasites, currently plays a limited but significant role in agriculture. In many cases, the use of these organisms has been integrated with selective use of chemical pesticides. In the future, more opportunities to combine genetic, chemical, biological and cultural control strategies will emerge, changing the control of pests. These developments will further reduce the use of pesticides in our food system.

India and the Asia-Pacific Region need to make further advances in many areas of pesticide-residue management. This region could make rapid progress if it followed the recently developed priorities of the 1998 International Workshop held in Yogyakarta, Indonesia, on Seeking Agricultural Produce Free of Pesticide Residues (Kennedy et al. 1998). The four main areas for implementation are: (1) monitoring of pesticide residues in produce and environment; (2) promoting additional research on developing simple affordable test methods for pesticide detection and using IPM; (3) training and extension on pesticide risk reduction strategies, field tests linked to quality assurance, better methods of pesticide application; and (4) harmonizing procedures for registration and regulation including the establishment of MRL with the aim of reducing costs and increasing trade.

REFERENCES

Agnihotri, N.P., S. Walia and V.T. Gajbhiye. 1999. Green Pesticides, Crop Protection and Safety Evaluation. Agnihotri N.P., S. Walia and V.T. Gajbhiye (Eds.). Society of Pesticide Science, India, Division of Agricultural Chemicals, Indian Agricultural Research Institute, New Delhi, India.. pp.1-20.

Atkin, J and K.M. Leisinger. 2000. *Safe and Effective Use of Crop Protection Products in Developing Countries*, J. Atkin and K.M. Leisinger (Eds.).CABI: Wallingford, U.K. 163p.

Australian Science and Technology Council. 1989. Health, Politics, Trade: Controlling Chemical Residues in Agricultural Products: A Report to the Prime Minister/ by the Australian Science and Technology Council (ASTEC). Canberra: Australian Government Publishing Service.

Bull D. 1982. *A Growing Problem: Pesticides and the Third World Poor.* Oxford, Oxfam, U.K. 192p.

Center for Science and Environment. 2000. URL: http://www.oneworld.org.

Codex Alimentarius Commission. 2000. Data base on maximum residue levels for pesticides. URL: http://www.fao.org/es*/codex/ also accessed 2001.URL: http://www.fao.org/ag/AGP/AGP/Pesticid/Default.htm..

Colburn T., D. Dumanoski and J.P. Myers. 1996. *Our Stolen Future: Are We Threatening our Fertility, Intelligence, and Survival? A Scientific Detective Story.* Penguin Group, N.Y. 306 p.

Economic Times. 2000. Pesticide residues found in wheat flour. June 24, Mumbai, India.

Environmental Working Group (EWG).1995. Forbidden Fruit: Illegal Pesticides in the U.S. Food Supply. Washington, D.C. URL: http://www.ewg.org/pub/home/Reports/fruit/Chapeter2.htm.

Federal Drug Agency (FDA). 2000. Detentions for OASIS for India. URL: http://www.fda.gov/ora/oasis/1/ora_oasis_c_in.html.

Federation of the Indian Chambers of Commerce and Industry. 2000. Agribusiness-Indian Agriculture. India. URL: http://www.agroindia.org/indiagri.htm.

Giovannucci D and T. Reardon. 2000. Understanding Grades and Standards and How to Apply Them. In World Bank Agribusiness and Markets Thematic Group. . 20p. URL: http://wbln0018.worldbank.org/essd/essd.nsf/Agroenterprise/grade_sd.

Harris, J. 2000. *Chemical Pesticide Markets, Health Risks and Residues.* CABI, Wallingford, Oxon, UK: 54p.

James, C. 2000. Global Status of Commercialized Transgenic Crops: 1999. ISAAA Briefs No. 17. ISAAA:Ithaca, New York. 65p.

Jiang-Lin and Ma-Cheng Zu. 2000. Progress of Researches on Biopesticides (review). In: *Acta Agrculturae Shanghai.* No. 16:73-77. (Supplement). Shanghai, China.

Kennedy, G. and T. B. Sutton. 2000. *Emerging Technologies for Integrated Pest Management: Concepts, Research, and Implementation.* APS Press. St. Paul, Minnesota.

Kennedy, I.R., J.H. Skerritt, G.I. Johnson and E. Highley., (Eds.). 1998. *Seeking Agricultural Produce Free of Pesticide Residues.* Proceedings of an International Workshop held in Yogyakarta, Indonesia, 17-19 February 1998. ACIAR Proceedings. No. 85:406p.

Krattiger, A.F. 1998. The Importance of Ag-biotech to Global Prosperity. ISAAA Briefs No. 6. ISAAA:Ithaca, New York.11p.

Mathur, S.C. 1999. The Future of Indian Pesticide Industry in the Next Millennium. Pesticide Information XXIV (4): 9-23.

Mistry, S. 1998. EU's pesticide-level norms on imported spices to be effective from September. In: *Financial Express,* Aug. 31, 1998. http://www.financialexpress.com/daily/19980831/24355454.html.

National Research Council (U.S.). 1987. *Regulating Pesticides in Food: The Delaney Paradox.* Committee on Scientific and Regulatory Issues Underlying Pesticide Use Patterns and Agricultural Innovation, Board on Agriculture, National Research Council. Washington, D.C., National Academy Press, 272 p.

New York State Integrated Pest Management (NYSIPM). 1997. M. Cowles, J. Tette, K. English-Loeb, C. Koplinka-Loehr, and J. Garlick (Eds.) The Marketplace Calls for Environmental Stewardship, New York State Integrated Pest Management Program Annual Report 4/97:4-5 p.

New York State Integrated Pest Management (NYSIPM). 2000. 1999-2000: *The Year in Review.* Woodsen and J. Shultz. (Eds.). New York State Integrated Pest Program Annual Report.

Pimentel, D., T.W. Culliney and T .Bashore. 2000. Public health risks associated with pesticides and natural toxins in foods. In: E.B. Radcliffe and W. D. Hutchison (Eds.), *Radcliffe's IPM World Textbook*. University of Minnesota, St. Paul, MN. URL: http://ipmworld.umn.edu

Seth, P.K., R.B. Raizada and R. Kumar. 1998. Agricultural Chemical Use and Residue Management in India. In: Kennedy, I.R., J.H. Skerritt, G.I. Johnson and E. Highley, (Eds.), *Seeking Agricultural Produce Free of Pesticide Residues*. Proceedings of an International Workshop held in Yogyakarta, Indonesia, 17-19 February 1998. ACIAR Proceedings. No. 85:46-53.

Unnevehr L. and N.Hirschron. 2000. Designing effective food safety interventions in developing countries. World Bank Agribusiness and Markets Thematic Group. 15p. URL:http://wbln0018.worldbank.org/essd/essd.nsf/Agroenterprise/grade_sd.

World Trade Organization Handbook. 2000.How to Apply the Transparency Provisions of the SPS Agreement. 52 p. http://www.wto.org/english/tratop_e/sps_e/sps_e.htm#handbook.

14 Reconciling Food Security and Environment Quality Through Strategic Interventions for Poverty Reduction

Ashok Seth

CONTENTS

POVERTY REDUCTION REMAINS THE MOST COMPELLING CHALLENGE

Since independence in 1947, India has continued to grow economically, poverty has declined and social indicators have improved. Despite this progress, however, India's poverty situation remains a serious concern, with the rural poor accounting for the largest numbers; in 1993–94 every third person in India still lived in conditions of absolute poverty (Datt, 1997). At the same time, the population has continued to grow at around 2.1% per annum as a result of India's high fertility rate, combined with a two-thirds drop in the death rate and a doubled life expectancy. These have led to substantial population increases, from 342 million in 1947 to over a billion people today. By 2050, India's population is projected to reach 1.5 billion, contributing to the formidable challenge of social welfare and poverty alleviation (World Bank, 1995).

India has some 470 million people living on less than US$1 a day. This is twice the number of poor in sub-Saharan Africa and accounts for some 40% of the total number of poor in the world (World Bank, 1999). Although significant progress has been made in improving the public social services, the non-income poverty indicators also remain stubbornly high. It is estimated that, between 20 and 30% of world's children not in school, the gender gap in education, under-5-year-old child deaths each year and maternal deaths each year are in India. Thus, poverty reduction remains India's most compelling challenge, a challenge of global significance (World Bank, 2000).

Interviews with India's poor (Consultations with the Poor India 1999, Country Synthesis Report by PRAXIS-Institute for Participatory Practices) revealed their vulnerability — to disease, crop failures, labor market fluctuations, domestic violence, natural disasters, floods and cyclones and their ensuing sense of insecurity. Any one of such events hit the poor particularly hard, and are generally important contributors to poverty.

WIDESPREAD POVERTY AND INCREASING POPULATION CONTRIBUTE TO DETERIORATING PHYSICAL ENVIRONMENT

India's physical environment is deteriorating in both urban and rural areas, due to poverty, increasing population pressure, weak management practices, poorly funded public services and inappropriate policy framework. Recent World Bank estimates of annual environmental degradation range from 6 to 8% of GDP. About 40% of this cost is related to the burden of diseases due to unsafe water and poor sanitation, 35% to air pollution, including both indoor and urban air pollution, 15% to soil degradation and 10% to other forms of natural resource degradation, such as range lands, forests and fisheries (World Bank, 2000). Thus, the search for solutions to the environmental problems needs to find an important place in the overall development process that benefits the poor. Such a process would need to adopt a holistic approach addressing the social, economic and technological needs of the poor, support adoption of sustainable management practices and reform institutions as well as policies impacting on the physical environment.

PRIORITIES FOR ACCELERATED ECONOMIC DEVELOPMENT

The overriding objective of India's Poverty Reduction Strategy today is growth with social justice and equity. To improve the enabling environment for growth, the strategy calls for greater public and private investment in infrastructure, the restoration of fiscal balance and continued liberalization, especially in agriculture, foreign trade and financial markets. The government gives the highest priority to provision of basic minimum services for all, including education, health, nutrition, water and sanitation. The second and third priorities are the interventions aimed at generating productive employment and restoring regional balance in growth and prosperity.

In a recent report on policies to reduce poverty and accelerate sustainable development, the World Bank noted that economic reforms introduced in the 90s propelled the rate of economic growth to 7% during 1993–97, declining to around 6% since then (World Bank 2000). To sustain the rate of growth at this level or above, the report highlighted the following areas for immediate action.

- Further deregulation of the economy along with reduction in high tariff rates.
- Increase in foreign direct investment. Although an order of magnitude above the levels of the early '90s, it is still incredibly low compared with what other large developing countries have achieved.
- Reform of the financial sector. Capital markets are deep but lack transparency, provide limited access to credit for the rural poor and lack long-term debt market.
- Reduction in fiscal deficits (9.2% of GDP for the general government, and 10.4% of GDP for the consolidated public-sector deficit) at both the central and state levels. State-level fiscal deficits have approximately doubled as a percentage of GDP from their level of 3 years ago. As a result, public borrowings to balance the budget have led to high interest rates that have crowded out private investment. At the same time, the heavy burden of interest payments and deteriorating composition of expenditure have resulted in a weaker development role for the government.

INTERVENTIONS TO PROMOTE PRO-POOR RURAL GROWTH

Fostering sustained agricultural growth, which has been critical for rural growth and poverty reduction, remains an important priority for central and state governments. In this endeavor, the World Bank is India's most important foreign financier. The government–bank partnership takes an integrated approach to rural development. More specifically, this includes investments in rural infrastructure (irrigation, rural roads and markets, drinking water supply and sanitation), natural resource management and agricultural support services (technology generation and dissemination) that are linked to key institutional and policy reforms (World Bank, 1999).

To ensure sustainable management of natural resources, different strategies are being adopted in irrigated and rain-fed areas. In irrigated agriculture, sustainability is linked to more equitable use of scarce water resources through: (1) institutional reform for effective intersectoral planning and allocation of water using a river basin approach, (2) improved delivery of services through greater user participation in irrigation and drinking water systems and (3) enhancing financial sustainability through greater cost recovery and increased allocations to system operations and maintenance.

Resource degradation problems represent a major challenge to rain-fed areas. In a review of the evidence Kerr (1996) identified soil erosion, falling groundwater levels and the degradation of community forests and grazing lands as the most widespread problems. Increasing population pressure, poverty and landlessness further compound the situation. In a recent paper, Hazell and Fan (1999) identified the following key elements for an improved paradigm for sustainable development of rain-fed areas:

- Promote broad-based agricultural development, including facilitation to improved technologies, credit and farm inputs, with a special focus on small farms and women farmers
- Improve technologies based on multidisciplinary location-specific research and farming systems incorporating a holistic approach to resource management practices
- Ensure property rights and effective institutions for managing natural resources
- Ensure that risks are managed effectively through agricultural research, provision of effective safety net (e.g., drought insurance, area-based insurance based on rainfall) by the government
- Investment in rural infrastructure and people
- Provision of the right policy environment and strong public institutions

For the development of rain-fed areas, the government strategy in India now adopts a decentralized watershed-based planning of interventions with strong community participation. The new strategy covers many of the actions listed above and incorporates lessons learned from the older programs. As a result, these interventions are increasing productivity through both intensification and diversification of the production systems in a sustainable manner and are successfully reaching some of the poorest communities engaged in rain-fed agriculture.

Creation of nonfarm employment through greater involvement of the private-sector investment in industry and services is critical in reducing poverty. This will require liberalization of rural economies, including deregulation of controls on trade and processing of agricultural commodities and reform of labor laws. Resultant expansion of the private-sector activities will help to generate jobs, especially by small- and medium-sized enterprises (SMEs). Attention is also being given to the reform of rural financial institutions to improve access to credit by the rural poor, especially women.

TECHNOLOGICAL INNOVATIONS

Technological innovations and their adoption by the farming communities are critical to improving and sustaining productivity growth and to alleviating rural poverty. In the recent past, the research system in India has been very successful, especially during the '60s and '70s, when adoption of Green Revolution technologies resulted in rapid increases in productivity. However, these gains have been limited to a handful of crops and are predominantly in irrigated areas. Basic problems of food and nutrition insecurity, poverty, employment and equity persist. Both growth and regional development objectives call for strengthening of agricultural research and extension systems, attention to diversification of production systems, better exploitation of rain-fed areas and sustainable increases in yields of crops in irrigated areas. In this endeavor, integration of "new" sciences, e.g. biotechnology, with conventional disciplines will be critical, especially to address hitherto unresolved or emerging problems.

The government, with assistance from the World Bank, is addressing the weaknesses of the national agricultural research system through the National Agricultural Technology Project (World Bank, 1998). The project is working to:

- Shift the focus of research on poverty and sustainability issues, increase community ownership of and participation in setting of the research agenda
- Avoid technology vacuum and productivity gaps
- Improve research management systems, research quality and research-extension linkages
- Enhance public–private cooperation
- Finance critical research themes including:
  - Sustainable management of natural resources
  - Conservation of agro-biodiversity
  - Integration of frontier sciences, especially biotechnology, to address new as well as older problems related to biotic and abiotic constraints to higher productivity
  - Postharvest management systems, especially for fruits and vegetables
  - Technology dissemination using participatory methodologies
  - Wider use of latest developments in information technology in support of research and extension activities

REFERENCES

Consultation with the Poor, India (1999). Country synthesis report by PRAXIS - Institute for Participatory Practice, Patna, Bihar, India.

Datt, G. (1997). Poverty in India and Indian States: An update. International Food Policy Research Institute, Washington D.C.

Hazell, P. and Fan Shenggen (1999). Balancing Regional Development Priorities to Achieve Sustainable and Equitable Agricultural Growth. Paper prepared for AAEA International Pre-Conference on Agricultural Intensification, Economic Development and the Environment, July 31–August 1, 1998, Salt Lake City, UT. Paper revised August 14, 1999.

Kerr, J. (1996). Sustainable Development of Rain-Fed Agriculture in India. Environment and Production Technology Division Discussion Paper No. 20, International Food Policy Research Institute, Washington D.C.

World Bank (1995). India's Family Welfare Program: Toward a Reproductive and Child Welfare Approach. Report No, 14644-IN, Washington D.C.

World Bank (1998). National Agricultural Technology Project, Project Appraisal Document. Report No. 17082- IN, World Bank, Washington D.C.

World Bank (1999). India: Toward Rural Development and Poverty Reduction, Vol. 1 (Summary)and II (Main Report and Annexes). Report No. 18921-IN, World Bank, Washington, D.C.

World Bank (2000). India: Policies to Reduce Poverty and Accelerate Sustainable Development. Report No. 19471-IN, World Bank, Washington D.C.

Part Three

Technological Options

15 Ensuring Food Security and Environmental Stewardship in the 21st Century*

S.K. De Datta

INTRODUCTION

One of the greatest achievements of the 20th century was harnessing agricultural sciences with policy interventions to challenge hunger, poverty and food insecurity in the developing regions of the world. International agricultural research centers, national programs, U.S. Land Grant Universities, government entities, non-governmental organizations (NGOs) and the private sector all contributed to enhanced food security in developing regions. It was obvious that this need, combined with vision, determination, commitment and resources, made it possible to tackle these challenges.

The most urgent need in most developing countries was simple and focused: increased food production and the transformation of food-deficient countries into ones self-sufficient in food production. There was remarkable success with cereal production in developing countries, most notably in India. These successes were brought about by focused agricultural research and policy instruments instituted by the governments of many developing countries. As a result, cereal production stayed ahead of the population increase. These successes were tempered by the contention that some segments of the population did not quite benefit from the enhanced cereal production because of poverty and lack of purchasing capacity. A debate continued

---

* Paper presented at the workshop Reconciling Food Security and Environmental Quality in Industrializing India, Ohio State University, Columbus, Ohio. March 7–8, 2001.

1-5667-0594-0/02/$0.00+$1.50
© 2002 by CRC Press LLC

as to whether the Green Revolution helped only the well-endowed farmers. The answer to that debate might be that the well-endowed farmers benefited more because of their capacity to invest in high inputs, which have been one of the ingredients for higher production. The low-income farmers did not benefit very much from the Green Revolution because of their inability to invest in the new technologies.

IMPACT OF THE GREEN REVOLUTION

The debate then shifted to a discussion of whether the high input technologies in the Green Revolution era might have reached a plateau in production and had the unintentional effect of challenging the environmental security of developing nations. In this debate, the issue of enhanced pesticide and inorganic nutrient use was intertwined and gave mixed signals, to the detriment of farmers' understanding of the issues. All inputs were lumped together and branded as "high input technology." However, there is no question that indiscriminate use of pesticides caused health and environmental problems. Similarly, pumping more water from the ground than the rechargeable capacity of the land to replenish it resulted in soil problems that included increased salinity and alkalinity (De Datta et al., 1993). The fact remains that, although we have developed rain-fed agriculture with some success, irrigated agriculture will continue to provide the most stable food production source.

It is widely recognized that the yield ceiling in developing nations has plateaued, and, in some cases, the yields have declined over time, particularly in areas where cereals have been grown intensively (Evans and De Datta, 1979; Flinn et al., 1982). With increased demand worldwide to sustain growth in food production and increased food security, a concerted effort in research that will enhance the yield ceiling is urgently needed. In the case of rice in the tropics, for example, basic research using physiological parameters suggests that at least a 15% and, in an ideal situation, up to a 20% yield increase is possible by modifying plant type and cultural practices (Dingkuhn et al., 1991; Dingkuhn et al., 1992). Unfortunately, the prototype of such higher yielding rice cultivars with fewer but longer panicles has not been found agronomically acceptable because of its lack of pest resistance. In this regard, modern tools such as genetically engineered plants should provide some additional opportunities for a breakthrough in the yield ceiling. However, research on the yield ceiling is time consuming and expensive.

At the same time, donor communities plagued with their domestic agendas are falling behind in supporting agricultural research. In fact, the role of agriculture in the international development agenda has been reduced significantly. New agenda items such as the environment, natural resources, poverty, democracy and governance, health, disease and population issues dominate development agendas. It appears that agriculture does not generate as much energy in the development debate as all the other issues mentioned above. The obvious synergy between child survival and food production is often not understood in the policy arena. And, food security and increased food production are intimately linked together with environmental management.

In fact, many of our concerns and projections about food security have been based on a simplistic judgment calculation of calories and protein intake. The importance of other nutritional bases such as vitamin A, iron and zinc as essential for the physical development of children is not widely recognized, although the recent news of "golden rice" in Asia has generated worldwide attention. International Rice Research Institute scientists, in cooperation with laboratories in Europe, have joined hands with the private sector to develop rices with enhanced Vitamin A production. This public–private-sector collaboration is critical for taking on the complex research topics involved in this urgent technology gap in food security that requires new tools, resources and commitments. The recent approved revision of Title XII entitled Famine Prevention and Freedom from Hunger Improvement Act of 2000, demonstrates the U.S. congress' support for such an approach. Its stated goals include (1) "improved human capacity and institutional resource development for the global application of agricultural and related environmental sciences," and (2) "providing for application of agricultural sciences to solving food, health, nutrition, rural income, and environmental problems, especially such problems of low income, food deficit countries."

ROLE OF BIOTECHNOLOGY

Now let us focus attention on the role of biotechnology and bioinformatics to address food security and environmental issues. There is a heated debate, particularly in Europe, about learning more about genetically modified organisms (GMOs) before placing food items on the supermarket shelves. This is a fine idea, and science should unfold some of the unresolved issues. Unfortunately, the discussion in the developing region of the world is on the urgency of food security; in some instances, food security is directly linked to national security. In this debate, the choice of developing countries is to use whatever tools are available in the pursuit of food and environmental security, including the use of biotechnology and hybrid seed programs. There is consensus that developing countries are moving forward with these new tools to speed up generating new crop varieties that are superior in production, with some specific attributes that will minimize pesticide use and environmental degradation. Detailed issues on the potential role of biotechnology in solving food problems in developing countries have been summarized by Herdt (1993). In an acceptance speech for receiving the Indira Gandhi Prize for Peace, Disarmament and Development, Dr. M.S. Swaminathan said, "while we should admire the prospects of progress and prosperity promised by the virtual world, it would be foolish to overlook the state of poverty, hunger, malnutrition and environmental degradation prevailing in the real world." We therefore need to pursue a research agenda which will touch upon all of the issues mentioned by Dr. Swaminathan.

INTERNATIONAL COLLABORATION

In the pursuit of global efforts on agricultural growth, food and environmental security, and rural development (Hazell and Lutz, 1998), international collaboration is not a choice but a requirement, producing a shared agenda with win–win results. The USAID Global Bureau has supported Collaborative Research Support Programs (CRSPs) and other research support programs led by the United States'

Land Grant Universities. These and other eligible universities are engaged in research programs, institution and policy development, extension, training and other programs for global agricultural development, trade and the responsible management of natural resources.

One such CRSP project is the Integrated Pest Management (IPM) managed by Virginia Tech. The IPM CRSP conducts participatory and collaborative integrated management programs to develop and implement economically and environmentally sound crop protection methods. The program strengthens global IPM capacity in both the United States and developing-country institutions. IPM CRSP research is currently under way in eight host country sites in Asia, Africa, Latin America, the Caribbean and Eastern Europe.

The IPM CRSP goals are to develop improved IPM technologies and institutional changes that will reduce crop losses, increase farmer income, reduce pesticide use and residues, improve IPM research and education program capabilities, improve the ability to monitor pests and increase the involvement of women in IPM decision-making and program design. Achievement of these goals should improve environmental quality, reduce poverty and enhance human health across the globe.

Central to IPM CRSP methodology is the use of a participatory process that includes participatory appraisals (PAs) conducted to identify local problems and the needs of farmers and other stakeholders. Research, training and information exchange activities are developed based on PAs and other information gathering and sharing. Local scientists in IPM CRSP host countries collaborate with U.S. scientists to implement interdisciplinary research, education and training. Most research is conducted on farms with farmer cooperators.

Eight prime sites in developing regions of the world have been strategically selected to create a regional fold from which IPM CRSP technologies can be effectively disseminated to neighboring countries. This enables IPM CRSP to promote the development and adoption of IPM technologies in a variety of cropping systems around the world. The result is higher income, greater food security and greater food safety in collaborating countries. In Asia, for instance, vegetables in rice-based systems are the targeted crops for IPM research. Collaborative IPM research is conducted at two sites, one in the Philippines (for southeast Asia) and the other in Bangladesh (for South Asia). But it is our expectation that results from the IPM of vegetables in rice-based systems at these two sites will benefit people across South Asia, including India and the rest of the southeast Asian countries.

In Bangladesh, IPM CRSP research activities are targeted for vegetable crops because they account for about 10% of the total pesticide use — a disproportionately large share. The research agenda was developed and initiated through a PA process in August 1998 for three intensive vegetable growing areas: Gazipur (Kashimpur), Commilla (Sayedpur) and Narasingdi (Shibpur). A large number of farmers and other stakeholders participated in the PA process. Following that, a planning workshop was held in Dhaka to identify and prioritize the research agenda. Four targeted vegetables, i.e., eggplant, cabbage, tomato, and okra and their pests were prioritized for IPM CRSP research. Each January collaborating scientists and other stakeholders in Bangladesh and the United States, the AVRDC and IRRI review the research progress and prepare the workplan for the following year, keeping in mind the IPM needs and problems faced by farmers.

The IPM CRSP in Bangladesh has had promising initial results in farmers' fields, including:

- A number of eggplant varieties have been identified as resistant to fruit and shoot borers, bacterial wilt, root-knot nematodes and jassids.
- Two eggplant varieties that are resistant to bacterial wilt are now being used for grafting with cultivated eggplants.
- Tests of synthetic pheromones and locally prepared insecticide-impregnated, smashed-sweet-gourd traps were highly effective for attracting and suppressing the cucurbit fruit fly population.
- An economic impact assessment procedure was developed for IPM CRSP research at the Asia site in Bangladesh that draws on Geographic Information Systems (GIS) and economic models. The models were tested for a soil-borne disease control strategy on eggplant and weed control in cabbage. Results from the test project reported several million dollars in net welfare gains given its projected adoption by farmers in Bangladesh over the next 30 years. This information was summarized and demonstrated in a field day organized by the IPM CRSP team in Bangladesh (IPM CRSP Bangladesh, January 2001).

Details on worldwide programs for IPM CRSP are summarized in the report IPM CRSP Annual Highlights For Year 7 (1999–2000), published by the Office of International Research and Development in November 2000.

POLICY INTERVENTION

In the new century, we face a population growth of about 86 million persons a year, mostly in the developing regions, which will contribute significantly to environmental degradation. Policy interventions are needed to mitigate these environmental problems while increasing yields substantially (Pinstrup-Anderson, 1997). Yet the World Bank Report of 1999, as quoted by Ismail Seregeldin (1999), suggests that doubling the yields of complex farming systems in an environmentally sound manner is a difficult challenge. Biotechnology and the associated bioinformatics for a food security and environmental stewardship program may be extremely useful in speeding up the technology development, which allows for fewer pesticides and other purchased inputs.

Over the past 5 years, areas planted with transgenic crops have shown dramatic and continuing increases. From 2.8 million hectares in 1996, this area increased to 27.8 million hectares in 1998 (James 1997 and 1999). The United States alone accounted for 74% of the area devoted to transgenic crops. Developing countries have been late in starting research using biotechnology and there is a lot of catching up to do with limited resources. The scientific tools are fast evolving and capital intensive. Here again, strong and targeted collaboration between developing and developed countries will be beneficial to both regions. The promises are great.

In developed regions, the private sector has invested and reaped the benefits of developing seeds of transgenic crops. However, the public sector has played an important catalytic role. Seregeldin (1999) argues in favor of public–private-sector collaboration to identify and put to work priority areas of technology development that will benefit developing countries while allowing the private sector to recover its investment. Recently, the Swiss company Syngenta and its partner, Myriad Genetics, a U.S. biotechnology company, revealed that they have not only decoded the rice genome, but have also found the location of most of the 50,000 genes it contains as well as the regulatory regions that control them. The map will give a big push to efforts to create new rice germplasm to feed the developing world's population. Syngenta has promised to work with research institutes to pass the benefits of the rice genes on to subsistence farmers (Firn, 2001).

With the revolution of information technology and the potential marriage between biotechnology and information technology (IT), bioinformatics is also becoming an important tool. It promises to speed up the process of developing crops and livestock that are genetically altered for higher productivity while remaining safe for the environment and consumers in both developing and developed regions.

CONCLUSIONS

In conclusion, I again partially quote Dr. Swaminathan, who advocates Gross National Happiness in addition to an increase in the Gross National Product. The major components of this index are: environmental protection, economic growth, cultural promotion and good governance. These are the covenants we must pursue for the 21st century.

REFERENCES

De Datta, S. K., H. U. Neue, D. Senadhira and C. Quijano. 1993. Success in Rice Improvement for Poor Soils. *Proceedings of a Workshop on Adaptation of Plants to Soil Stress,* August 1-4, 1993, pp. 248-268. INTSORMIL Pub. No. 94-2, University of Nebraska, Lincoln.

Dingkuhn, M., H. F. Schnier, S. K. De Datta, K. Dörffling, and C. Javellana. 1991. Relationships between ripening phase productivity and crop duration, canopy photosynthesis, and senescence in transplanted and direct seeded lowland rice. *Field Crops Research.* 26. 327-345.

Dingkuhn, M., H. F. Schnier, C. Javellana, R. Pamplona, and S. K. De Datta. 1992a. Effect of late season nitrogen application on canopy photosynthesis and yield of transplanted and direct seeded tropical lowland rice. I. Growth and yield components. *Field Crops Research* 28. 223-234.

Dingkuhn, M., H. F. Schnier, C. Javellana, R. Pamplona, and S. K. De Datta. 1992b. Effect of late season nitrogen application on canopy photosynthesis and yield of transplanted and direct seeded tropical lowland rice. II. Canopy stratification at flowering stage. *Field Crops Research* 28. 235-249.

Evans, L. T., and S. K. De Datta. 1979. The relation between irradiance and grain yield of irrigated rice in the tropics, as influenced by cultivar, nitrogen fertilizer application and month of planting. *Field Crops Research.* 2(1):1-17.

Firn, D. 2001. International Economy:Syngenta Wins the Race to Publish Rice Genome—Food Crop Sequencing Project. *Financial Times*, Jan. 26, 2001.

Flinn, J. C., S. K. De Datta, and E. Labadan. 1982. An analysis of long-term rice yields in a wetland soil. *Field Crops Research*. 5: 201-216.

Hazell, P.B.R. and E. Lutz, (Eds). 1998. Agriculture and the Environment:Perspectives on Sustainable Rural Development. A World Bank Symposium, World Bank, Washington, D.C.

Herdt, R. 1993. The Potential Role of Biotechnology in Solving Food Production and Environmental Problems in Developing Countries. Paper prepared for the ASA-CSSA-SSSA Annual Meetings, Cincinnati, Ohio, 7-12 November 1993, 28 pp.

International Rice Research Institute, the Rockefeller Foundation, and Syngenta AG. 2001. News about Rice and People: Golden Rice Arrives in Asia.

International Service for the Acquisition of Agri-biotech Applications (ISAAA). 1999. Global Review of Commercialized Transgenic Crops. 1998. Brief No. 8.

IPM CRSP Annual Highlights for Year 7 (1999-2000). 2000. Office of International Research and Development, Blacksburg, Virginia, 55 pp.

IPM CRSP in Bangladesh, an Overview. 2001. Horticulture Research Center, Bangladesh Agricultural Research Institute, 8 pp.

James, C. 1997. Global Status of Transgenic Crops in 1997. ISAAA Brief No. 5.

James, C. 1999. Global Review of Commercialized Transgenic Crops. 1998. ISAAA Brief No. 8.

Office of International Research and Development (OIRD). IPM CRSP in Bangladesh. In:*IPM CRSP—An Overview*. Project managed by OIRD/Virginia Tech and funded by the USAID Global Bureau. It is a collaborative project between U.S. Land Grant Universities (Virginia Tech, Pennsylvania State University, and Ohio State University) and National Systems in Bangladesh (BARC, BARI, BRRI, BSMR Agricultural University, DAE-Plant Protection wing and CARE/Bangladesh). International Collaborators are IRRI, AVRDC, and NCPC Philippines.

Pinstrup-Andersen, P., R. Pandya-Lorch, and M.W. Rosegrant. 1997. The World Food Situation:Recent Developments, Emerging Issues, and Long-Term Prospects. 2020 Vision Food Policy Report, International Food Policy Research Institute, Washington, D.C.

Seregeldin, I. 1999. Biotechnology and Food Security in the 21st Century. In: *Science*, 285. July 16, 1999. AAAS (with assistance from Stanford University's Highwire Press).

Swaminathan, M.S. 2000. Indira Gandhi Prize for Peace, Disarmament and Development Response Speech.

U.S. House of Representatives Title XII Legislation. Famine Prevention and Freedom from Hunger Improvement Act of 2000. H. R. 4002. Sec. 1.

16 Water Harvesting and Management to Alleviate Drought Stress

Gary W. Frasier

CONTENTS

INTRODUCTION

Many parts of the world's land surface are too dry for intensive agriculture without supplemental water. Traditionally, supplemental water has been in the form of irrigation using surface water diversion or pumped groundwater. There are many locations in arid and semiarid areas where surface or groundwater for irrigation is inadequate, unavailable or unsuitable. Yet, many of these lands, in the past or currently, support some form of cultivated agriculture, even in areas that receive less than 200 mm of rainfall per year (Evenari et al., 1961). How can there be intensive agriculture in areas where annual rainfall is less than 200 mm? The answer is that crops are grown using a technique of water supply called water harvesting. In most arid lands, even with limited precipitation, relatively large quantities of water are potentially available if the rainwater can be concentrated, collected, and stored until needed.

1-5667-0594-0/02/$0.00+$1.50
© 2002 by CRC Press LLC

WHAT IS WATER HARVESTING?

Water harvesting is a technique of water supply that collects precipitation from a specific land area for some beneficial use. Precipitation runoff is collected from a relatively large area and stored or concentrated onto a smaller area. This provides a multiplication factor for maximizing the benefits of the limited precipitation. The water collection area can be a natural undisturbed hill slope or some type of prepared impermeable surface. The collected water can be used for growing crops, drinking water for humans and animals, or other domestic uses. It can be used immediately by placement in the soil (infiltration) or stored in an appropriate container for later use.

The term water harvesting has several meanings describing a multitude of methods for collecting and concentrating runoff water from various sources for a variety of purposes. The term is frequently used interchangeably with rain-fed, dry-land or irrigated agriculture (Reij et al., 1948). This chapter will use the meaning that water harvesting is a method of water supply entirely dependent upon local rainfall (overland flow or ephemeral streamflow). Water harvesting for crop production is an intermediate point between rain-fed farming (dry-land agriculture) and standard irrigation from wells or rivers.

Water harvesting as a means of water supply is not a "new" technique. There is evidence of water harvesting structures being used over 9000 years ago in the Edom Mountains of Southern Jordan, and the people of Ur practiced water harvesting as early as 4500 B.C. Studies have shown that extensive agricultural systems using water harvesting techniques existed in several areas 3000 to 4000 years ago in what we now refer to as the Middle East. There is evidence that similar techniques were used over 400 years ago in the southwestern United States, where Mesa Verde National Park is located (Frasier, 1984). Many of these ancient systems were located in areas where the annual precipitation was 200 to 500 mm per year. For these early systems to function satisfactorily, not only did the people effectively collect and store the limited rainfall, they also developed water management techniques to maximize the benefits of the limited water.

POTENTIAL OF WATER HARVESTING

A common concept is that water harvesting has been used only in, or is most suitable for, arid lands. In reality, water harvesting can be used almost anywhere where other water sources are inadequate or unavailable. If all the water that falls as precipitation on a given piece of land can be collected and put to beneficial use, there is usually adequate water to sustain life and support some form of agriculture. This can be illustrated using an example from the Negev Desert of Israel. Current yearly records show that precipitation ranges from 28 to 168 mm per year, with an average of about 86 mm per year. Most of the precipitation occurs during the winter months, November to March, with about 16 rainy days per year, 12 days with precipitation greater than 1 mm, 3 days with precipitation greater than 10 mm and with only a single storm greater than 25 mm per day every 2 years. Average hourly intensities are relatively low, less than 5 mm/hour, but for short periods of 5 to 10 minutes, precipitation intensities up to 20 to 50 mm per hour have been recorded (Anonymous,

1967). Even with low annual precipitation in a very few storm events, considerable water can be collected. One millimeter of precipitation per square meter is equal to 1 liter of water. In this example, if all the annual precipitation (86 mm) occurring on 10 square meters of land can be collected and used to irrigate 1 square meter, it is the equivalent of 850 mm of precipitation.

TYPES OF WATER HARVESTING FOR CROPS (RUNOFF FARMING)

Water harvesting for crop production is commonly referred to as runoff farming. Runoff farming techniques can range from direct water application on the fields during the precipitation event to collecting the precipitation runoff and storing it in a suitable container for later application to the cropping area by some form of irrigation system. There are almost as many types of runoff farming systems as there are installations. These systems can be grouped into four or five general types based on degree of complexity.

One of the simplest and maybe the oldest method of runoff farming is called floodwater farming. The precipitation runoff flowing down an ephemeral channel or watercourse during a storm event is directed or diverted onto a field or cropping area. Sometimes the water is spread (water spreading) onto the banks of the channel by low dams that hold back a portion of the water, allowing it time to infiltrate into the soil. Excess water flows through constructed spillways in the dams to the next lower "spreading dam" (Figure 16.1). In these systems, the first spreading area will

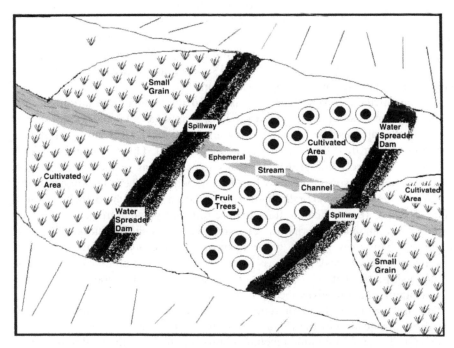

FIGURE 16.1 Water spreading using spreader dikes in an ephemeral stream channel.

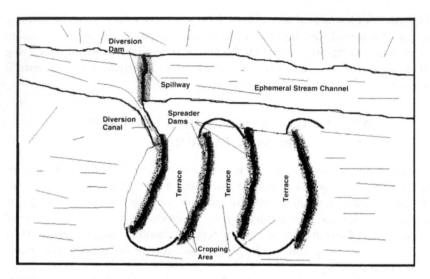

FIGURE 16.2 Water diversion onto contour terraces from an ephemeral stream channel.

potentially receive the most water, with decreasing amounts downstream, depending on the runoff quantities in the channels. A second method of floodwater farming involves the diversion of a portion (or in some instances all) of the water onto a series of contour terraces designed to pass the water from one level to another in a controlled flow regime (Figure 16.2). Again, excess water is allowed to flow downstream through spillways in the diversion dam. In some instances, the runoff water is diverted into some storage container or pond during the storm event. The stored water is then applied to the lower-lying plants or fields at a later date by some form of gravity irrigation. In India, this is referred to as "tank irrigation" (von Oppen, 1983). If the fields are located upslope of the storage, the water can be applied with some form of sprinkler or drip irrigation system.

Another method of runoff farming is called microcatchment farming. With microcatchments, each plant or small group of plants has a small runoff contributing area directly upslope of the growing area. Typically, the runoff area is five to 20 times larger than the cropping area (Figure 16.3). This technique has been used very extensively for growing various trees such as pistachio, olives and almonds (scientific names of plants listed in Appendix). These techniques apply the water to the cropping area during the precipitation event (Photo 16.1). In some instances, water from a hillside flows onto a terraced planted area (Figure 16.4). In Tunisia, a combination form of microcatchment areas called "meskats" is used for various fruit trees. Runoff water from an upslope area is directed onto a cropping area. Any excess water passes over a small spillway into another planted area downslope (Reij et al., 1948) (Figure 16.5). Again, as in floodwater farming, the first planted area receives the most water.

The most complex form of runoff farming encompasses a combination of both direct application of the runoff water and later irrigation with excess water from a stored source. A common technique involves forming the land into a series

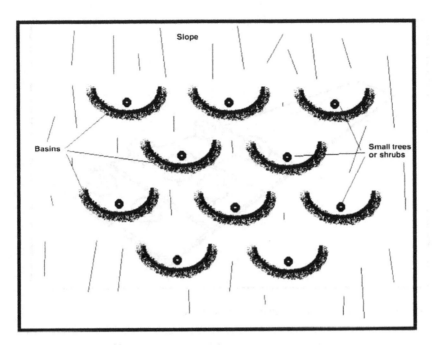

FIGURE 16.3 Microcatchment basin water harvesting.

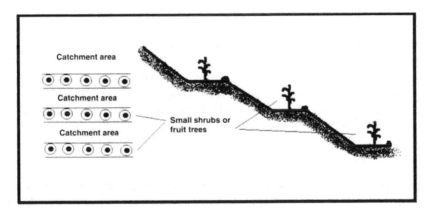

FIGURE 16.4 Microcatchment terraces.

of large ridges and furrows. Crops such as fruit trees or grapes are planted in the bottom of the furrows. Runoff water from the side slopes of the ridges drains onto the crop area in the bottom of the furrows (Photo 16.2). Excess runoff water that is not directly infiltrated into the planted area continues down the center of the furrow into some storage pond or container. At some later date, the water is pumped back onto the crop area as needed, using some form of sprinkler or drip irrigation system.

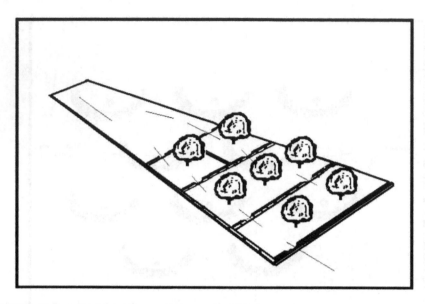

FIGURE 16.5 A Tunisian microcatchment called Meskats.

FACTORS THAT INFLUENCE RUNOFF
FARMING SUCCESS

Each site has unique characteristics that must be considered in the design, installation and operation of a successful runoff farming installation. These characteristics include timing of precipitation with respect to when the water is needed, storm quantities and intensities, soil type and slope, availability of land, labor and materials, potential crops and socio-economic acceptance. Many of these factors are interrelated and must be simultaneously considered (Frasier and Myers, 1983).

Following is a list of a few of the more important factors that must be considered for a successful runoff farming installation.

SOIL TYPE

The soil in the cropping area should allow for good infiltration with a high water holding capacity that will retain the collected water within the plant rooting depth. On the water collection area (catchment) the soils should have an impermeable surface that reduces water infiltration and maximizes runoff. These two opposing conditions can frequently be obtained near the edges of small drainage areas. The hill slopes have relatively shallow soil and steeper slopes, while the bottom areas have deeper soils with flatter slopes.

PRECIPITATION

The quantity of precipitation occurring during a given time interval is one of the most difficult parameters to accurately depict. Monthly averages obtained from long-term records are the most common database. Short-term random fluctuations from the mean can significantly affect the performance of the runoff farming system. To minimize the effect of precipitation variations, it is desirable to use a minimum of 10 years of record. If variations in the precipitation quantities are extreme, data from the two wettest years should be eliminated to maximize the probability that there will be sufficient water when needed (Frasier, 1983). Even harder to estimate are precipitation intensities. There must be some period during the storm events when the precipitation intensity is greater than the infiltration rate on the catchment area. Otherwise there will never be any runoff to collect.

Maximum benefits of water harvesting are achieved if the precipitation occurs during the cooler weather when evapotranspiration rates are the lowest. There is an added benefit if the precipitation occurs during the cropping season. This reduces the period of time necessary to store the collected water and usually permits smaller water harvesting systems.

CROP TYPE

For water harvesting to be most effective, the crop species must be adapted to withstand droughts and effectively utilize water when it is available. Cropping practices must include plant species or cultivars that are capable of utilizing the available water efficiently yet can withstand prolonged time intervals when water may be limited or nonexistent. Cropping practices must also recognize that water requirements for plant establishment are frequently different from the water requirements for mature established plants. During the establishment phase, plant rooting depths are usually shallow, which necessitates that the water be available in the upper layers of the soil profile. Under these conditions, there is the potential for significant losses of the soil water by evaporation from the unprotected (nonshaded) soil surface.

The total water quantity and seasonal distribution requirements will vary for each crop type. Table 16.1 lists the total consumptive water use for selected crops (Erie et al., 1982) that are potentially suited for runoff farming applications. The information was developed under extensive irrigation practices and will probably be higher than needed for many runoff farming applications, but can be used as relative guidelines.

Of equal importance to the total water requirement is the timing of the water needs. Figure 16.6 is an example of the seasonal distribution of water needs for a crop of barley. Total water requirement is 635 mm, with most of the water required in March and April when the grain is in the stage of maximum growth and seed development (Erie, 1982). This water requirement pattern must be satisfied by the design of the runoff farming system. The required water must be either stored for use (in the soil or some storage container) or collected during the critical times of the growing season.

The extra water supplied by a water harvesting system usually improves the yield of crops over what would be obtained by conventional dry-land farming. Studies at ICRISAT Center near Hyderabad, India showed yields of pearl millet, sorghum, and groundnut could be increased with applications of water during

TABLE 16.1
Total Consumptive Water Use for Selected Crops in Mesa, Arizona

Crop	Period of growth	Total seasonal use (mm)
Cash or oil crops		
Castor beans	Apr–Nov	1130
Cotton	Apr–Nov	1050
Flax	Nov–Jun	795
Safflower	Jan–Jul	1150
Soybeans	Jun–Oct	560
Sugar beets	Oct–Jul	1090
Lawn or hay crops		
Alfalfa	Feb–Nov	2030
Bermuda grass	Apr–Oct	1100
Blue panic grass	Apr–Nov	1330
Small grain crops		
Barley	Nov–May	635
Sorghum	Jul–Oct	645
Wheat	Nov–May	655
Fruits		
Grapefruit	Jan–Dec	1215
Grapes (early maturing)	Mar–Jun	380
Grapes (late maturing)	Mar–Jul	500
Oranges	Jan–Dec	990
Vegetables		
Broccoli	Sep–Feb	500
Cabbage (early)	Sep–Jan	435
Cabbage (late)	Sep–Mar	620
Cantaloupe (early)	Apr–Jul	520
Cantaloupe (late)	Aug–Nov	430
Carrots	Sep–Mar	420
Cauliflower	Sep–Jan	470
Lettuce	Sep–Dec	215
Onions (dry)	Nov–May	590
Onions (green)	Sep–Jan	445
Potatoes	Feb–Jun	620
Corn (sweet)	Mar–Jun	500

Source: Erie et al., 1982

short droughts that might occur during the rainy season. Yields of pigeon pea, castor and cowpea could almost be doubled by additional water applications in the post rainy season (Table 16.2) (El-Swaify et al., 1983). This water could be obtained using water harvesting techniques and storing until it is needed (von Oppen, 1983).

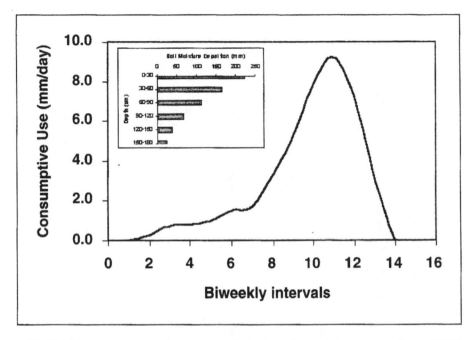

FIGURE 16.6 Mean consumptive water use for barley at Mesa, Arizona, according to USDA data, for years 1952–53, 1969-70 (Erie et al., 1982).

The amount of yield increase with additional water varies by crop. Figure 16.7 shows a typical response of sorghum and groundnut to additional water (Willey et al., 1983). While there is a general increase in yield with increased water availability, there is also a need for increased fertility management in the form of fertilizer application. The danger is that if the fertilizer is applied and there is no rain to collect, the cost and effort of applying the fertilizer are lost. There is an economic maximum of crop yield vs. size of water harvesting system that must be determined for each site. The risk factor of not having sufficient rain to collect must also be considered in these decisions. In many places, the maximum benefit of water harvesting is not realized with increased yields, but better exemplified as getting *some* crop when there would have been none without the additional water. Some of the most successful water harvesting systems have been obtained using plants that are hardy (capable of surviving drought periods) and long lived (olives, pistachios, and almond) or annual plant species that can produce a harvestable crop with one application of water (wheat, pearl millet and barley).

ACCEPTANCE AND NEED AS VIEWED BY USER

The user of the system must be involved in the design and construction as much as possible. The performance and success of the system will depend on the user for proper operation and maintenance. All runoff farming systems will require periodic maintenance. If the user cannot provide the necessary maintenance, the system will fail. In some areas, runoff farming may not be acceptable because of various social

TABLE 16.2
Yield of Selected Crops with and without Supplemental Water during the Growing Season at ICRISAT Center Near Hyderabad, India

Period	Period	Treatment	Crop			
			Pearl Millet	Sorghum	Groundnut	Tomato
Drought during rainy season	1[a]	Control	2100	2820	690	9600
Supplemental water (4 cm)			2700	3218	1050	14400
	2	Control	1630	—	686	23200
		Supplemental water (4 cm)	1725	—	890	13100
			Pigeonpea	Castor	Cowpea	
Post rainy season	1[b]	Control	660	715	310	17500
		Supplemental water (4 cm)	790	928	665	29300
		Supplemental water (8 cm)[c]	1120	1280	725	
	2	Control	850	795	500	
		Supplemental water (4 cm)	910	870	685	
		Supplemental water (8 cm)[c]	1185	1335	795	

[a]Period 1 1981, Period 2 1982
[b]Period 1 1981–82, Period 2 1982–83
[c]Two irrigations of 4 cm each
Source: El-Swaify et al., 1983

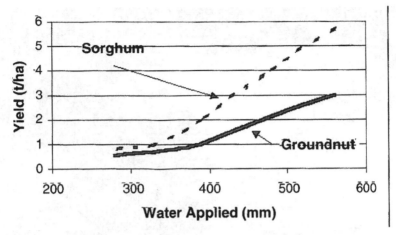

FIGURE 16.7 Effects of additional water on yield of sorghum and groundnuts at ICRISAT Center, Hyderbad, India (Willey et al., 1983).

or economic factors. These factors are not always evident to outsiders. The user must believe that the system is the best for the local purpose or situation. Otherwise, there will be problems in the operation and maintenance of the system. In areas where the concepts of runoff farming are not known or fully accepted, the first installation must be constructed using techniques and materials that will have minimum maintenance requirement and maximum effectiveness. If the user has been shown that the ideas are valid, the user will expend the extra effort to properly operate and maintain the system.

ADVANTAGES AND DISADVANTAGES OF WATER HARVESTING

Water harvesting has the potential to supply water in most areas. It should not be considered an inexpensive means of water supply. Costs of preparing runoff areas (catchments) and water storage facilities can be appreciable. Maximum runoff efficiency is obtained by sealing or covering the soil surface. This may not be cost effective in some areas. An alternative is to increase the size of the catchment area to compensate for lower runoff efficiency. In these instances, it may require higher rainfall quantities to initiate runoff. At sites where land area and labor are relatively inexpensive and readily available, smoothing of the soil surface may be the most effective means of collecting the required quantities of water.

In many locations, the cost of constructing the water storage facility can represent the major expense of a water harvesting system. In these instances, it may be desirable to design the storage to meet the water needs only during the critical growing periods even if there is excess water during part of the year (Frasier and Myers, 1983).

For maximum long-term effectiveness, water harvesting systems must have scheduled, timely maintenance and repair. Many systems have been adequately

FIGURE 16.8 Microcatchment water harvesting for growing jojoba near Phoenix, Arizona, in a 230-mm annual precipitation zone.

FIGURE 16.9 Ridge-and-furrow water harvesting system for growing pistachios near Saltillo, Coahuila, Mexico. Excess precipitation runoff is collected in a storage pond at the lower edge of the field for later application to the trees by a drip irrigation system.

designed and constructed and yet have failed to supply the anticipated quantities of water within a relatively short period of time because of inadequate maintenance. Usually, the required maintenance or repair can be accomplished in a relatively short period of time without a lot of expense. Other systems have failed, despite proper materials and design, because local social and economic factors were not adequately integrated into the systems (Renner, 1993). These systems failed because of personnel changes, water was not needed, or because of communication failures. Word-of-mouth publicity of one failure will often be more widespread than all the publicity from 10 successful units.

A successful system must be:

- Technically sound, properly designed, and maintained
- Socially acceptable to water users and their method of operation
- Economically feasible in both the initial cost and maintenance at the user level

System failure is more likely when funds are available for construction at no obligation to the user, unless there is a clear understanding of how the maintenance is to be performed, by whom, and when.

There is no universally "best" system of runoff farming or water management. Some type of system will be the best for a given location. Each site has unique characteristics that will influence the design of the most optimum system. All factors — technical, social and economic, must be considered (Renner and Frasier, 1995a, b).

The available literature describing techniques for runoff farming is usually not widespread and readily accessible. Much of the information was developed by trial and error, with only brief overviews and descriptions of the successful installations presented in proceedings of meetings. Very little information reaches the scientific journals.

REFERENCES

Anonymous. 1967. Ancient and modern water harvesting in the Negev Desert. Dept. of Botany, Hebrew University, Jerusalem, Israel. mimo.: 17.

El-Swaify, S.A., S. Singh and P. Pathak 1983. Physical and conservation constraints and management components for SAT alfisols. In *Alfisols in the Semi-Arid Tropics. Proc. Consultants' Workshop on the State of the Art and Management Alternatives for Optimizing the Productivity of SAT Alfisols and Related Soils*. P. Prabhakar, S.A. El-Swaify and S. Singh (Eds.), 1-3 December 1983, ICRISAT Center, India, International Crops Research Institute for the Semi-Arid Tropics, Patancheru, Andhra Pradesh 502 324, India: 33-48.

Erie, L.J., O.F. French, D.A. Bucks and K. Harris. 1982. Consumptive use of water by major crops in the southwestern United States. *Conservation Rpt. No. 29*, U.S. Dept. Agriculture, Agricultural Research Service, Washington, D.C.: 40.

Evenari, M., L. Shanan, N.H. Tadmor and Y. Aharoni. 1961. Ancient agriculture in the Negev. *Science*. 133(3457):979-996.

Frasier, G.W. 1983. Water harvesting for collecting and conserving water supplies. In *Alfisols in the Semi-Arid Tropics. Proc. Consultants' Workshop on the State of the Art and Management Alternatives for Optimizing the Productivity of SAT Alfisols and Related Soils*. P. Pathak, S.A. El-Swaify and S. Singh (Eds.), 1-3 December 1983, ICRISAT Center, India, International Crops Research Institute for the Semi-Arid Tropics, Patancheru, Andhra Pradesh 502 324, India: 67-77.

Frasier, G.W. 1984. Water harvesting, including new techniques of maximizing rainfall use in semiarid areas. In *Proc. of the Fourth Agriculture Sector Symposium.* Ted J. Davis Editor, The World Bank, 1818 H. Street, N.W., Washington, D.C. 20433: 46-71.

Frasier, G.W. and L.E. Myers. 1983. Handbook of water harvesting. *Agriculture Handbook No. 600.* U.S. Dept. of Agriculture. Agricultural Research Service: 45.

Reij, C., P. Mulder and L. Begemann. 1948. Water harvesting for plant production. World Bank Technical Paper No. 91, The World Bank, Washington D.C.: 120.

Renner, H. 1993. The potential of microcatchment water harvesting for agricultural production in sub-Saharan Africa: physical, technical and socio-economic design considerations. M.S. professional paper, Colorado State University, Fort Collins, Colorado.

Renner H. and G. Frasier. 1995a. Microcatchment water harvesting for agricultural production: part I: physical and technical considerations. *Rangelands* 17(3):72-78.

Renner H. and G. Frasier. 1995b. Microcatchment water harvesting for agricultural production: part II: socio-economic considerations. *Rangelands* 17(3):79-82.

von Oppen, M. 1983. Tank irrigation in southern India: adapting a traditional technology to modern socioeconomic conditions. In *Alfisols in the Semi-Arid Tropics. Proc. Consultants' Workshop on the State of the Art and Management Alternatives for Optimizing the Productivity of SAT Alfisols and Related Soils.* P. Pathak, S.A. El-Swaify and S. Singh (Eds.), 1-3 December 1983, ICRISAT Center, India, International Crops Research Institute for the Semi-Arid Tropics, Patancheru, Andhra Pradesh 502 324, India: 89-93.

Willey, R.W., M.S. Reddy and M. Natarajan. 1983. Conventional cropping for alfisols and some implications for agroforestry systems. In *Alfisols in the Semi-Arid Tropics. Proc. Consultants' Workshop on the State of the Art and Management Alternatives for Optimizing the Productivity of SAT Alfisols and Related Soils.* P. Pathak, S.A. El-Swaify and S. Singh (Eds.), 1-3 December 1983, ICRISAT Center, India, International Crops Research Institute for the Semi-Arid Tropics, Patancheru, Andhra Pradesh 502 324, India: 155-163.

Appendix 16A:
Scientific Names of Plants

Common Name	Scientific Name
Alfalfa	*Medicago sativa*
Almonds	*Prunus amygdalus*
Barley	*Hordeum vulgare*
Bermuda grass	*Cynodon dactylon*
Blue panicgrass	*Panicum antidotale*
Broccoli	*Brassica* Spp.
Cabbage	*Brassica oleracea capitata*
Cantaloupe	*Cucumis melo cantalupensis*
Carrots	*Daucus carota*
Castor	*Ricinus communis*
Cauliflower	*Brassica oleracea botrytis*
Chickpea	*Cicer arietinum*
Corn	*Zea mays*
Cotton	*Gossypiuia* Spp.
Cowpea	*Vigna sinersis*
Flax	*Linum usitatissimum*
Grapefruit	*Citrus paradisi*
Grapes	*Vitis* Spp.
Groundnut	*Apios tuberosa*
Jojoba	*Simmondsia californica*
Lettuce	*Latuca sativa*
Olives	*Olea europaea*
Onions	*Allium cepa*
Oranges	*Citrus narang*
Pearl Millet	*Pennisetum glaucum*
Pigeon pea	*Cajanus cajan*
Pistachio	*Pistacia vera*
Potatoes	*Solanum tuberosan*
Safflower	*Carthamus tinctorius*
Sorghum	*Sorghum vulgare*
Soybean	*Glycine max*
Sugar beet	*Beta vulgares*
Wheat	*Triticum aestivum*

17 Postharvest Food Losses to Pests in India

David Pimentel and K.V. Raman

CONTENTS

INTRODUCTION

The current world population numbers more than 6 billion (Population Reference Bureau, 2000). The population in India is about 1.1 billion, with a growth rate of 1.8% per year. Based on this growth rate, India's population is projected to double in less than 40 years (Population Reference Bureau, 2000). As India's population continues to grow, the country's serious shortages of cropland, water resources, forests, and energy resources are exacerbated. India's population of 1.1 billion currently exist on about one third of the land area of the United States.

Worldwide, the food situation is critical. The World Health Organization (www.who.int/nut/malnutrition-worldwide.htm) reports that more than 3 billion people are malnourished. This is the largest number and proportion ever in history. India, since its independence in 1947, has undergone a significant transformation from a food grains importer to an exporter. While it is reported to be the third largest producer of food grains in the world after China and the USA, Ramesh (1998) predicted that the country will have to import food grain at the rate of 45 million tons per year by 2000. In spite of major advances in food production in India, it continues to have serious food problems, especially for the poor.

Pre- and postharvest losses vary greatly by crop, by country and by climatic region, partly because there is no universally applied method of measuring losses. As a consequence, estimates of total postharvest food loss are controversial and range widely — generally from about 10% to as high as 40% (www.wri.org/wr-98-99/foodloss.htm). Preharvest food losses to pests (insects, weeds and plant pathogens) are estimated to be more than 40% worldwide, despite the application of more than 2.5 billion kg of pesticides (Pimentel and Greiner, 1997). In India, the estimate is that about 50% of potential food production is lost to pests before harvest, with insects destroying 20%, weeds 15% and plant pathogens 15% (Pimentel and Hart, 2001).

In this chapter, we examine the losses of food to pests during postharvest. The major emphasis will be on grains, because grains make up about 80% of the world's food and are often stored. Recently, India has emerged as an important tropical-fruit and vegetable producer, ranking second after Brazil. India's fruit production is estimated at 32.8 Mt of fruit annually (Roy, 1996). Detailed information on post-harvest losses for fruit and vegetable production in India is not well documented. Our review for this sector is therefore limited.

POSTHARVEST FOOD LOSSES

Worldwide postharvest food losses, primarily grains, to pests (insects, microbes and rodents) are estimated to be about 25% (FAO, 1998; Pimentel and Greiner, 1997; Cao et al., 2001a, b). Postharvest food losses added to preharvest food losses suggest that approximately 52% of all potential food produced in the world are lost to pests, despite all pesticide use and other pest controls employed worldwide.

In India, the estimate is that postharvest food losses to pests are about 30% (Cao et al., 2001a, b). The losses to insects and mites are estimated to be about 5% and microbes are also approximately 5% (Cao et al., 2001a).

The major insect pests of grain are beetles and caterpillars (Metcalf and Metcalf, 1993). These pests infest the grain usually from other infested grain stored in the same building or nearby. In India, the insect species of most importance damaging the food grains including the pulses (legume grains) belong to the order Coleoptera and Lepidoptera. Trained inspectors, with the help of recognition charts and some excellent keys, can identify these insects. However, such service is available at only a few limited locations. A majority of the farmers and extension agencies storing grains lack such information, and, as a result, insects continue to be a problem for stored grains in the hot climates of India.

Microbe infestations occur mostly in the field. Both the microbe and insect infestations require relatively high levels of moisture in the grain for the pests to multiply — about 20% moisture or higher is needed. Insects feeding and metabo-lizing the grain ingested will release moisture and, as this moisture increases, the environment for insects improves and the insect population infestation increases. With high levels of moisture, the microbe populations also increase rapidly.

No one favors consuming grain heavily infested with insects or microbes. Although eating insects in grain has little or no health threat (Pimentel and Greiner, 1997), some of the microbe infestations are a serious threat to public health. In particular, the aflatoxin produced by the fungus *Aspergillus flavus* will poison people and also cause physiological abnormalities resulting from ingestion of secondary metabolites or mycotoxins produced by this fungus. Ingestion of these mycotoxins causes a disease commonly referred to as "mycotoxicosis" (Busby and Wogan, 1979). The Protein Advisory Group of the United Nations has recommended con-suming less than 30 ppb aflatoxin in food rich in protein. In India, the governmental agencies responsible for procuring food grains try to create quality consciousness among farmers through education. They are encouraged to adopt scientific methods of food-grain storage with a view to minimizing the qualitative and quantitative. The quality control teams within these governmental agencies are responsible for monitoring the quality of food grains. In spite of these monitoring mechanisms, India continues to have its exports rejected due to high levels of aflatoxins. Climatic conditions in most regions of India are also conducive to mould invasion, proliferation and production of mycotoxins in grains. Rains and flash floods are common in India and the high moisture content of the grain makes them more vulnerable to fungal attack.

Rodents, especially rats, are a major threat to grains in storage. Three major reasons that rats and mice are considered pests are:

1. They consume and damage human foods in the field and storage. In addition, they spoil food in storage by leaving urine and droppings, thus reducing the sales value.
2. Through their gnawing and burrowing habit, they destroy many articles (packaging, clothing, furniture) and structures (floors, buildings). By gnawing through electrical cables they can cause fires.
3. They are responsible for transmitting diseases dangerous to man. In India, the estimate is that grain losses to rats range from 20% to 30% (Cao et al., 2001b).

In India and Pakistan, individual rats have been reported to consume or contaminate with urine and feces as much as 700 kg of grain per year (FAO/INPHO, 1998). Rats are a particular problem for stored grain because of the ease with which they can invade it. In contrast to insects and microbes, rats can gnaw through plastic, wood and some metals, such as aluminum, to invade grain. Once they have gained entrance to the stored grain, the rats multiply rapidly, each female producing 30 young rats each year.

Rats are also a major problem pest for rice production in India. For instance, rats are reported to consume and destroy approximately 25% of the rice in the field before it can be harvested (Cao et al., 2001b). An individual grown rat is estimated to consume or destroy about $15 worth of grain per year (Pimentel et al., 2000). With an estimated 1.25 billion rats in the United States and assuming the $15 cost per rat, the total damages from rats per year is reported to be US$19 billion (Pimentel et al., 2000). Equally important, rats are implicated as reservoirs and vectors for about 50 diseases, including salmonellosis, leptospirosis, plague, and typhus, to mention just a few (Cao et al., 2001b).

PROTECTION OF STORED GRAINS

Most grains in India are harvested and stored on farms before they are sold and stored in commercial facilities. Most of the traditional methods for storing grains are not insect-, microbe- and rodent-proof. The wooden, burlap and plastic storage facilities are easily invaded by rats and other pests. In addition, the grain usually has a high level of moisture (about 20% or higher), which makes the grain an ideal environment for insects and microbes. To prevent rapid insect and microbe growth, the grain should contain no more than 13% moisture when placed in storage.

With a low level of moisture and uninfested with insects and mites, the grain is generally safe from insects and microbes if stored in heavy plastic bags. However, the grain in a plastic bag is not safe from the invasion of rats and other rodents. To protect the grain from rodents, it must be placed in metal garbage cans with tight lids or in heavily screened areas. Heavy, thick types of tight wooden containers, lined with plastic, might provide sufficient protection from rats and insects.

Once infested, a few methods can control insect and mite pests. High temperatures of about 120° C for an hour will kill most insect and microbe pests. If the grain has already been infected with aflatoxins, the high temperatures will not rid the grain of the toxin. If the grain has a high level of the toxin, the only option is to destroy the grain.

Insect-infested grain can be fumigated with several different pesticides such as cyanide and methyl bromide, but these are dangerous materials that are highly toxic to humans and other animals. These chemicals and other hazardous materials require professionals for treatment of the grain.

In India and other developing countries, it is not uncommon for various insecticides to be added to grains and other stored food products (Cao et al., 2001a). In India, about 98% of the foods purchased have detectable residues of pesticides and 25% of the foods have levels of pesticides above the acceptable tolerance level. This widespread use of pesticides is now responsible for pesticide resistance developing in pest insects.

Insecticides are often added to foods by wholesalers and retailers who desire to protect their resources. Farmers also may treat their grain to protect it from insects and mites. In India, a natural botanical insecticide, such as neem (*Azadrichita indica*), has been added to grain (Cao et al., 2001a). The safety of neem and other botanicals added to grains and in turn eaten by humans remains to be determined.

PROTECTION OF FRUITS AND VEGETABLES

The government of India places high emphasis on the use of postharvest management to prevent postharvest losses in fruits and vegetables. Total losses of fruits and vegetables vary by crop and region. Those due to inadequate postharvest handling, transport and storage of fruits and vegetables vary from 20–40% (Maini, 1997; Mehrotra et al., 1998). Major postharvest diseases of fruits and vegetables in India have now been identified and control measures are being developed (Roy, 1989). To reduce postharvest losses, fruits and vegetables require treatments such as curing, pre-cooling, washing, grading, sorting, packaging, transport, storage and irradiation. Maturity indices including harvesting techniques are now described for many vegetables (Mehrotra et al., 1998). Similarly, new developments in packing and cooling systems are now being developed for fruit crops, and new approaches such as solar drying, pickling and fermentation are reducing postharvest losses of India's fruit (Maini, 1997).

Proper use of postharvest techniques developed within India when effectively implemented in fruit and vegetable production will lead to (1) more availability, (2) benefits for farmers and consumers, (3) better nutrition, (4) more raw material for industry, (5) fewer pesticides used; (6) employment opportunities and (7) improved quality of life.

Realizing the importance of this sector, the government of India has placed great importance on horticultural development during the 8th plan by approving a budget of Rs. 1000 crores (US $250 million) (Maini, 1997).

CONCLUSION

With more than 3 billion people malnourished in the world and food production per capita declining since 1983, greater efforts are needed to reduce losses of food to pests, both pre- and postharvest. Preharvest food losses are estimated to be more than 40% and postharvest food losses are estimated to be 25% worldwide. In India, food losses to pests are estimated to be nearly 50% preharvest, and postharvest, to be about 30%.

Reducing postharvest food losses has priority because, once the food is produced, it should be protected and utilized. In addition, the cost per kilogram of food protected in storage in general is less than the costs of protecting a kilogram of crop food under preharvest conditions.

Grains, which make up about 80% of the world's food, are more easily protected postharvest than many other types of food, such as fresh vegetables and fruit. Although insects and microbes are not easily controlled, a wide array of relatively simple storage units, like heavy plastic bags, can be used to store grains, The grain placed in storage must have less than 13% moisture and be free of insect pests when placed in the heavy plastic bags for storage.

Protecting grains from rats and other rodents is a more difficult problem than insects and microbes because of the ability of rats to gnaw through plastics and many other materials to attack the stored grain. Clean metal garbage cans or heavy metal screening are required to keep rats and other rodents from gaining entrance to grain-storage facilities.

Postharvest losses of fruit and vegetables in India and other countries of Asia are high because of the inherent difficulty of collecting and transporting quantities of produce from numerous small farms and trying to collect these into a large enough quantity for efficient domestic marketing or for export. Even if large shipments can be collected, the produce is often highly variable in size and quality, so it is difficult to apply standardized grading and storage procedures. The warm, humid weather in many fruit- and vegetable-producing regions of India accelerates the decay of tropical produce. Postharvest losses of fresh produce are high, ranging from 20 to 50%. There is, therefore, a great deal of research and training needed to prevent losses in this sector.

Postharvest losses in India continue to be high in many rural areas, primarily because of the lack of proper information, distribution, marketing, postharvest treatment and packaging. Many of these losses could be avoided if some of the relevant recommendations developed in the Caribbean (CARICOM) countries were implemented in India (http://www.fao.org/docrep/x0046e/x0046e00.htm). The recommendations of relevance to reducing postharvest losses in India are:

- The development of commercial enterprises through the introduction of small-scale processing to help reduce postharvest losses and to generate employment in rural areas.
- The postharvest activities conducted by Indian Research Institutions within the umbrella of the Indian Council of Agricultural Research (ICAR) could be expanded, with the objective of disseminating information to existing cottage industries in the rural areas of India. To facilitate this process, extension booklets, show-and-tell activities, farm and postharvest Internet portals in local languages could be developed for use at the village level.
- Additional training of farmers and agroprocessors is required in all aspects of cottage industries, i.e., production, packaging, labeling, marketing and postharvest techniques.
- The training should be carried out at ICAR institutions as well as at other appropriate institutions. Current existing training courses conducted by local Indian institutions should be expanded to include small-scale agroprocessing. Funds for training should be provided by both national and international agencies such as FAO.
- Cottage industries should operate based on sound business principles. Relevant local agencies, such as industrial-development corporations and development banks, should be encouraged to provide business counseling and extension services to cottage industries to promote sustainable business operations in India.
- Bearing in mind that the availability of reasonably priced packaging is a constraint in India, there is need for central, local or regional facilities for importing and selling a variety of packaging materials to small processors. This would be an interim measure aimed at facilitating the availability of packaging where no manufacturing of packaging material exists. This should stimulate the development of packaging industries within the private sector in India.
- Where packaging is unavailable due to lack of appropriate technology, e.g., package molds, efforts should be made to standardize and produce packaging efficiently for the different regions of India.
- A postharvest network could be developed within India and later expanded to include other regions in Asia to provide for the proper exchange and dissemination of information on successful cottage agro-industries. The onus should be placed on the national governments to ensure the success and viability (long term) of this network. Specific technology developed in various regional institutions should be exchanged via an identified network representative in each country. Newsletters should be exchanged on a regular basis. Exhibitions might be held annually in different countries to aid in developing successful cottage industries.

- Various institutions within ICAR, Central Food Technology Research Institute (CFTRI) should be accessed to provide information on research that has been conducted in food technology. Other regional and international organizations should also be accessed for relevant information on food technology
- Appropriate, proven and inexpensive technology should be disseminated via the press, media and Internet connections.
- An inventory of available small-scale processing equipment — where such can be purchased and other general information on technology — should be made available.
- Continuity in the transfer of technology is necessary, so that the different regions of India can be kept informed of the available technology. Teachers need to become involved in agro-industry extension. Agroprocessing should be worked into the school curriculum (via the food and nutrition or home economics programs). Nongovernment organizations (NGOs) should also be involved in this extension service.
- More private and public partnerships in the postharvest sector are needed. Several large and medium-size private firms now regularly acquire food products from farmers in India. The private sector, in several instances, is unaware of the appropriate grades and standards that need to be applied to grains, fruits and vegetables (see Chapter 13 for more details).

REFERENCES

Busby, W.F. Jr. and G.N. Wogan. 1979. *Food Borne Infections and Intoxicants*, Academic Press, New York: pp. 519.

Cao, D., D. Pimentel and K. Hart. 2001a. Postharvest food losses (vertebrates), In *Encyclopedia of Pest Management*, Marcel Dekker, New York, in press.

Cao, D., D. Pimentel and K. Hart. 2001b. Postharvest food losses (invertebrates). In *Encyclopedia of Pest Management*, Marcel Dekker, New York, in press.

FAO. 1998. Food balance sheet, http://armanncorn:98ivysub@faostat.fao.org/lim...ap.pl

FAO/INPHO. 1998. *Proceedings of the Roundtable on the Reduction of Postharvest Fruit and Vegetable Losses through the Development of the Cottage Industry in Rural areas in the Caribbean Countries*, Nassau, Bahamas, 6-8 November, 1991, FAO Regional Office for Latin America and the Caribbean, Santiago, Chile, 1998: pp.127.

Maini, S.B. 1997. Present status and future prospects of postharvest technology of vegetables, *Agricultural Marketing*, 40:21.

Mehrotra, R.S., A. Aggarwal and S. Khanna. 1998. Management of postharvest diseases of fruits and vegetables, In *Pathological Problems of Economic Crop Plants and their Management*, S.M.P. Khurana (Ed.), Scientific Publishers, Jodhpur, India: 431-442.

Metcalf, R.L. and R.A. Metcalf. 1993. *Destructive and Useful Insects: Their Habits and Control*, Academic Press, New York, Chap. 19.

Pimentel, D. and A. Greiner. 1997. Environmental and socio-economic costs of pesticide use. In *Reducing Pesticides: Environmental and Economic Benefits*, D. Pimentel (Ed.), John Wiley & Sons, Chichester, UK: 51-78.

Pimentel, D., L. Lach, R. Zuniga and D. Morrison. 2000. Environmental and economic costs of non-indigenous species in the United States, *BioSci.* 50: 53.

Pimentel, D. and K. Hart. 2001. Pesticide use: ethical, environmental and public health implications. In *New Dimensions in Bioethics: Science, Ethics and the Formulation of Public Policy,* W. Galston, E. Shurr (Eds.), Academic Publishers, Boston:79-108.

PRB 2000. World Population Data Sheet; Population Reference Bureau, Washington, D.C.

Ramesh, A. 1998. Priorities and constraints of postharvest technology in India, In *JIRCAS International Symposium Series No. 7,* 5th JIRCAS International Symposium, Tsukuba, Ibraki, Japan, 9-18 Sept. 1998: 33-43.

Roy, S.K. 1989. Role of postharvest technology of horticultural crops in India, In *Trends in Food Science and Technology,* Proceedings of 2nd International Food Convention, Mysore, India, February 18-23, 1988, Association of Food Technologists, Mysore (India), 1989.

Roy, S.K. 1996. Potentiality of processing and export of tropical fruits from India. *J. of Appl. Hort.*(Navsari) (2): 34.

WHO. 2000. Malnutrition worldwide,http:://www.who.int/nut/malnutrition-worldwide.htm, July 27,2000.

WRI. 1999. Disappearing Food: how big are postharvest losses?http://www.wri.org/wr-98-99/foodloss.htm.

18 Storage and Processing of Agricultural Products

Judith A. Narvhus

CONTENTS

INTRODUCTION

The problem of food insecurity is complex, but one aspect that has received insufficient attention is how to reduce the destruction and spoilage of food following harvesting, gathering or butchering. Some estimates put the loss of food through spoilage as high as 35% in developing countries (Argenti, 2000). It may be surprising that the estimate for food losses in industrialized countries is as high as 25%. However, the reasons for these losses are not the same. The loss of food in industrialized countries is due largely to wastage and poor utilization of raw materials — to a great extent a problem of an affluent society. In developing countries, on the other hand, the loss of food is mainly due to spoilage by microorganisms or to being eaten and sullied by insects or larger animals, especially rodents.

In a drive toward increased food security and food safety in developing countries, several important aspects need to be addressed. The provision of enough food must include preservation (in general terms) of the food that is produced. To grow more food when 35% is destroyed before it can be eaten is not good economy and is certainly ecologically indefensible. The available food should be made safe and free from both pathogenic microorganisms and poisonous chemicals. The food that is available to a population must provide a balanced diet.

The small-scale farmer in developing countries is likely to stay poor unless radical changes are made to food production systems. The poor farmer cannot get rich by selling the excess from the farm directly to the world market. It is questionable, however, whether a policy that advocates and promotes production only of low-price commodities and self-sustainability offers these farmers much of a future.

If a farmer produces exactly enough food for the family, and nothing else in the way of saleable items, their financial situation will worsen, because there will be no available cash to purchase any more of life's necessities. Production of food in excess of requirements gives the potential of earning money by selling the surplus. How successful this is depends on the demand from the local market and also on whether the local people have sufficient buying power. At best, such income will be spasmodic. At times, the farmer may have no surplus to sell, at other times, a glut of a commodity may make sale difficult or unprofitable.

Food is an essential commodity that plays a crucial part in raising the standard of living. The development of a country must go hand in hand with the development of a food processing industry. However, the question of whether such industry should be small- or large-scale needs to be assessed in each situation.

FOOD LOSSES

Much food is lost due to spoilage during storage. In tropical countries, the hot climate is conducive to rapid deterioration due to the growth of microorganisms. Pest control is also more difficult than in temperate climes, where harsh winters exert a certain seasonal control. With the problems and expense of creating a system of cooled food transport, it may not be possible to get the food to larger markets in good condition. Trucking on poor roads may cause considerable bruising of fruit and vegetables, thus hastening decay and reducing their sale value. Transporting milk in uncooled tankers over long distances to a large dairy results in growth of microorganisms that, at best, make the milk a poor raw material for further processing, at worst, unsuitable for use at all. In a recent study in Zimbabwe (Gran et al., 2002), the number of microorganisms in milk produced by rural farmers increased approximately fivefold during uncooled transport to the dairy. The numbers were found to be positively correlated to the distance between the farm and the dairy. Similar results have been found in a study of rural milk production in India (Wetlesen, 2001). Raw meat and fish suffer similarly during transport. Thus, the transportation of raw foods to a distant market may result in an inferior product.

Much of the total profit in food production lies, not in the actual growing of crops or the rearing of animals and their sale, but in the processing of food raw materials into value-added food products. Transport of raw materials into the towns will mean that the potential additional profit from food processing is moved from the rural to the urban communities. The income of rural populations can be meaningfully increased only if processing of raw materials is done in the rural areas so that the profits of this processing are returned to the local producers. This can be achieved by the setting up of small-scale cooperative food processing units (Galun, 1996). It is important to point out that this sequence of events has taken place in the past in most industrialized countries, in particular within dairying. Surplus processed food products can either be sold locally or transported to urban areas. The latter can result in a much-needed translocation of revenue from urban to rural areas that may also help to slow down urbanization. An additional advantage with local processing is that processed foods often have an extended shelf life compared with the raw materials and they are therefore easier to transport. Thus, a significant advantage can be gained by processing a highly perishable product locally, and thereby reducing losses through early treatment. Wastewater can be used for irrigation. Waste from food processing in rural areas can be composted or fed to animals; in urban areas, this waste constitutes a pollution problem and is expensive to dispose of properly (Cybulska, 2000).

Large-scale production of processed food products may bring the benefits of more cost-effective processing and improved food quality. However, the establishment of large factories in urban areas is unlikely to benefit rural farmers but will benefit the investors and the factory workers and thereby possibly increasing the influx of rural people to the towns. In addition, the present shortage of expertise in food science in developing countries would make large-scale units dependent on foreign management.

WHAT CAN FOOD TECHNOLOGY OFFER?

The processing of food usually results in an extension of its shelf life. However, this is not the only advantage. Food that has been processed is often safer from pathogenic microorganisms. Several different processes may have this effect and, of these, heat treatment is probably the most important. If food processing can be done near the area of raw material production, the opportunity for growth of unwanted microorganisms in unprocessed food is reduced due to the shorter time that transpires before processing can take place. This reduction in time also reduces the possibility of development of microbial toxins in raw materials.

Many food-processing techniques change the nutritional value of the product, a factor that can be either positive or negative. For example, heat treatment of a food may result in a more digestible product, but may also reduce the amount of vitamins or the availability of amino acids.

Processing of food almost always results in changes in sensory attributes. In some processes, this change is not desirable and every effort is made to reduce such changes. An example of this is heat treatment for milk, which results in the least possible change in taste while still achieving the desired reduction in the number of microorganisms. However, in many other cases, food-processing technology results in changes that are necessary to attain the desired taste or texture. Foods that have been dried and salted do not taste the same as the original raw material and, as such, may be regarded as another food. Fermentation processes result in production of important flavor compounds that are characteristic for the product (Steinkraus, 1996). Many food processes have the additional advantage of providing a wider variety of foods.

CHALLENGES FOR FOOD TECHNOLOGY IN DEVELOPING COUNTRIES

Many of the foods eaten in developing countries are not those that figure in food technology textbooks. The raw materials may be uncommon or even unknown in industrialized countries and the technology used in traditional processing may never have been published. However, these foods are an important part of the people's heritage and culture. The raw materials for traditional foods are usually produced in sustainable agricultural systems that are suited to the area's climate and soils, whereas the introduction of alternative foods or technologies based on raw materials used in industrialized countries is not necessarily going to be a success story.

More advanced technologies for food processing may be dependent on the availability of electric power or other fuels and this can, at present, be a problem in remote areas. The use of wood for fuel cannot be recommended as part of the development of local food processing due to the negative environmental impact.

Central to many food processes is the availability of plenty of potable water, which can present a problem for the introduction of food processing in rural areas. Water may be mixed with the food during the process and will most certainly be used for cleaning of equipment. However, it should be remembered that local water supplies may not have sufficient capacity for even small-scale food processing units. Pinstrup-Anderson and Pandya-Lorch (1998) advocate that water policies should be reformed to make better use of existing water supplies. Agriculture, the single largest user of fresh water, accounts for ~75% of current human water use (Wallace, 2000). If efficiency can be improved in the agricultural sector, local water resources may be sufficient to supply a local food processing industry.

Vagaries of climate are also a challenge for the development of small-scale food processing industries. High ambient temperatures are a particular problem because this promotes spoilage. Heavy rains and poor roads can also hinder transport of products away from the local production areas to small or large towns. The need for pest control is greater in tropical areas than in temperate countries. This introduces a further problem if the food product is destined for export, as the purchasing country may impose maximum allowable levels for pesticide residues that are difficult to attain if the pests are to be controlled.

Distribution, sales and marketing are unfamiliar concepts in areas that have previously based their food production and consumption on self-sufficiency. However, these aspects must be addressed when developing systems for local food processing.

TRADITIONAL FOOD PROCESSING

The use of heat to treat food is one of the most ancient of food technologies. Many raw materials change in taste, consistency and digestibility when subjected to heat. Simultaneously, the food becomes safer as pathogenic microorganisms are destroyed. Heat treatment at the household level is a fairly uncontrolled process, with neither even nor constant temperatures and times being employed.

Sun drying of foods is an economical way of preserving some foods, for example some fruits, grain, nuts, fish and meat. Dried foods, due to their low water content, are less prone to microbial degradation. Nevertheless, dried foods may be spoiled by yeasts and moulds and are not necessarily free from pathogenic organisms if they have been dried under unhygienic conditions.

Some fruits and vegetables can be processed to extract juice, to be drunk as fresh juice or fermented to alcoholic brews. Food can also be preserved by salting or by adding sugar.

TRADITIONAL FERMENTED FOODS

In most countries of the world, fermented foods of various types are consumed. We are all familiar with dairy products such as yogurt and cheese, and with olives and coffee. These, along with many other everyday foods, are, in fact, produced using fermentation techniques. In tropical countries, many foods undergo spontaneous fermentation, resulting in new products with new properties of flavor and consistency. In these countries, the range of fermented foods is often greater, a natural consequence of high ambient temperature and lack of cooling facilities.

Fermentation of food is caused by the selective growth of specific microorganisms, in many cases lactic acid bacteria and yeasts. These microorganisms may be naturally present in the food raw material, or they may be purposely added as starter cultures. During fermentation, microorganisms grow in the food and their metabolism of particular components produces compounds that bring about specific changes in the taste and consistency of the original raw material. Certain compounds, such as lactic acid and ethanol, when produced in high concentrations during the fermentation exert a preservative effect that may prevent the growth of pathogenic organisms. Virtually all types of foods can be subjected to fermentation processes — vegetables, fruits and cereals (Battock, 1998; Haard et al., 1999), milk, fish and meat. In the case of spontaneous fermentation, the microorganisms that cause the desired changes are those present in or on the raw materials. Which microorganisms will dominate in the fermented food can be influenced by various technological procedures. In most fermented products, the desirable organisms have been found to be various specific species of lactic acid bacteria or yeasts. However, in such an uncontrolled production system, the chance of other less desirable organisms also being present represents a threat to both health and food quality. An unsuccessful fermentation can therefore result in wastage of large amounts of raw materials.

UPGRADING OF TRADITIONAL FERMENTED FOOD TECHNOLOGY

Few of the changes, or their causes, that occur during fermentation of the majority of traditional tropical fermented foods have been documented. If the food traditions of developing countries are to be preserved, these processes must be researched so they are not forgotten and replaced by unfamiliar foods introduced from industrialized countries. Such research requires detailed documentation of the traditional production technology, including local variations. The microorganisms responsible for these fermentations must be isolated, characterized and selected according to their desirable contribution in the fermentation process. They can subsequently be added as starter cultures to new batches of raw material to promote the desired fermentation. The development of suitable small-scale processing equipment is also necessary. This can facilitate the preparation of the raw material before the fermentation step or contribute to the actual fermentation by providing an environment in which the fermentation can proceed under controlled conditions of temperature and humidity and also be kept free from contamination by unwanted microorganisms or pests. Control of fermentation processes produces safer foods of consistent and better quality, because the fermentation is no longer a matter of chance (Steinkraus, 1996). Implicit in the potential for this improvement is the availability of safe water.

SELECTION OF FOODSTUFF AND TECHNOLOGY

When developing traditional food technologies for small-scale processing, the selected raw material and intended product should be familiar to those who carry out the processing. The target market must also be defined. The scale of production and the requirements for distribution and packaging are, to a large extent, dependent on whether the product is destined for local markets, urban areas or export. The introduction of small-scale processing of raw materials to foods that are known to have a sustainable production and a stable market is more likely to be successful than introduction of, for example, a nonindigenous plant that is to be processed into an unfamiliar product with unknown long-term appeal. When selecting a raw material or food product, the type of storage or distribution network necessary for an acceptable shelf life must also be assessed.

The establishment of small-scale food manufacturing systems must also consider the seasonal variation in the availability of raw materials. Production based on raw materials that have a limited keeping quality or that are harvested during only 2 months of the year is not likely to be an economic success and will at best provide spasmodic income.

Small-scale technology has an advantage over large-scale manufacturing because the equipment required can be kept relatively simple and may even be based on man or animal power. This reduces the chances for stops in production due to breakdown of equipment in areas where it may be difficult to obtain spare parts quickly and where qualified technical assistance may be hard to come by.

The economic aspects of upgrading traditional food technologies cannot be ignored and there is a need for experts in this field to assess the market potential for products before significant investments are made.

THE GENDER ISSUE

In many developing countries, women are responsible for the production and processing of food for the family, possibly also for sale. These women have an inherent understanding of the food processes and the procedures necessary for promoting the products' safety. The production of food is part of women's cultural heritage. The making of saleable commodities for the market not only gives women social contact and status but also money in their hands that they can use according to their priorities — usually for the improvement of the family's well being.

However, women unfortunately have less access to improved technology, training and extension programs and credit (Paris, 2002) and this may prove to be a hurdle in the development of small-scale production systems within the present social structure. If food-manufacturing businesses become dominated by men, women will lose an important source of income and contact. This threat to the social structure of rural communities can be mitigated by assistance in the setting up of women's cooperatives and by making special credit facilities available.

THE POTENTIAL EFFECT OF LOCAL FOOD PROCESSING ON POVERTY AND HUNGER IN RURAL AREAS

The manufacture of value-added products at the local level could bring much-needed revenue to rural or semi-rural populations. The cost of transport of raw materials is reduced and the raw materials can also be processed at a time that is optimum for achieving the best quality end product. Reduction in spoilage of raw materials and products makes for better economy for the producer. An improvement in food safety will reduce the incidence of food-borne diseases and the manufacture of food under controlled conditions will result in better and more stable quality.

A more organized and effective production and processing of food, where the economic gains are returned to the primary producers, will contribute to increasing the income of these people and also their standard of living. Development of processes that are not radically affected by seasonal availability of raw materials gives the farmer a steady income compared with yearly harvesting of a cash crop.

WHAT MORE IS NECESSARY?

In developing countries, much emphasis has — rightly — been on increasing the production of food by improved agricultural systems, the use of fertilizers and the introduction of new varieties of crops. Ecological aspects such as controlling soil erosion and reducing deforestation have also been in focus. Some developing countries are becoming increasingly aware that they must move from being primary producers to also becoming processers of food (Anon, 2001). However, this paradigm shift requires competency within the field of food science, both in the industry and also in educational institutions. In many countries, university departments of food science are in their infancy.

By increasing the competency of staff at educational establishments, knowledge can be passed on to future students destined for the country's food industry. The institution can also become a source of help for the industry on a consultancy basis, thus building on these ties for mutual benefit. Funding aimed at building this type of competence must also include the provision of the necessary "hardware" (for example pilot plants and laboratories) to give the necessary practical experience.

Production hygiene is an aspect that must receive special attention. If the necessary precautions are not taken, the shift from home processing to small-scale or even large-scale processing creates the possibility of widespread food poisoning or mass food spoilage. It is imperative that knowledge and understanding of the principles of production hygiene are conveyed to all workers in food and catering industries.

In rural areas, outside investment in processing equipment can give a much-needed head start. Advice about and investment in transport and distribution systems and also infrastructure such as buildings is necessary. As a national food industry develops, the need for control of food quality and safety becomes more pressing. It may not be possible to export processed foods if certain safety and quality standards cannot be documented (Henson and Loader, 2001). Central control laboratories are indispensable if it is to be ensured that the food produced is of acceptable quality.

REFERENCES

Anon. 1996. Small-scale food processing in rural development. Spore. Information for Agricultural development in ACP countries. No. 65, 1-3.

Anon. 2001. Government of India, Press Information Bureau. Report on Tasks Before National Task Force on Food Processing. Symposium, September, 2001.

Argenti, O. 2000. Achieving Urban Food and Nutrition Security in the Developing World. Feeding the cities: Food Supply and Distribution. IFPRI. 2020 Focus 3, Brief 5 of 10. Http://www.ifpri.cgiar.org/2020/focus

Battock, M. 1998. *Fermented Fruits and Vegetables. A Global Perspective.* FAO Agricultural Services Bulletin No. 134. FAO, Rome.

BOSTID, 1992. Applications of biotechnology to traditional fermented foods. Report of an *ad hoc* panel of the Board of Science and Technology for International Development. National Academy Press.

Cybulska, G. 2000. *Waste Management in the Food Industry: An Overview.* CCFRA Key topics in Food Science and Technology – No. 3. Chipping Campden, Gloucestershire, U.K.

Galun, E. 1996. Rural population helped by "partnerships" in processing industry? *Biotechnology and Development Monitor,* September issue.

Gran, H.M., Mutukumira, A.N., Wetlesen, A., Narvhus, J.A. 2002. Smallholder dairy processing in Zimbabwe: hygienic practices during milking and the microbiological quality of the milk at the farm and on delivery. *Food Control:* 13, 41-47

Haard, N.F., S.A. Odunfa and C-H Lee. 1999. *Fermented Cereals. A Global Perspective.* FAO Agricultural Services Bulletin No 138. FAO, Rome.

Henson, S. and R. Loader. 2001. Barriers to agricultural exports from developing countries: the role of sanitary and phytosanitary requirements. *World Development* 29: 85-102

Paris, T.R. 2002. Crop-animal systems in Asia: socio-economic benefits and impacts on rural livelihoods. *Agricultural Systems.* 71, 147-168

Pinstrup-Anderson, P. and R. Pandya-Lorch. 1998. Food Security and sustainable use of natural Resources: a 2020 vision. *Ecological Economics* 26: 1-10.

Steinkraus, K.H. 1996. Introduction to Indigenous Fermented Foods. In: *Handbook of Indigenous Fermented Foods.* Marcel Dekker, New York.

Wallace, J.S. 2000. Increasing agricultural water use efficiency to meet future food production. *Agriculture, Ecosystems and Environment.* 83: 105-119.

Wetlesen, A., 2001. Personal Communication.

19 Postharvest Food Technology for Village Operations

Poul M.T. Hansen and Judith A. Narvhus

CONTENTS

INTRODUCTION

The trends in food production and population in the developing countries have shown that parallel growth has been approximately maintained over recent decades. In quantitative terms, the increases in world food production have kept pace, even increased, with population growth for this period of history. The apparent success is due to the introduction of improved agricultural technologies with respect to plant breeding programs, fertilizers and irrigation, but has also required placing additional land under cultivation. There is, however, a growing awareness that reducing food waste may increase food supplies. According to estimates, the loss of some food commodities may approach 30–40% of the total production, but the question of how much of the world harvest really is lost remains. There is no solid information on the precise amount and nature of loss because losses vary greatly by crop, by country and by climatic region. Furthermore, there is no universally applied method of measuring losses. As a consequence, estimates of total postharvest food loss are controversial and range widely — generally from about 10% to as high as 40% (Miller, 2000).

Along with improving economies, dietary habits have changed dramatically from predominantly grain-based foods to those of animal-based products and higher-value fruits and vegetables (Delgado et al., 1999). Such foods are inherently perishable and, therefore, food losses are likely to increase due to spoilage (Rajorhia, 1999). For example, the seasonal nature of fruits and vegetables and their poor keeping quality prevent their use throughout the year. When in season, much wastage may occur when the fruits and vegetables are cheap and abundant (Nagi and Bajaj, 1996). In a similar fashion, milk production is also highly seasonal and, when the abundant season coincides with higher temperatures, much of the milk may sour and spoil (Shaikh, 1999) unless chilling facilities are made available. Thus, food processing and preservation become one of the important strategies for future food security.

However, in developing countries, the introduction of postharvest food technology on a broad scale must be approached with some caution. In a study of food wastage at an open-air food market in Sumatra, Van Giffen, (1985) made these observations:

"If we define food wastage as *'every piece of food being thrown away,'* there seems to be very little food wasted at the Pasar Raya food market. Usually, the food, which is too ripe to be stored until the next day, is sold at the end of the market at a discount. For an economist, however, this practice may represent wastage, i.e., *a loss of economic value* that could have been prevented by better storage and display facilities.

"For the large group of buyers that usually enter the market at late afternoon, this practice is of direct importance to their existence. It means food at a price they can afford, which is not the case with the better quality, fresh products sold at higher prices in the morning. Any processing that will do away with this practice will have disastrous effects on significant parts of the urban population."

The central question is not whether we can produce enough food, but whether people can afford to buy it or grow it (Lean, 1978). We must ask if introduction of food processing technology will increase the cost of food and put it further out of reach of the poor. We must hope that the long-term effects will be a reduction in food prices as we have seen in industrialized countries.

FOOD PROCESSING

The most efficient way to increase per capita food availability in the Third World is to prevent food spoilage and increase channels of distribution by adopting processing methods that will insure the preservation of food that cannot be consumed in the fresh state. As food production increases, so does the need for food preservation. The food industry employs a number of different methods for food preservation depending on the kind of food and the market available. Some of these methods are based on ancient traditions for food preparation that have been adapted for larger-scale operations. Other methods are based on technological innovation and require sophisticated and expensive equipment. In the industrialized countries, the food industry has become highly mechanized and automated because of high labor costs and because of a need for maintaining uniformity of quality.

In other parts of the world, substantial quantities of food are still processed by traditional methods; for example, olive oil and wine production in the Mediterranean countries remain labor-intensive industries, yet the products compete effectively on the export market. Table 19.1 lists processing methods for food preservation.

TABLE 19.1
Methods for Food Preservation

A. Temperature Control
 1. Cold storage (milk, meat, fruits and vegetables)
 2. Frozen storage (meat, fish and vegetables)
 3. Modified atmosphere (fresh fruit and vegetables)
B. Heat Processing
 1. Blanching (to control biochemical changes)
 2. Pasteurization (to reduce microbial content)
 3. Sterilization
  • Canning (batch process for vegetables, meat and fish)
  • Ultra-High-Temperature-Short-Time (UHTS)
  • Radiation preservation (still in experimental stage)
C. Water Removal, Drying and Dehydration
 1. Concentration (tomato paste, puree)
 2. Vacuum Evaporation (milk, orange juice and tomato juice)
 3. Solar and wind drying (fish, fruits, fruit slices)
 4. Oven drying (cereals, potato chips, protein isolates)
 5. Freeze drying (high-cost items e.g. spices, coffee and tea)
 6. Roller drying (process for low-cost items, e.g. animal feed)
 7. Spray drying (typical industrial process for large volume processing of dry milk, baby formula, instant coffee and many powdered food ingredients, incl. powdered eggs)
D. Microbiological preservation
 1. Yeast fermentation (beer and wine)
 2. Lactic acid fermentation (yogurt, cheese, pickled vegetables)
 3. Complex fermentations (coffee, tea, cocoa, soy sauce)
 4. Indigenous fermented foods (thousands of them)
E. Chemical preservation
 1. Addition of sugar (jams and jellies)
 2. Addition of salt (salted meat or fish)
 3. Addition of acids (pickles, soft drinks etc.)
 4. Addition of preservatives (soft drinks)
 5. Smoking (smoked fish, smoked meat)
F. Packaging
 1. Protection against contamination
 2. Protection against deterioration
  • Loss of moisture (meat, vegetables, cheese)
  • Gain of moisture (breakfast cereals, potato chips)
  • Loss of aroma compounds (coffee, tea, cocoa)
 3. Barrier against oxygen permeation (frying oil, ghee)
 4. Odor barrier (cheese, fish, some fruits)

TABLE 19.2.
Reasons for Processing Food

A. Preservation of perishable/seasonable foods
  - Decrease food waste
  - Stopping unwanted biochemical changes
  - Stopping growth of spoilage organisms
  - Destruction of pathogenic organisms

B. Generated changes
  - Taste
  - Flavor
  - Consistency

C. Improved food distribution
  - Long shelf life
  - Improved packaging

D. Reduction in transportation costs
  - Reduction in volume and weight
  - Efficient loading and unloading

·BENEFITS

The different methods used by the food industry for preserving food against spoilage or for prevention of physical deterioration serve to extend the storage life of food. But there are additional advantages. Some of the benefits to be gained by food processing technology are listed in Table 19.2.

Decrease of food waste is high on the list. Waste is the avoidable loss of something valuable, and, for food, principally represents sullying caused by insects and rodents and also spoilage due to microorganisms and inherent biochemical changes that take place in all unprocessed food. According to Ewards (1979), food spoilage can be regarded as any change in the nature of a fresh or processed food material whereby changes in chemical, physical or organoleptic properties of the food take place leading to its rejection by the consumer. If spoilage can be controlled, the storage life of food can be extended with opportunities for improved distribution and marketing.

CONSTRAINTS ON FOOD PROCESSING

In postharvest food technology, constraints must also be considered in applying a technical fix to the problems of food preservation. Some of these are listed in Table 19.3. These relate to the altered characteristics of processed foods and to the high costs for capital expenditure and energy associated with establishing a food processing industry. Considerations of the sociological and environmental effects on the communities, which must host the incoming industry, cannot be ignored.

TRANSFER OF FOOD TECHNOLOGIES TO VILLAGES

The processing technologies and plant designs to be introduced into a region in transition from traditional processing to advanced industrial manufacture may

TABLE 19.3
Constraints on Food Processing

- Loss of nutrients
- Loss of flavor (e.g. loss of aroma from orange juice)
- Altered taste (e.g. staleness, rancidity, cooked flavor)
- Capital investment (cost of plant and equipment)
- Availability and cost of potable water
- Availability and cost of electricity
- Availability and cost of detergents and sanitizers
- Cost of waste disposal
- Sociological and environmental effects

initially be focused on small-scale operations to serve the rural communities. While small-scale food processing operations stand in sharp contrast to the trends for the food industry in the industrialized countries, proceeding in this manner has its justifications. The expenditure on large-scale automated equipment may not be justified in countries where labor costs are low, and the creation of new jobs for food handlers and technicians will provide an economic boost to the community. Centralized food processing depends on a well-developed infrastructure of road and railway systems and reliable supplies of water and electric power. If these provisions are not in place, the manufacturing plant will fail. Transportation is a crucial issue, both for product quality and for the economy of operation; delays due to congested roads may cause food spoilage and the maintenance of vehicles driving on poorly maintained roads is a formidable expense. Such considerations are contained in reports from an Indian dairy factory, Baroda, Inc, which is planning construction and operation of satellite chilling centers in rural communities (Shaikh, 1999). By locating the chilling centers within reach of the milk producers, the time between milking and effective cooling is greatly reduced and the quality of the milk is preserved (Table 19.4).

TABLE 19.4
Proposed Baroda Dairy Chilling Centers

Operation
- Self-sufficient units
- Veterinary doctors and services
- Processing facilities for flexible pouch filling

Economics
- Capital cost: Rs 1.85 crore/chilling centre
- Advantage: Elimination of 22 tank trucks
- Transport savings: Rs 8,600 per day.
- Amortization of capital cost: 9 months

Source: The Times of India, Dec. 1999

CENTERS FOR FOOD PROCESSING

The concept of establishing milk-chilling centers can be expanded to operation of similar village or community centers for small-scale food processing and storage of selected products. The goal of such centers would be for local farmers to find improved temperature- and storage conditions for produce until sale at the markets and to convert surplus milk into products with long shelf life and marketing potential. Such operations could be integrated into biovillage schemes as launched by the M.S. Swaminathan Foundation in southern India in 1999-2000 (UNDP, 2000).

Modified atmosphere packaging or storage has become a valuable technique for prolonging the storage life of fresh fruit and vegetables (Thompson, 1998) and of fish and fishery products (University of California, 2000). Modified atmosphere is a condition created by enclosing the products within sealed plastic film that is slowly permeable to the respiratory gases. During respiration of the fresh fruit or vegetable, the surrounding atmosphere will change in composition with lower oxygen content and higher concentrations of CO_2. The altered atmosphere with respect to oxygen- and CO_2 levels slows the rate of respiration and the aging processes of the food. A somewhat different technology is used in controlled atmosphere storage. Here, the composition of the surrounding atmosphere is monitored and controlled by injecting gas mixtures at the required composition. Controlled atmosphere conditions have made possible truck transport of fruit and vegetables over large distances, yet still assure consumers of high-quality products (Poulsen and Cowley, 1989).

Making ghee is important in many cultures, but perhaps nowhere more than in India. Ghee is an anhydrous butterfat product that represents a valuable portion of milk with extended keeping quality. The current knowledge of ghee making has been reviewed (Sserunjogi et al., 1998). The authors stress that homemade ghee is a commodity of variable quality due to different technological production procedures and the uncontrolled microbiological status of the raw milk or butter. The keeping quality of ghee is governed by such factors as the ripening of the cream, method of manufacture, clarification temperature and the permeability of the packing material to air and moisture. Several of these factors are difficult to control at the homestead level. To produce a more uniform product of assured composition for marketing purposes, it may be possible to reprocess ghee at a central facility.

Dry acid-casein: Casein is an important protein found in skim milk (2–3%). When properly prepared and purified, it finds use as a food ingredient but also has technical uses in nonfood items. Casein may be prepared as a curd from skim milk by methods similar to those traditionally used for making curd from yogurt. After the curd has been separated, it is washed carefully in clean water to remove all traces

---

TABLE 19.5
Village Food Processing Centers

- Modified atmosphere and controlled temperature storage for vegetables and fruits
- Reprocessing of ghee
- Casein manufacture

of whey; it is then broken into small pieces and thoroughly dried in the sun. In this form, the product is called dry acid-casein and has no direct use. However, it is an intermediate product for making *sodium-* or *calcium-caseinate* (Tetra Pak, 1996). In this form, the products are water soluble and are used throughout the world as an ingredient in manufactured foods. The making of dry acid-casein can be done on a small scale without need for expensive equipment and would be a suitable shelf-stable product for a village cooperative. The conversion of acid-casein into sodium- or calcium-caseinate can be done only by industrial scale equipment.

UNIVERSITY ROLE FOR POSTHARVEST FOOD TECHNOLOGY TRANSFER

INTERDEPARTMENTAL COOPERATION

Postharvest technology is the integrated application of science, engineering and technology for improved utilization of agricultural commodities. Special provisions may be needed to promote integration of postharvest technology among the university departments. Also, institutional cooperation with the national food research centers and government agencies is a needed element for formulating strategies and implementing new programs. Postharvest technology is closely identified with the disciplines of food science, food technology and agricultural engineering. There are compelling reasons to link these programs with those of other departments within the agricultural university system. High among the priorities is a combined effort for technology transfer to improve the quality of agricultural outputs from the farms. It is necessary that technology transfer be based on sound economic policies and with consideration for possible impact on the social structure and the environment.

ORGANIZATION OF VILLAGE PROCESSING CENTERS

Modern food processing is rapidly gaining ground in India. The development of the cooperative dairy system, Amul, has resulted in a network of large dairy plants supplying abundant milk to all of the metropolitan centers of the country. Similar large food processing plants dealing with other foods are in evidence and responsible for supplying cities with processed foods, including bakery and vegetable products. India has a large reserve of professional food technologists and engineers who spearhead the development of the food industry for urban areas, which represent a ready market for the surrounding farming areas. In contrast, there are vast rural areas, remote from urban population centers, where the farmers do not have a ready outlet for surplus agricultural products. By reason of poor infrastructure, the various products of milk, produce and meat may not reach marketing or processing centers in acceptable condition. A suggested strategy for introduction of food processing to minimize postharvest losses would be to focus on establishing village processing centers, possibly as cooperative ventures.

The food technology departments at Indian universities are in a position to support design and development of such enterprises (Table 19.6). By taking leadership in an educational program for training individuals for the food industry at all

---

TABLE 19.6
University's Role in Postharvest Technology

A. Education
   1. Provide workers for an established food industry
   2. Provide entrepreneurs for small-scale food production
   3. Food control authorities
B. Interaction with the food industry
   1. Pilot food processing plant
      • Product development
      • Equipment development
      • Practical teaching for entrepreneurs and employees
   2. Consultancy
      • Problem solving
      • Product development
   3. Further education
      • Short courses on specific products and technologies

---

levels, the university departments become active participants in developing resources for future food. As part of this strategy, the food science and technology departments may consider consolidating the teaching of fundamental principles of food processing and food preservation under a postharvest technology program, and conduct an aggressive training program in the practical aspects of food processing. With the availability of resource personnel and needed food processing facilities, the food disciplines will be in a position to lead efforts to reduce postharvest losses through the development of value-added products (Wheatley et al., 1995) and to improve food quality and safety (FDA, 1999) to generate a better return for farmers.

REFERENCES

Delgado, C., M. Rosegrant, H. Steinfeld, S. Ehu and C. Courbois. 1999. Livestock to 2020: The Next Food Revolution. (Discussion Paper 28), International Food Policy Research Institute, Washington D.C.

FDA. 1999. Annex 6: Food Processing. Food Code 1999. p. 414-425. Department of Health and Human Services, Public Health Service, Food and Drug Administration, Center for Food Safety and Applied Nutrition, Office of Seafood, Washington, DC.

Lean, G. 1978. *Rich World, Poor World*. George Allen & Unwin, London.

Miller, S. 2000. Disappearing Food: How Big Are Postharvest Losses? World Resources Institute, 10 G Street, NE (Suite 800), Washington, DC

Narvhus, J., Research and Development Of Indigenous Fermented Foods For Small Scale Commercial Processing, In *Successes in Rural Development*, R. Haug and J. Teurlings, Eds. Noragric, Aas, Norway, pp 88-91.

Nagi, M. and S. Bajaj. 1996. *Home Preservation of Fruits and Vegetables: A Manual*. Communication Centre, Punjab Agricultural University.

Poulsen, K.P. and M. Cowley. 1989. *Guide to Food Transport: Fruit and Vegetables*. Mercantile Publishers, DK—2100, Copenhagen.

Rajorhia, G.S. 1999. Analysis of Aga Studies on Peri-Urban Livestock Production Systems in Asia, West Asia and Near East, Africa and Latin America. Food and Agriculture Organization. Rome, Italy.

Sserunjogi, M.L., R.K. Abrahamsen and J. Narvhus. 1998. A review paper: Current Knowledge of Ghee and Related Products. *International Dairy Journal*, 8: 677-688.

Shaikh, S. 1999. Baroda Dairy to set up two chilling stations. *The Times of India*. (20 December 1999) http://www.timesofindia.com/201299/pagemahm.htm.

Tetra Pak. 1996. Dairy Processing Handbook. Tetra Laval Marketing Services, Tetra Pak Processing Systems AB Lund, Sweden.

Thomsen, A.K. 1998. *Controlled Atmosphere Storage of Fruits and Vegetables*. CAB International, New York.

UNDP 2000. Harnessing Women, Nature and Biotechnology to Fight Poverty! http://rbapintra.undp.org/RBAP/INDIA1.htm.

University of California 2000. Chapter 8: Vacuum and Modified Atmosphere Packaged Fish and Fishery Products. http://seafood.ucdavis.edu/haccp/compendium/Chapt08.htm.

Van Giffen, D.F. 1985. Research on Food Waste-Some Preliminary Observations. Symposium on Future Prospects for Food Processing in West Sumatra. Padang, 29-30 July, 1985.

Wheatley, C., G.J. Scott, R. Best and S. Wiersema. 1995. Adding Value to Root and Tuber Crops: A Manual on Product Development. CIAT ISBN 958-9439-14-4.

20 Reconciling Animal Food Products With Security and Environmental Quality in Industrializing India

Herbert W. Ockerman and Lopamudra Basu

CONTENTS

INTRODUCTION

Based on current technologies, the world is now approaching the limit of global food production capacity. New technologies are going to be needed, and a reordering of world priorities will also be necessary to solve the world's, as well as India's, food problems.

1-5667-0594-0/02/$0.00+$1.50
© 2002 by CRC Press LLC

World population grew slowly over most of recorded history but started accelerating in 1900 (Horiuchi, 1992). Population had reached 5.5 billion by 1992 and 6 billion in 2000 with a growth rate of 1.7% per year. At this rate, the population will double every 40 years. Recently, fertility in a number of countries has gradually decreased (United Nations Population Fund, 1991; UNFPA, UN Population Fund, 1991).

The world's food production must increase at a rate greater than the population growth if future generations are going to have an adequate diet (Kindall and Pimentel, 2001). In the 1960s, many countries had an adequate supply of food, but today, only a few countries fall into that category (Kindall and Pimentel, 2001). High-yielding crops and energy-intensive agriculture led to the Green Revolution, which expanded world grain production at the rate of 2.7% per year between 1950 and 1984 (State of the World, 1990; World Resources Institute 1990). Today, per capita production has slowed (Moffat, 1992) and may even be declining.

Some of the major food problems (in addition to production) are inadequate distribution, spoilage after harvest, and the adverse economic situation in many countries. Added to this is the soil degradation that has also become a major threat to the world food supply (Lal and Pierce, 1991). Water is also becoming a limiting factor for food production because 16% of the total land (producing one third of crop production) is under irrigation and it consumes 70% of the fresh water used by humans (Leyton, 1983; Batty and Keller, 1980; Ritschard and Tsao, 1978; World Resources, 1992-93).

India, located in southern Asia, has a total area of 3.3 million km², of which 2.9 million km² is land area and 0.31 million km² is water (Crosswalk, 2001). This makes it slightly more than one third the size of the United States. Its climate ranges from tropical monsoon in the south to temperate in the north; its terrain from upland plain (Deccan Plateau) in the south to flat to rolling plain along the Ganges, deserts in the west, and Himalayas in the north. Natural resources include coal (fourth largest reserves in the world), iron, manganese, mica, bauxite, titanium, chromate, natural gas, diamonds, petroleum, and limestone (India, 2001). Of the land area, 56% is arable, 1% is in permanent crops, 4% in permanent pastures, 23% in forest and woodlands and 16% other land uses (Crosswalk, 2001). Irrigated land was 480,000 km² in 1993. As of July 1999, India was inhabited by more than 1 billion people. India's coastline is 7,516 km long and nearly 20% of the population lives in the coastal areas. The age structure is 34% between 0–14, 61% between 15 and 64 and 5% over 65 (Crosswalk, 2001). Many highly populated and industrialized cities such as Bombay, Madras, Calcutta, Cochin and Visakhapatnam are located near the coast (ICM Country Profile, 2001). Religions in India are 80% Hindu, 14% Muslim, 2.4% Christian, 2% Sikh, 0.7% Buddhist and 0.5% Jains (Crosswalk, 2001). In India, 67% of the population works in agriculture, which contributes 25% to the country's gross domestic product (Crosswalk, 2001). Agricultural products are primarily rice, wheat, oilseed, cotton, jute, tea, sugarcane, potatoes, cattle, water buffalo, sheep, goats, poultry and fish (Crosswalk, 2001).

The human fertility rate in India is declining, but is still rather high (Figure 20.1), and progress is being made in most of the parameters normally used to indicate progress. Life expectancy in India is increasing, so the total population is still expanding in this

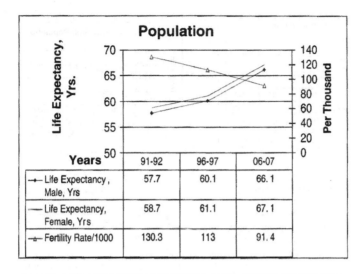

Years	91-92	96-97	06-07
Life Expectancy, Male, Yrs	57.7	60.1	66.1
Life Expectancy, Female, Yrs	58.7	61.1	67.1
Fertility Rate/1000	130.3	113	91.4

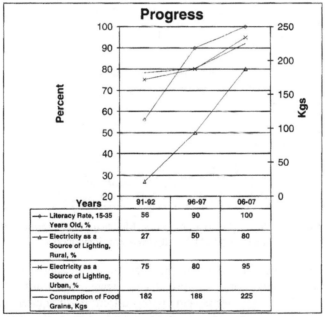

Years	91-92	96-97	06-07
Literacy Rate, 15-35 Years Old, %	56	90	100
Electricity as a Source of Lighting, Rural, %	27	50	80
Electricity as a Source of Lighting, Urban, %	75	80	95
Consumption of Food Grains, Kgs	182	188	225

FIGURE 20.1 Indicators of social development (2001a). Modified from Department of Agriculture & Cooperation: Statistics at a Glance. http://www.nic.in/agricoop/statistics/growrh9.htm.

country. Population growth remains a basic problem because it requires approximately 0.5 ha of cropland to feed a human with extensive agriculture and 0.07 with intensive agriculture (Lal, 1991). In 1990, only 0.27 ha per capita of cropland was available in India. A nation with over 2% population growth has almost no hope of improving its per capita food supply (Kindall and Pimentel, 2001) in the future.

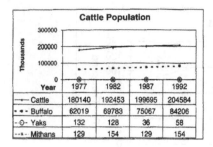

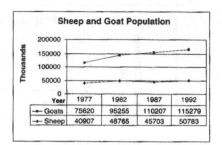

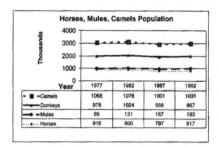

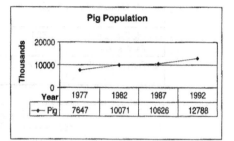

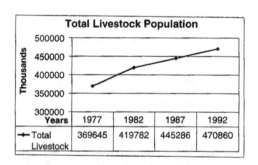

FIGURE 20.2 Changes in Indian livestock production. Modified from Department of Agriculture & Cooperation: Statistics at a glance (2001b). http://www.nic.in/agricoop/statistics/prod1.htm.

Irrigated land is often prone to salinization and water logging, which are serious problems in India. If this continues, 30% of the irrigated area will be lost by 2025 (Anonymous, 1992). Also, water is being pumped from fossil aquifers in excess of the recharge rate. In southern India, the underground water level is dropping at the rate of 2.5–3 m per year (State of the World, 1990; World Resources Institute, 1990). In the world as a whole, livestock use plant growth as a source of food, so, for example, these animals graze on about one half of the total land area and one quarter of the cropland is planted in grain for livestock (Durning and Brough, 1992). Therefore, anything that affects crops also affects livestock. Livestock are important because they convert cellulose (the most abundant plant material), which cannot be converted by humans into nutritious food items, and they also produce manure, which is useful for crop fertilization. The data in Figures 20.2 and 20.3 show India's animal production and animal products perspective. These data indicate that cattle,

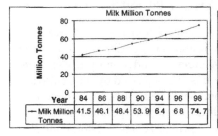

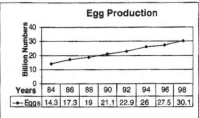

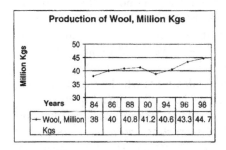

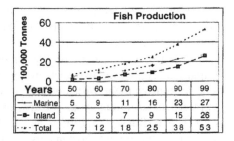

FIGURE 20.3 Animal products production. 2001 Modified from Department of Agriculture & Cooperation: Statistics at a glance. http://www.nic.in/agricoop/statistics/prod2a.htm. and http://www.nic.in/agricoop/statistics/fish1.htm.

sheep, goats, pigs and total livestock production have been increasing while yaks, mithuns (wild ox in northeast India), horses, mules and camels have remained about constant, suggesting that the important food producing animals are increasing in numbers. This is also indicated by the dramatic increase in milk, eggs and fish production, and the more gradual increase in wool production.

India's gross domestic product (1993–1994 constant price) in 1999–2000 was Rs. 11,51,991 billion, which was a growth rate of 6.4%, and, for 2000–2001, was Rs. 12,21,174 billion or a 6% growth rate (1 US $ = 45 Rs in 2000). The per capita income for 1999–2000 was Rs.10,204, a 4.8% increase; for 2000–2001 was 10,654, a 4.8% increase; for 2000–2001 it was 10,654 billion, which was a 4.4% increase (Press Information Bureau Government of India, 2001). Agriculture accounts for 34% of the gross domestic product (India, 2001). The percentage of undernourished people decreased from 38% in 1979–81, to 26% in 1990–92 and 22% in 1995–97, still placing India in the moderately high category of developing countries (FAO, 1999; World Food Programme, 2001). In spite of this progress and due to its large population, India has the largest number (204 million) of undernourished people of any country or even groups of countries in the developing world (FAO, 1999; World Food Programme, 2001).

The agriculture, forestry and fishing sector of India had a quarter (July–Sept., 2001) increase (constant prices) of 1.0% compared with 6.0% gross domestic product for the same period. This 1.0% was the lowest of the eight sectors listed (Statistics India, 2001).

LIVESTOCK DIVERSIFICATION IN INDIA

The Himalayan area makes up a unique geographical and geological region of India with diverse agro-economic and environmental conditions. It occupies 591,000 km², which is approximately 18% of India (Chander and Harbola, 1996) but contains only about 6% of the population. It is characterized by little cultivable land, lack of economic diversity, out-migration, low accessibility, low productivity, little infrastructure, little employment, and little social and political articulation (Chander and Harbola, 1996). The economy is primarily agro-pastoral and very livestock dependent. Cattle account for 47.5%, buffalos 12.3%, goats 15.9%, and sheep 10.4% of the livestock in the region (Chander and Harbola, 1996). Pigs and poultry are found in the eastern Himalayan, and sheep and yaks in the alpine areas. Horses are often used for transportation. Many attempts have been made to improve the livestock production units in the area but most have failed because livestock was seen in isolation and there was inadequate follow-up and lack of farmer participation. Most of the livestock are low in productivity, which stems from lack of genetic improvement with no clear direction, lack of adequate quantities of food and balanced rations, poor pastures in forest land, lack of management, poor housing, harsh climatic conditions and lack of marketing and disease control in this difficult environment. Fortunately, indigenous cattle are well acclimatized to many of these conditions. Livestock diversification has also been recommended for this area (Chander and Harbola, 1996). Species such as Angora and broiler rabbits, poultry in the lower altitudes and Pashmina goats (there is some concern from a conservation standpoint with sheep and goats), which thrive in the higher altitudes. However, local markets, processing facilities, and a strong extension service would be necessary to make these species successful. Things that are needed (Chander and Harbola, 1996) in this area to improve livestock production and consequently a firmer economic base are: greater coordination and collaboration of international institutions that are trying to help, genetic improvement of livestock, adequate feed and fodder, improved veterinary service with artificial insemination capabilities, diversification of livestock, promotion of cooperatives, greater role of women in management, nongovernment organizations (NGOs) working closer with farmers, and an efficient extension service. People-centered programs need to be a priority.

PROGRESS IN FOOD PRODUCTION

World cereal and meat product constant prices declined from the early 70s through the mid 90s (World Bank, 1998). It now appears that there is a "livestock revolution," which is driven by demand for animal food products, which increased in the same areas where increased consumption occurred. In developing countries, between 1982 and 1994, it grew at the rate of 5.4%, which is five times the rate of growth in developed countries (FAO, 1998a).

India is the largest producer of fruits and the second largest producer of vegetables in the world. It almost has a monopoly on spices and condiments (ICAR, 2001). According to the Indian Council of Agricultural Research (ICAR, 2001),

India has achieved a breakthrough in milk (world's largest producer), meat, eggs (28 billion per year), poultry (300 million broilers in 1966) and wool production (ICAR, 2001). The animal industry also supplies approximately 73 million draft animals that cultivate about 55% of the cropland (ICAR, 2001). Fish is another area of success and India is now in seventh place in the world, producing 4.3 million tonnes of fish from marine and fresh water and exporting $1 billion worth of fish and fish product in 1994–1995 (ICAR, 2001).

Despite the progress, a lot remains to be done. Undernutrition is still a problem. It is often not due to the lack of food but the inability of needy groups to purchase it. Governments alone cannot solve these problems of food security, and civil organizations such as trade unions, self-help associations, cooperatives, women's groups, NGOs and informal groups can be tremendously helpful (FAO, 1998b). Cooperatives, with many examples of failures and problems (a major one is low participation of women), are usually organizations of the poorer segments of society and play an important role in obtaining social capital, which is an important ingredient of food security and sustainable agriculture (FAO, 1998b). India is probably the best example of cooperative empowerment of milk producers. For example, in the Amul district, there were 180,000 producers in 700 cooperatives in 1970, but this number had increased to 544,000 producers in 950 cooperatives in 1995. There were 9 million producers in India in 1995 (FAO, 1998a; Department of Agriculture & Cooperation, 2001a, b, c).

For developing countries, total meat consumption is growing at the rate of 5.4% and milk consumption is increasing at the rate of 3.1%. Even small increases in per capita consumption are magnified in many developing countries that have rapidly increasing populations. Under the most realistic set of assumptions (Rosegrant et al., 1997) it is estimated that the annual consumption of meat in India will increase by 2.9% per year and milk consumption by 4.3% per year for the period 1993 to 2020. This will result in total consumption in 2020 of 7.25 million tonnes of meat and 145 million tonnes of milk, which translates into 6 kg of meat and 125 kg of milk per capita per year in 2020 (Rosegrent et al., 1997).

CATTLE

Cattle are distributed throughout the world and, of the 1.3 billion head, about 31% are found in Asia. India has more cattle than any other country, with 20% of the world's cattle and buffalo (FAO, 1998a) in 1993. The market value of these animals is estimated at Rs. 25.6 billion and their output (milk, meat, dung and draught power) is estimated at Rs. 22.5 billion per year (Trivedi, 1990). Approximately 50% of the households own milk animals (Trivedi, 1990). From 1982 to 1994, growth rate per year for cattle was 3.6% and the growth in percent of animals slaughtered was 2.2% per year (FAO, 1998a). *Bos taurus* originated in Europe and are responsible for most of the modern beef and dairy breeds of cattle. *Bos Indicus* and Brahman cattle originated in India and can be recognized by a hump at the withers. They are usually white, with large droopy ears and a large dewlap. This breed can now be found in Africa, Asia, the southern parts of North America,

Central America, and the northern and central parts of South America. Today about 274 cattle breeds exist and continue to evolve.

Dairy cattle have been developed to produce milk, and the major dairy breeds of the *Bos Indicus* found primarily in India include Gir, Hariana, Red Sindhi, Sahiwal, and Tharparker (Bupps, 2000). The National Dairy Development Board has initiated progeny test programs and the Dairy Herd Improvement Programme Action (DIPA) maintains bulls and supplies semen for artificial insemination (A.I.) use. In 1990, 20% of the dairy cooperatives were using A.I. (Trivedi, 1990). Embryo transfer, cryopreservation, cloning and embryo sexing have also been initiated (Trivedi, 1990).

Beef cattle have been developed primarily for the production of meat. Of these, the Brahman and some of its crosses such as Brangus and Santa Gertrudis are popular and are used in various arid parts of the world. In India, there is virtually a ban on the slaughter of cattle (Jul and Padda, 1984) for religious reasons.

BUFFALO

There is no religious belief associated with the buffalo that prevents it from being slaughtered for human consumption. However, this operation is still somewhat of a clandestine industry (Jul and Padda, 1984). From chemical, physical and organoleptic standpoints, there is little difference between beef and buffalo meat, with the exception that buffalo is slightly lower in fat content. The buffalo also has the advantage of consuming cheap roughage (e.g. wheat straw, paddy byproducts, maize stubble, *jowar*, and sugar cane tops) and performs much better than cattle on this diet.

Buffalo milk recording is also under way and some animals are producing 3000 kg of milk in 300 days (Trivedi, 1990).

MILK PRODUCTION

Approximately 80% of Indians are Hindus, and a significant number (approximately 210 million) of them are lacto-vegetarians (Hull et al.1993). Milk and dairy products are important to the cultural life of Hindus, with *mithais* (dairy sweets) being offered at weddings, birthdays and religious occasions (Bhaskaran, 1996). Therefore, milk is an important source of protein to a large percentage of the Indian population. India's number of cattle increased by 6.4% per year and those milked by 1.6% per year from 1982 to 1994 (FAO, 1998a). Milk production was 973 kg per head in 1992–1994, having grown at the rate of 4.8% per year (FAO, 1998a).

Increase in the number of middle- and lower-middle-class families, significant economic achievements, liberalization of economy and identification of public policy dictated by food security and food self-sufficiency caused a tremendous increase in India's milk output ("white revolution"). India, being the world's largest milk producer, is one of the world's real success stories. India's total milk consumption increased from 31 million tonnes in 1983 to 47 million tonnes in 1993 (FAO, 1998a). For example, the Indian government approved more than 250 milk processing and dairy manufacturing licenses between 1991 and 1993 (Abichandani, 1994) and multinational companies such as Nestlé and Unilever have increased capacity and product offerings. Consequently, milk production in India increased from 15.5 million tonnes in 1951 (Acharya and Yadav, 1992) to 61.2 million tons in 1994 (FAO, 1995). Milk was ranked as the most important rural product in India in 1993 (Aneja, 1994). Per capita milk consumption in India increased from 39 kg in 1985 to 66 kg per annum in 1990, by far the highest growth rate in Asia.

With current dietary habits and significant cultural changes as the result of an increase in income, the demand for milk and milk products will increase. However, some barriers to this exist. Dairying is a secondary activity for most farmers. Inefficient dairy farming methods, limitations of feed and fodder (25% of world's population and 2.5% of land mass and 0.5% pastureland; Doornbos and Gertsch, 1994), overstocking and opposition to culling of nonproductive animals are always major impediments to growth of the dairy industry in India.

MEAT CONSUMPTION

History suggests (Figure 20.4) that meat consumption goes up with increase in prosperity (Jul and Padda, 1984; Alexandratos, 1998). This is true even in countries that have religious limitations to meat consumption. The data in Figure 20.4 show an exponential relationship between national per capita income and per capita meat consumption. People are reaching their meat saturation point in the higher income countries. Also, within this trend, countries differ according to culture, and India lies below the general trend because of religious preference against meat. In general,

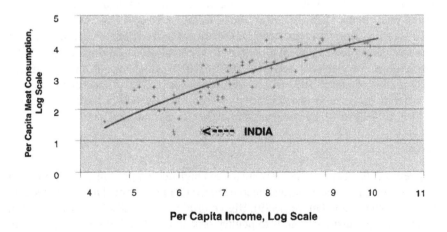

FIGURE 20.4 Relationship between meat consumption and income. Each plot represents a developing or a developed country examined. The trend line is statistically significant. Replotted from data by Jul and Padda, 1984, and Alexandratos, 1998.

developed countries consume three to four times as much meat and five to six times as much milk products as developing countries (Delgado et al. 1998). However, this pattern is changing, with consumers in developing countries obtaining a greater percentage of their calories and protein from animal products in 1993 ("red revolution") compared with 1983 (FAO, 1997c). Lal (1977) indicated that the most acceptable meat in Indian homes is from goats and sheep and states, "among Indians, meat eaters are no longer in a minority." In 1984, there were 2800 organized and 9000 unorganized slaughterhouses in India (Jul and Padda, 1984). In addition to organizing and modernizing slaughterhouses, many individual companies are offering fresh and frozen meat materials and poultry products (India Mart, 2001).

India's income growth rate is increasing, which is promoting increased food choices, more dietary diversity and increased animal food product consumption. Meat consumption increased 3.6% annually (FAO, 1998a) between 1982 and 1994. Although this is significant growth, production will exceed domestic consumption for many years to come. Cattle beef productivity was 103 kg per head in 1992–1994, which was an increase of 1.4 % per year between 1982 and 1994 (FAO, 1998a). Canned buffalo meat is an acceptable export item for many countries. Almost all the meat is sold in retail butcher shops and under no chilling process (Jul and Padda, 1984). Growth in human and livestock populations is higher in developing countries than in the developed world (Branckaert and Gueye, 1999). Per capita meat consumption is 120 kg per year in the USA, 90 kg per year in Europe, 40 kg per year in China and 4 kg per year in India. China's meat consumption increased considerably during the last two decades of the 20th century.

The Green Revolution provided many more calories in developing countries than the increase in meat consumption did during the same time period, but the additional meat consumption was worth almost three times as much at constant world prices. During 1971–1995 the additional consumption of meat, milk and fish in developed countries was greater than that of cereal in terms of weight and value. The most realistic set of assumptions governing international and national economics would suggest that the projected annual growth of total cereal used as animal feed in India will increase at 5% per year between 1993 to 2020, which would mean a change from 2.7 million tonnes in 1993 to 12.7 million tonnes in 2020. This change suggests that per capita increase in cereal utilized as feed for livestock in India will change from 4 kg in 1993 to 11 kg per person in 2020 (Rosegrant et al., 1997).

POULTRY

Poultry represents an appropriate system to feed high quality protein to a fast growing population and to provide additional income to small, low-income farmers, especially women (Branchaert and Gueye, 1999). High mortality, especially Newcastle disease, constitutes the greatest constraints to poultry production. In addition to high quality protein from meat and eggs, poultry can provide skin, feathers, manure for fertilizer, fuel and feed, as well as a means of capital accumulation. Although all categories of livestock have increased in numbers in the developing world, the increase in poultry is much greater than in pigs and ruminants (Branchaert and Gueye, 1999). Over the last decade, poultry has increased 76% in the developing countries. Poultry production in India has increased sixfold in 10 years (Branchaert and Gueye, 1999). In 1993, India had 2% of the world's chickens (FAO, 1998a). From 1982 to 1994, India's chicken population increased 11.9% per year and the number of animals slaughtered increased by the same amount (FAO, 1998a). The productivity of India's poultry was 0.9 kg per head, which did not change between 1982 and 1994 (FAO, 1998a). All over the developing world, these low-input–low-output husbandry systems have been a component of small farms and will probably continue into the future (Branchaert and Gueye, 1999). Eighty percent of the poultry population is found in traditional family-based production involving almost all ethnic groups, and this contributes about 20% of the protein (meat and eggs) in the developing world (Branchaert and Gueye, 1999). Four management systems are usually employed, including free range, backyard, semi-intensive and intensive systems. Sonaiya et al. (1998) estimated that 70% of total poultry production in developing countries comes from farm poultry flocks.

An FAO technical cooperation project (FAO, 1997b) was started in 1994 in Sikkim, India, which involved an improved chicken breed, Rhode Island Red, that was introduced from another part of India. Twelve chicks were distributed to each participant. Extension workers trained the participating women. The women organized themselves, sharing experiences and good roosters (for better breeding), and even agreeing on the selling price of their eggs. The participants were discouraged from eating the chickens, which would have put an end to the project. Instead, the participants passed on ten chicks from their growing flock to the next group of villagers.

SEAFOOD PRODUCTS

During the last decade, the worldwide catch of wild fish from both inland and marine waters had an average growth rate of less than 2%, but the contribution to human nutrition declined by 10%, because the percentage of wild species decreased in value and more of the fish were used to produce fish meal for feed and fertilizer (World Bank, 1998). This marine catch is shared by a handful of nations. India did not make the top 10 in 1938, was seventh in 1970 with 2.5% of the catch and in 1990 with 3.9% of the catch, and by 1991 was up to 4.2% of the world catch (FA0-CRO, 1994; Weber, 1998). That 4.2% is approximately 2% of India's GNP. Almost 90% of the export value is from prawns, lobsters and cuttlefish (Marine Products Export Development Authority, 1993). During this same period, aquaculture (the farming and husbandry of aquatic organism such as fish (68%), crustaceans (7%), mollusks (25%) and seaweed) increased at an average rate of 10% This was dominated by China, which produced 57% of the world's supply (FAO, 1997a; National Fisheries Institute, 2001). India is the second leading country (referred to as the "blue revolution") with 9% of the catch (FAO, 1997a), and the second leading country in noncarnivorous fish (carp, tilapia, milkfish, striped bass, perch and others). The rest of the world doubled production between 1984 and 1995 (World Bank, 1998, National Fisheries Institute, 2001). India's aquaculture production (excluding aquatic plants) in 1992 was 1.25 million tonnes (National Fisheries Institute, 2001). The average world consumption of fish in 1995 was 14 kg, and, if population projections are correct, the current aquaculture production will have to double by 2010 to maintain the same level of consumption, which would require an approximately 10% annual rate of increase per year. In the year 2000 in India, 70% of the population were nonvegetarians, eating fish if it was available at an affordable price (Weber, 1998).

India, including islands, has a 7,516 km coastline, with an exclusive economic zone of 2.02 million km^2 and 0.41 million km^2 of continental shelf, which contains 90% of the marine life (Weber, 1998). Water to 50 m in depth has the greatest concentration of fish and it is estimated that 2 million tonnes (out of a total of 3.5 million tonnes) maximum yield can be found at this depth (Weber, 1998). Brackishwater areas include marshes, backwaters, mangroves, inter- and sub-tidal areas of 1,416,300 hectares that act as feeding and nursery areas for a variety of fish, prawns and crabs (ICM Country Profile, 2001). Indian mangrove forests often function as spawning, breeding, and nursery grounds for fish, crabs, prawns and molluscs such as *Mugil cephalus, Hilsa ilisha, Lates calcarifer, Scylla serata, Meretrix casta, Crassostrea grephoides* and *Penaeus* spp. (ICM Country Profile, 2001). The fisheries sector employs more than 1 million people, of which 450 are fishermen and the rest are working as fish vendors or in fish processing plants (Weber, 1998).

In addition to contributing to protein in the human diet, aquaculture has benefits such as less spoilage if fish can be raised close to the consumption area, decentralized employment, more management by women, generation of foreign exchange, and the possibility of integration with other farming systems. These include fertile pond water used to irrigate crops; crop residue used to feed fish and other livestock; fish used in rice fields to control insect pests; bivalves (clams, oysters, scallops), seaweed and sea cucumbers raised with finfish to utilize fish waste and improve water quality; and production of freshwater and marine pearls. The major problem with aquaculture is that devastating diseases and overstocking can lead to fish waste, causing anoxic conditions that can kill plants. New tools for monitoring water quality are now available. Also needed are adequate postharvest fish handling facilities.

Most fish are consumed in the areas in which they are caught, but there is a growing international trade in fishery products. In 1960, only 10% were consumed in nations other than where they were caught; however, by 1987, this had risen to 39% (*Yearbook of International Trade Statistics*, various editions). Today, developing countries play a major role in catching and exporting of fish to developed countries (Weber, 1998).

GOATS

Goats can grow on low-grade feed and fodder and survive in diverse agroclimatic conditions. They can be raised for the sale of live goats, meat and milk with low investment and therefore are attractive to landless, marginal and small farmers. A 75% increase in the goat population was reported (Pai, 1996) from 1967–1971 to 1993 (117 million head). India has the world's largest goat population and goat meat is acceptable to both Muslims and Hindus. In India, there is a positive correlation between wasteland and the goat population of the area. In 1993, India had reported 10% of the world's sheep and goats (FAO, 1998a). There are 20 recognized goat breeds in India but 75% of the animals fall into the nondescript category. The common management system is grazing on natural rangeland and migration during lean periods of feed. In 1993, 47 million goats were slaughtered, which yielded 0.47 million tonnes of meat, 12% of the total meat production of India (Pai, 1996). Even though meat production increased from 0.22 million tonnes in 1969–71 to 0.47 million tonnes in 1993, the relative contribution to total meat production has decreased from 37% to 12%. An average Indian goat carcass weights 10 kg compared with 15 kg or more in other Asian countries (Pai, 1996). Goat meat is usually cooked in the form of curry and only 1.5% is processed into meat products. Composition of goat meat is 74% moisture, 21% protein and 3.6% fat (Pai, 1996). Goat meat accounts for 65–75% of the total meat exported from India.

SHEEP

Domesticated sheep (*Ovis ammon aries*) come in large and small sizes, with and without horns, and with and without wool. India ranks fifth in the world in sheep production. Sheep are sometimes criticized for cutting the grass much closer to the ground than cattle. Sheep produce 2–4 kg wool per year and is usually cut twice per year. Lamb meat today is the most important product of sheep and the quantity is usually 15 to 20 kg per slaughtered lamb. A milk sheep can produce 600 kg of milk per year.

PIGS

In 1993, India had 2% of the world's pig population (FAO, 1998a), which was up from 1% in 1983. From 1982 to 1994, India's pig output increased by 2.8% per year and the number of pigs slaughtered also increased by 2.8% per year (FAO, 1998a). In 1992–1994 India's pork was produced at the rate of 35 kilograms per each slaughtered pig (FAO, 1998a).

CAMELS

The one-humped camel (*Camelus dromedarius*) is raised in the western part of India. The camel population in India is approximately 1.5 million (Khanna and Rai, 1995). Traditionally, camels were used exclusively for transport and the farmers did not sell female camels, camel wool or camel milk. Slaughtering of camels and consumption of camel meat was an absolute taboo (Kohler-Rollefson, 1992). However, these attitudes seem to be rapidly changing. A recently (1984) discovered off-white, small, short-leg breed, the Malvi camel, is largely located in the Madhya Pradesh. It is estimated that there are between 2500–3000 of these animals. This breed is a dual-purpose animal, with the males used for work and the females for milk. With adequate forage and milked twice a day, the average milk yield is about 2 kg. Lactation length averages 1 year and calving is approximately every 2 years (Kohler-Rollefson and Rathore, 1996).

LEATHER

Indian leather, one of the prime exports to 120 countries, is valued at US$1.80 billion per year (Indian Leather, 2001). Major leather products are hides and skins such as cow and buffcalf, sheep nappa, goatskin, kid leather and wet blue. Footwear products include shoes, shoe uppers, and soles. Leather garments include gloves, saddlery, travel bags and totes, purses, wallets and briefcases. The major leather production centers are in Chennai, Ambur, Ranipet, Vaniyambadi, Trichy, Dindigul in Tamil Nadu, Calcutta, Kanpur, Jallandhar, Delhi, Hyderabad, and Bangalore (Indian Leather, 2001).

SUMMARY

In the present scenario, it is evident that reconciling animal food products with security and environmental quality in industrializing India has both positive and negative potentials. Population growth is a major problem, because per capita land and water resources are rapidly decreasing. Fertility rate is decreasing (but still too high) but life expectancy is increasing. Both of these are positive factors but the latter, even though desirable, increases the total population. There are no acceptable short-term solutions to population control in India. Long-term solutions that would help are a dependable retirement system, improved education, acceptable employment opportunities for women, and marriage at a more advanced age.

India's primary animal food sources are cattle (milk and meat), water buffalo (milk and meat), sheep (meat and milk), goats (meat and milk), poultry (eggs and meat), and fish. Recent statistics indicate that all of these are increasing in numbers and increasing the quantity of product available for the consumer. Problems encountered after production include distribution, spoilage, soil degradation, poor economic base to purchase food, insufficient fresh water supply, salinization, overstocking and insufficient land area. In spite of these limitations, milk, poultry including eggs, and fish production have increased rather remarkably. Per capita income has been increasing and the percentage of undernourished people has been improving, but they still need much more progress. Increases in milk production in the animal food product category have been accomplished through cooperatives and with increasing numbers of cattle. However, this cannot increase indefinitely due to limited feed and fodder. Genetic improvement can help, but a selection factor that must also be included is lifetime milk production because there is a strong limitation to culling nonproductive animals. Seafood farming has also been successful, but better technology is needed to prevent environmental degradation. Poultry production is growing and also supplies a significant amount of protein to the Indian diet. It also has the added advantage of being useful for small farm operations often managed by women. Cooperatives have been successful when managed properly and need to be encouraged. Women need to be given a greater role in livestock management. New technologies and reordering of governmental and personal priorities are needed to solve India's animal food problems.

REFERENCES

Abichandani, H. (Ed.). 1994. Industrial Entrepreneurs Memorandum (IEM) for milk and milk products filed with the Government of India. In *Indian Dairying: Global Opportunities, proceedings of XXV Dairy Industry Conference*, pp. 59-116. New Delhi. Indian Dairy Association.

Acharya, S. and R. K. Yadav. 1992. *Production and Marketing of Milk and Milk Products in India*. New Delhi. Mittal Publications.

Alexandratos, N. 1998. Quoted in: *Poultry Times of India*. http://www.poultrytimesofindia.com per issues per 1998/december/news.html.

Aneja, R. P. 1994. Traditional milk specialities: a survey. In *Dairy India*. pp 259-275. New Delhi: Indian Dairy Association.

Anonymous. 1992. Salt of the earth. *The Economist* 323,34.

Batty, J.C. and I. Keller. 1980. Energy requirements for irrigation. In: *Handbook of Energy Utilization in Agriculture*. Pimentel, D. (Ed.) CRC Press. Boca Raton, FL. P 35-44.

Bhaskaran, 1996. Culture's consequences: dairy market opportunities in India. Marketing Bulletin, 7,39-50. http://marketing-bulletin.massey.ac.nz/article7/article5b.asp.

Book of Sheep, 2001. The domestic sheep. http://hem.passagen.se/egilize/nine.htm.

Branckaert, R.D.S. and E.F. Gueye. 1999. FAO's programme for support to family poultry production. http://www.husdyr.kvl..dk per htm/tune99/24-Brackaert.htm.

Bupps, P. T. 2000. Cattle. Encyclopedia article. http://encarta.msn.com/index/conci...05813000.htm?z=1&br=1.

Chander, M. and P. C. Harbola. 1996. Livestock management in the Himalayan environment: constraints, opportunities and strategies with reference to India. hddp://www.mtnforum.org/resources/library/chanx96a.htm.

Crosswalk. 2001. Country profiles. http://religiontoday.crosswalk.com/CountryProfiles/in.html

Delgado, C., C. Courbois, and M. Rosegrant.1998. Global food demand and the contribution of livestock as we enter the new millennium. Paper presented at the British Society of Animal Science. *Kenya Agric. Res. Inst.* Nairobi.

Department of Agriculture & Cooperation: Statistics at a Glance. 2001a. http://www.nic.in/agricoop/statistics/growrh9.htm.

Department of Agriculture & Cooperation: Statistics at a glance. 2001b. http://www.nic.in/agricoop/statistics/prod1.htm.

Department of Agriculture & Cooperation: Statistics at a glance. 2001c. http://www.nic.in/agricoop/statistics/prod2a.htm.

Doornbos, M.M. and L. Gertsch. 1994. Sustainability, technology and corporate interest: resource strategies in India's modern dairy sector. *Journal of Development Studies*, 30(3),916-950.

Durning, A.T. and H.B. Brough. 1992. Reforming the livestock economy, In: *State of the World*. Brown, L.R. (Ed.). W.W. Norton and Company, N.Y. pp. 66-83.

FAO, 1995. FAO statistics series number 125. Rome: Food and Agricultural Organization of the United Nations.

FAO, 1997a. The status of fisheries and aquaculture–1996. Rome, Italy.

FAO, 1997b. Women work together to improve livelihoods in rural India. http://www.fao.org/NEWS/1997/971211-e.htm.

FAO, 1997c. FAO statistics database. http:faostat.fao.org/default.htm.

FAO, 1998a. FAO statistics database. http:faostat.fao.org/fefault.htm.

FAO, 1998b. Cooperatives and food security FAO's perspective. http://www.copacgva.org/faoidc97.htm.

FAO. 1999. The state of food insecurity in the world. Rome, Italy.

Press Information Bureau Government of India. 2001. Advance estimate of national income, 2000-01. Press Note, http://www.nic.in/stat/t1.htm

FAO-CRO, 1994. Commodity review and outlook (FAOCRO), Rome

Horiuchi, S. 1992. Stagnation in the decline of the world population growth rate during the 1980s. *Science*, 761-765.

Hull, R.R., A. Evans, and D.A. Gupta. 1993. India–food opportunities, Department of Industry Technology and Regional Development, Canberra, Australia.

ICAR. 2001. Indian Council of Agricultural Research. http://www.nic.in/icar/ICAR3.HTM.

ICM Country Profile, 2001. India. http://icm.nooa.gov/country/india/india.html

India. 2001. India. http://user.cholliant.net/~iksoo/in.htm.

Indian Leather. 2001. Indian leather and leather products. http://www.webinda.com/india/leather.htm.

India Mart, 2001. Meat and poultry food, manufacturers, exporters and suppliers from India. http://www.indian.com/indianexporters/ag_meat.html.

Jul, M. and G.S. Padda. 1984. Meat production in India: The potential of buffalo beef *World Animal Review*. 50, 36-44.

Khanna N.D. and A.K. Rai. 1995. Perspective and strategic plan for camel improvement research programs in NRC on camels at Bikaner (1995-2025). Bikaner National Research Center on Camels, Bikaner, India.

Kindall, H.W. and D. Pimentel. 2001. Constraints on the expansion of the global food supply. http://dieoff.org/page36.htm.

Kohler-Rollefson, I. 1992. The Raika dromedary breeders of Rajasthan: a pastoral system in crisis. *Nomadic Peoples* 30: 74-83.

Kohler-Rollefson, I. and H. S. Rathore. 1996. The Malvi camel: A newly discovered breed from India. League for Pastoral People, Animal-Genetics Resources-Information. (18) 31-42.

Lal, R. 1977. Meat dishes, Bombay India Book House Pvt. Ltd.

Lal, R. 1991. Land degradation and its impact on food and other resources. In: *Food and Natural Resources*, Pimentel, D. (Ed.) Academic Press, San Diego. pp. 85-104.

Lal, R. and F. J. Pierce.1991. The Vanishing Resources. In: *Soil management for Sustainability*. Soil and Water Conservation Soc. Ankeny, Iowa. pp. 1-5.

Leyton, L. 1983. Cropwater use: principles and some considerations for agroforestry. In: *Plant Research and Agroforestry*, Huxley, P.A. (Ed.), International Council for Research in Agroforestry, Nairobi, Kenya.

Marine Products Export Development Authority, 1993. Statistics of Marine Products Export 1991. Cochin 1993.

Moffat, A.S. 1992. Does global change threaten the world food supply? *Science* 256: 1140-1141.

National Fisheries Institute, 2001. Commercial fishing statistics. http://www.nfi.org/industr7.html.

Pai, U. K. 1996. Goat: promising meat animal in India. *Asian Livestock*, XXI, (9), 97-101.

Press Information Bureau, Government of India. 2001. http://pib.myiris.com/textfile/search/article.php3?&filename=../press/000d929161640,htm&srch=gross.

Ritschard, R.L. and K. Tsao. 1978. *Energy and Water Used in Irrigated Agriculture During Drought Conditions*. Lawrence Berkeley Lab., Univ. of Calif. Berkeley.

Rosegrant, M.W., M. Agcaoili-Sombilla, R.V. Gerpacio and C. Pinger.1997. Global Food Markets and U.S. Exports in the 21st Century. Paper presented at the Illinois World Food and Sustainable Agriculture Program Conference, Meeting the Demand For Food in the 21st Century: Challenges and Opportunities for Illinois Agriculture. Urbana.

Sonaiya E.B., R.D.S. Branckaert and E.F. Gueye. 1998. Research and development options for family poultry. Introductory paper to the First INFPD/FAO Electronic conference on family poultry. http://www.husdyr.kvl.dk/htm/php/tune99/24-Branckaert.htm.

State of the World. A World Watch Institute Report. 1990. Washington D. C.

Statistics India. 2001. India: economic and financial data. http://www.inc.in/stat/sdrsum.htm.

Trivedi, K.R. 1990. Animal breeding programmes in India in the dairy cooperative sector. *Animal Science Papers and Reports.* 6:113.

UNFPA, UN Population Fund. 1991. The state of the World population 1991. UN. New York.

United Nations Population Fund. 1991. Population, resources and the environment. United Nations, New York.

Weber, E. 1998. Fisheries development in India: who benefits? http://www.uni-freiburg.de/gradwald/fisher.htm.

World Bank. 1997. World Bank projections and the manufacturing unit value indes used for expressing values in constant 1990 US dollars. World Bank.

World Bank. 1998. Agriculture Technology Notes. http://wbln0018.world-bank.org/essd...sf/rural%20development/aquaculture.

World Food Programme, 2001. Hunger map. Rome, Italy.

World Resources. 1992-93. A report to the World Resources Institute. Oxford University Institute, Oxford University Press. New York.

World Resources Institute. 1990. World resources 1990-1991.Oxford University Press. New York.

Yearbook of International Trade Statistics, various editions.

21 Sustainable Agriculture on a Populous and Industrialized Landscape: Building Ecosystem Vitality and Productivity

Richard R. Harwood

CONTENTS

INTRODUCTION

Environmental quality in India's "evergreen revolution" requires a dramatic change in paradigm toward a vision of purposeful management of ecosystem processes. It is time to abandon the outdated and no-longer-useful concept of the "environment" as something external to human activity, and particularly to agriculture. With increasing human activity on, or influencing, every square meter of India's land and water base and the atmosphere above them, there are no longer "externals." Everything is becoming internal to human activity. There is no "away" to which materials can go, or be thrown. Outputs from one activity affect many others. Negative impacts become loadings, with costs being transmitted downstream. India of the future must regain much of its lost ecosystem health and vitality, even while human activity (including agriculture) is rapidly increasing.

The first corollary of the new paradigm is that ecosystem structure must serve the health and well-being of its people as well as its crops and animals. Supplemental structures and processes are then built into that environment. This turnaround in thinking is crucial to well-being, especially of economically disadvantaged people, in a resource-limited environment. The Chandigarh Eco-Declaration of November 1997 recognized this relationship most eloquently: "There is a strong inter-relationship between degradation of resources and poverty — a vicious circle — degradation of resources leads to poverty and poverty leads to degradation of resources. A converse relationship also holds true." Resource degradation, as discussed in this chapter, includes both geophysical endowments and the biological structure and processes that sustain and enrich the lives of people who live and work in the landscape.

A CHANGING CONTEXT

India's rapid industrialization will bring changes to the landscape at an accelerating pace. The productivity, resource constraints and environmental demands placed on agriculture change according to the contextual setting. Over the next two decades, with globalization of the marketplace for an increasing number of agricultural products, the true costs of many products, including most of the globally produced cereals, oilseeds, and many animal products, will either remain steady or continue to decline (Alexandratos, 1995; Pinstrup-Andersen et al., 1999). The global market-place will increasingly demand greater food safety with reduced tolerance of pesticide or antibiotic content. The marketplace in India will follow both the economic and food safety trends. The global market will increasingly move toward an industrial-commodity approach to plant protein, starch and basic oils, with ingredients for manufactured foods coming increasingly from the lowest-cost production environments. As the global economy evolves, these trends are increasingly debated and have become causes for alarm; their salience for trade and development will increase over time (Heffernan, 1997).

Many other commodities will remain specialized and local. Economic efficiencies, local preferences, demand for high-quality fresh produce, and, in some cases, demands for local or household food security, will bring pressure to preserve a diverse structure in our food systems. That diverse structure has influence on the degree to which farms are diversified and integrated (Harwood, 2001).

Over the next 50 years, India's population is expected to increase from 1.0 to 1.5 billion. Some significant portion of that increase will be geographically horizontal, that is, expanding across the landscape. Per capita incomes are expected to increase several fold, from the $400 level to, some say, the $6,000 range. The gross domestic product also will grow by several fold. Ruttan (1996) has given a concise summary of those relationships:

"A 1989 study at the International Institute for Applied Systems Analysis (IIASA) advanced what came to be known as the 2-4-6-8 scenario... a doubling of population, a quadrupling of agricultural production, a sextupling of energy production and an octupling of the size of the global economy by 2050."

For our purposes, the exact coefficients are not important. Whether the Indian GDP will continue to grow at a rate of 4% per capita per year remains to be seen.

The point is that, as the economy grows, some significant portion of the growth will also be geographically horizontal, competing with both domestic and agricultural use of the land. With economic growth, the material content per unit of GDP decreases for most materials (National Research Council, 1999), but total consumption of both carbon and water is likely to increase. This means that agriculture's share of the finite resources of land and fresh water will continue to decrease.

It is important for India's policy makers to continue to monitor shrinkage in the agricultural land base and to track the increasing price of agricultural land. Policy intervention in the marketplace for land must evolve together with development of long-term land use strategies. As incomes rise, demand for a continued increase in environmental quality and services will continue to grow. The growing economy and the increase in numbers of Indians will, in themselves, add to environmental loading. It is safe to say that agriculture's share of a shrinking acceptable level of environmental loading will have to continue to decrease. With increasing incomes, however, the demand for high-value, quality fruit, vegetables, animal products and ornamentals will continue to grow. India's agriculture will thus be increasingly constrained, on the one hand, and provide opportunities for economic growth on the other.

REQUIREMENTS FOR LANDSCAPE SERVICES

As demographic and economic activity on the landscape increases, many of the services provided by natural landscapes are diminished or drastically changed. Plant and animal diversity on the landscape decreases, often causing instability in pest and disease populations, population swings in wildlife, and reduced geophysical services. The ability to capture rainfall and to recharge groundwater is diminished. The demand for greenbelt areas, for trees and plant diversity will continue to increase. The true value of the multifunctionality of agriculture will increasingly be recognized (Anon., 1999).

As the economy grows, the marketplace evolves to place value on a changing array of services (Figure 21.1). In the early stages of farm and village progress, with only rudimentary market development, a high diversity of agricultural products and services is required for community stability, self-sufficiency and stability. As markets evolve, those services decline markedly. Rural people produce fewer marketable products and remove themselves from personal environmental interaction. As markets evolve (in a postmodern sense), the demand for landscape services increases. The values are internalized through pressures of social contract, regulation, direct subsidy or payment and, eventually, partitioning of the marketplace for land through land trusts, development restrictions or similar mechanisms.

With a highly evolved marketplace, landscape services and the emphasis on environmental quality become increasingly important relative to the value of agricultural product output. In such a context, the types of products usually shift toward higher-value, high-quality specialty products targeted for local and niche markets as well as for high-value global markets. Other services such as water harvesting, wind protection, shade and aesthetics become purposefully emphasized. The science of landscape ecology, as yet, has little to offer in identifying

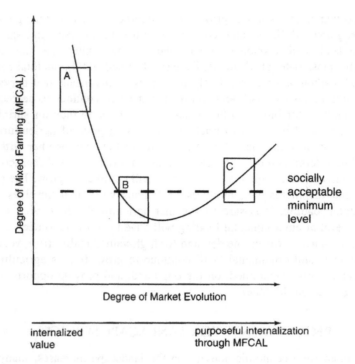

FIGURE 21.1 Changes in the purposeful development of the multifunctional character of agricultural land (MFCAL) as markets evolve. Adapted from L. Fresco, personal communication.

levels of biodiversity appropriate for such landscapes. It would seem likely that there will be eventual targeting of minimal levels of diversity for key functional groups in varied landscapes.

THE INCREASING DEMAND
FOR HYDROLOGICAL SERVICES

India's per capita freshwater availability is projected to decrease from just over 2000 cubic meters per person to around 1300 m³ per person in 2050, a condition generally considered to result in stress (Gardner-Outlaw and Engleman, 1997). Such gross national data do not reflect, however, the huge differences in water availability across India. As per capita income (or GDP) increases, withdrawals of water for both domestic and industrial use increases, but at different rates. Demands for water quality will increase concomitantly with greater demands for quantity.

The agricultural sector will remain, in most areas, a major user of the land. The agricultural landscape will increasingly be seen as a provider of hydrological services — rainfall capture, storage and groundwater recharge. Agriculture is not just a part of the problem, it is integral to the solution. Swaminathan (Chapter 1 in this volume) states that most of India's annual rainfall occurs during 100 hours of precipitation (Swaminathan, 2001). The capture and multiple use of this water will be critical both at a national economic level and for individual well-being. Its availability and purity to modest- and low-income people where they live, will be increasingly important.

The paradigm on national water planning and use has gradually changed. Unfortunately, consensus on a framework to sort out the interrelationships of scale, end users, pricing policy and control of water is not in sight. Global and regional models of renewable water availability and variability are rapidly improving, as are the geographical information systems (GIS) for their verification. At the macro level, the Climatic and Agricultural Atlas prepared by the International Water Management Institute (IWMI) provides reliable estimates of renewable water, at least at a river basin scale. It is linked to PODIUM, the Policy Interactive Dialogue Model, to assist in planning at a macro-scale (Seckler and Amarasinghe, 2000). Most national planning occurs at the river basin level, particularly for large-scale infrastructure and for macropartitioning of water resources (McKinney et. al., 1999). They discuss the concepts of conjunctive management of surface and groundwater. This is defined as "multiobjective, multipurpose and multifacility solutions at the river basin level." Pricing of water is often done at that level, if at all (Perry et al., 1997).

While several aspects of policy, infrastructure creation, management, and pricing occur at the river basin level or above, most of the critical interactions with agriculture and the agricultural structures and practices that influence the quantity and quality of ecosystem services occur below those levels. They are local, requiring knowledge and a sense of place for their understanding and management. India's essential problem is this: it receives 40 million hectare-meters (mham) of precipitation annually, which is supplemented by some 20 mham of river flow from neighboring countries. By the year 2025, India is expected to use 105 mham annually, up from 38 mham annually in 1974 (Nag and Kathpalia, 1975, as quoted in Agrawal and Narain, 1999).

Effective water harvesting, a key ecosystem service by agriculture on landscapes that are dominated by agriculture, occurs primarily at the farm and village level. It requires a wide range of technologies, most of which are traditional. Many have been abandoned or have fallen into disarray for a variety of reasons — social, political and economic. As water becomes scarcer under growing demand, local technologies must again provide a foundation upon which larger-scale availability is based. In today's India, numerous examples exist of the growth and development of such systems to replenish tanks, wells and groundwater and to harvest monsoonal downpours that would otherwise be largely lost to swollen stream and river flow (Agrawal and Narain, 1999). The concept of ecorestoration is used to describe the process. It requires many tiers of institutions at the state level, for policy coordination; at the district and miniwatershed level, for implementation and coordination; and at the village level to ensure that all villagers acquire an interest in the effort. For example, 1,748 women's groups, with 25,506 participants, were created in 374 villages of Jhabua.

But most importantly, serious efforts have been made to give local communities powers over decision making and control over resources. For instance, the villagers play an active role in managing the funds to implement the watershed program. Nearly 80% of the funds for the program are put in a bank account managed by the

Watershed Development Committees made up of village people. The Watershed Development Committee tries to bring together all the important interest groups in the village and thus replicates the concept of the *gram sabha* (village council).

These case studies show clearly that ecorestoration is possible even in highly degraded lands and that it can regenerate the local rural economy and thus help in poverty alleviation in a sustainable and cost-effective manner. In other words, helping the people to help themselves by improving their local natural resource base is a variable and effective strategy for poverty alleviation. The key to this ecorestoration lies in good management and use of the local rainwater, but the entire exercise must be underpinned by community-based decision-making systems and institutions, and enabling legal and financial measures that promote community action (Agrawal and Narain, 1999).

The 73rd Constitutional Amendment in India leading to decentralization of development work to subnational level — state, district, block, village — is an excellent conduit to effectively implement the programmes for effective ecological balance (Dhaliwal et al., 1998, p. xii).

The environmental issues relative to agriculture and water thus go far beyond the need to conserve or to partition available supplies. Agriculture, as a primary (and first) recipient of renewable water, will increasingly be required to harvest, use a portion and transfer water to groundwater and to constructed infrastructure in an acceptably pure form. This will become agriculture's primary ecosystem service. Appropriate marketplace mechanisms must evolve for farm compensation.

THE ROLE OF CARBON STOCKS IN ENVIRONMENTALLY SOUND AGRICULTURE

Much has been made of the global importance of atmospheric carbon balance with respect to greenhouse effect, with an abundance of literature on the subject. Carbon dioxide has been projected to increase from its 1990 level of 353 ppm by volume to between 700 and 800 ppm in the early decades of the 21st century (Engleman, 1998). The good news is that landscape stocks, on average, will be more easily accumulated, given adequate moisture, because of increased plant productivity. India's resource degradation includes an overharvest of its carbon resources. Soil organic matter has been reduced and trees have been harvested over time to well below the capacity of the landscape to be optimally productive of both goods and landscape services.

Numerous programs and technologies are suggested for reversing these trends (Dhaliwal et al., 1998). The Chandigarh Eco-Declaration addresses these issues. Having appropriate biological structure and living stocks of carbon on the landscape is an art of landscape design. Without appropriate policies and management at several levels, the stocks are nearly always deficient. The appropriate biomass of standing stocks depends on water, terrain, soil types and a plethora of social and economic factors. Scientific efforts need to go beyond the empirical site-specific nature of the carbon issue to understand processes and their management. Guidelines need to be developed for the broad range of Indian ecosystems to provide targets for farmers and villagers with respect to carbon stocks.

The effects of failure to do so — land and other resource degradation — are well known (Scherr and Yadav, 1996). The benefits of such action include more than just the provision of ecosystem services. They are linked to what has been called the "gross nature product" (Agrawal and Narain, 1999), derived from a community-based system of natural resource management. The net effect of improving the nexus among carbon stocks, agricultural practice and water harvesting is an increase in local productivity as well as a cascading of services to a larger scale (Figure 21.2).

This more holistic view of agricultural productivity to include ecosystem services is similar, in many ways, to the concept of the "multifunctional character of agricultural land" (MFCAL). This particular term has become entangled in political and trade issues and the debate over agricultural subsidies (see the Web page summaries of the 1999 Maastricht, Netherlands Conference on Sustainable Developments), but the underlying concept of agriculture's multifunctionality is perfectly valid and appropriate.

BIOCONTROL OF PESTS

Environmental and human health hazards will require limited and very targeted use of pesticides. The intensively interactive nature of India's agriculture with people and with the environment underscores that need. As a concentrated effort is made to use better agricultural land management for rainfall interception and groundwater recharge, the purity of water filtered through agricultural fields will become critical. Ecologically based pest management (EBPM) will be increasingly stressed. The U.S. National Research Council (1996) summarized EBPM as follows:

"EBPM promotes the economic and environmental viability of agriculture by using knowledge of interactions between crops, pests, and naturally occurring pest-controlled organisms to modify cropping systems in ways that reduce damage associated with pests. Ecologically based management relies on a comprehensive knowledge of the ecosystem, including the natural biological interactions that suppress pest populations. It is based on the recognition that many conventional agricultural practices recommended by EBPM will augment natural processes, supplemented by biological-control organisms and products, resistant plants, and targeted pesticides."

Integrated pest management (IPM) uses these same principles, but does not stress the management of biological relationships and processes as the foundation for pest management. The use of pesticides becomes supplemental. In such an approach, EBPM will be required to ultimately stabilize pest populations where genetic modification is used for crop or animal resistance. The holistic nature of tomorrow's pest management will utilize EBPM as a base, with the management practices, host-plant resistance, including transgenic pest-protected plants, and specifically directed pesticidal application serving in a supplemental role (Alberts, 2000). In tomorrow's agriculture, however, even host-plant resistance will come under scrutiny. In many, if not most cases, host-plant resistance is conferred through production by the plant of certain insecticidal or fungicidal compounds. While these are all biodegradable, they do have life expectancies in crop residues and soil. The great advantage of most modern plant breeding methods, and especially transgenic biotechnology methods, is that we know exactly what those compounds are and have very sensitive tests for them.

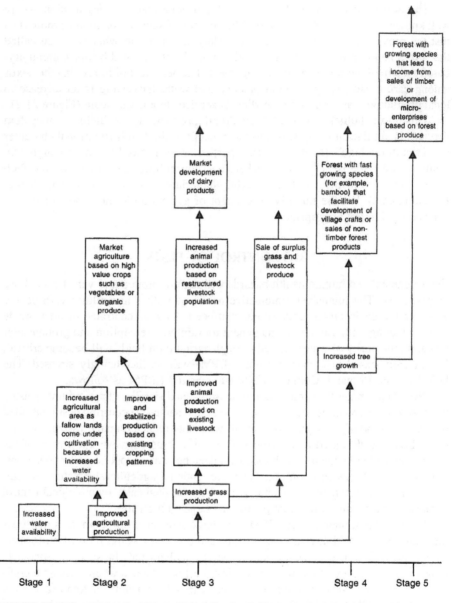

FIGURE 21.2 Improving the "Gross Nature Product": Ecological regeneration and its impact on a biomass-based village economy. Adapted from Agrawal and Narain, 1999.

Transgenic pest-protected crops have been planted in the U.S. since 1995. They are close to release in India. In 2000, 6.8 million hectares were planted in six countries to crops containing the *Bacillus thuringiensis* (Bt) gene, conferring resistance to certain insects (Adkisson, 2000; James, 2000). The specific toxin produced in these plants is the Cry 1A and 3A Bt endotoxins. These proteins have a mammalian toxicity somewhat greater than 4,000 micrograms per kg of body weight (about the same as that with the commonly used insecticides Sevin or Malathion). They are produced at a level of approximately 750 grams per hectare in an average corn crop. The proteins are water-soluble and seem to have modest to little effect on most nontarget organisms either above or below ground. Total global production of these insecticides in 2000 was approximately 5 million kilograms. No scientific data whatsoever exist as to what impact, if any, such broad ecosystem loading will have on biological processes, or on how much, if any, will appear in groundwater. In highly interactive environments where rainfall capture, soil infiltration and ground-water recharge are a focal point, possible contamination from all sources becomes important, and pest management in intensive agriculture will become both scientif-ically important and a publicly viable component of a sustainable agriculture.

Strategies for reduction of pesticide use in India are complex. While insecticide use in field crops has shown some decline in recent years, herbicide and fungicide use continues to increase (Pingali and Gerpacio, 1997). Host-plant resistance offers a major avenue for progress, subject to constraints from a different type of envi-ronmental loading. Biopesticide use in place of inorganic chemicals appears to have broad potential, but the new materials and their technologies are not widely available (Alam, 2000). Whatever combination of approaches is used, collective action at the community level will be needed, particularly for controlling migratory pest infestations.

DEVELOPMENT DIRECTIONS

It is clear that the major environmental problems of salinity buildup, soil loss, and aquifer overdraft must be addressed on a regional as well as a local scale. Water harvesting methods to increase water-use efficiency as well as to reduce downstream flooding need to be part of the agenda. Flood-tolerance mechanisms must be built into landscapes as appropriate. It is safe to assume that agriculture in an industrial-izing economy with high rural population pressure (700 to 1,000 or more persons per square kilometer) will have to be increasingly productive per unit area of land and per unit of water. As agriculture's share of acceptable environmental impacts decreases, it must increase nutrient flows from soil to crop or animal and back to soil with high efficiency. Nutrients, crop and animal residues and pesticides must be increasingly contained within field and farm boundaries and in the upper soil layers, rather than being allowed to leach and diffuse more broadly. These require-ments will dictate a need for developing plant and animal systems that optimize plant and animal health, including reduced or minimal requirements for pesticide use. They will minimize soil erosion and air contamination. They must intercept and infiltrate rainwater for groundwater recharge with minimal contamination, all within a production enterprise having high material flows.

Production ecology will provide the structural framework for this. Farmers will learn to manage key biological processes and flows (Harwood, 1998). These include careful management of the genetics of crops and animals, as well as key functional microbial groups in the soil. Farmers will have to manage the genetics of pest populations as well as their overall abundance. The sciences and technologies for doing this are evolving rapidly.

It is extremely important that a significant part of public-sector scientific research shift to a process-level understanding of hydrological processes, of carbon stocks and flows, of landscape-level biogeochemical processes, and of the dynamics of pest populations and pest-predator relationships. This will build up a framework of understanding that will permit us to move from empirically derived site-specific solutions to purposeful designs based on principles. Many U.S. and other developed country universities and laboratories are at the cutting edge of such process-based cutting edge research.

We need to have a clearer sense of the need for a farm-and-community-based pyramid of institutions and communities to put together viable multi-level solutions to sustainable resource use. Most importantly, we need to view agriculture as part of the solution, not just a contributor to the environmental problem. The science to accomplish this is on the horizon. Unfortunately, the evolution of an enabling and supportive marketplace does not appear to be so visible.

REFERENCES

Adkisson, P.L. 2000. Preface to *Genetically Modified Pest-Protected Plants: Science and Regulation*. National Academy Press, Washington, D.C.: p. xi.

Agrawal, A., and S. Narain. 1999. *Making Water Management Everybody's Business: Water Harvesting and Rural Development in India*. Gatekeeper Series No. 87. International Institute for Environment and Development, brochure, London: p 12.

Alam, G. 2000. *A Study of Biopesticides and Biofertilisers in Harayana, India*. Gatekeeper Series No. 93. International Institute for Environment and Development, brochure, London: p24.

Alberts, B. 2000. Foreword to *Genetically Modified Pest-Protected Plants: Science and Regulation*. National Academy Press, Washington, D.C.: p. ix.

Alexandratos, N., (Ed.) 1995. *World Agriculture: Towards 2010*. FAO and Wiley and Sons, Rome, Italy. p 480.

Anonymous. 1999. In: *Sustainable Developments*. <http://www.iisd.ca/linkages> Summary of Conference on the multifunctional character of agriculture and land. Maastricht, Netherlands. September 12-17.

Dhaliwal, G.S., N.S. Randhawa, R. Arora, and A.K. Dhawan (Eds.) 1998. *Ecological Agriculture and Sustainable Development*, Vol. II. Indian Ecological Society, Punjab Agricultural University, Ludiana, India and Centre for Research in Rural and Industrial Development, New Delhi: pp. 712.

Engleman, R. 1998. Profiles in Carbon: An Update on Population, Consumption and Carbon Dioxide Emissions. Population Action International, Washington, D.C.: p 42.

Gardner-Outlaw, T., and R. Engelman. 1997. Sustaining Water, Easing Scarcity: A Second Update. Population Action International, Washington, D.C.: p20.

Harwood, R.R. 2001. Sustainability in agricultural systems in transition — at what cost? In *Sustainability in Agricultural Systems*, W.A. Payne, D.R. Keeney and S.C. Rao, (Eds.)American Society of Agronomy, Special Publication #64, Madison, WI: 7-30.

Harwood, R.R. 1998. In: Michigan Field Crop Ecology: Managing Biological Processes For Productivity and Environmental Quality, M.A. Cavigelli, S.R. Deming, L.K. Probyn, and R.R. Harwood, (Eds.). Michigan State University Extension Bulletin E-2646. East Lansing: Michigan State University, pp. 92.

Heffernan, W.D. 1997. Domination of world agriculture by transnational corporations. In *For All Generations: Making World Agriculture More Sustainable*, J. P. Madden and S. G. Chaplow, (Eds.) World Sustainable Agriculture Association, Glendale, CA: pp. 173-181.

James, C. 2000. Global Review of Commercialized Transgenic Crops: 2000. International Service for the Acquisition of AgriBiotech Applications, Ithaca, NY: pp. 15.

McKinney, D. C., X. Cai, M. W. Rosegrant, C. Ringler, and C.A. Scott. 1999. Modeling Water Resources Management at the Basin Level: Review and Future Directions. International Water Management Institute, Colombo, Sri Lanka, pp. 6-12.

Nag, B.N. and G.N. Kathpalia. 1975. Water and human needs. Paper presented at the Second World Congress on Water Resources. *Proceedings*, Vol. II. CBIP, New Delhi.

National Research Council Board on Agriculture. 1996. *Ecologically Based Pest Management: New Solutions for a New Century*. Washington, D.C.: National Academy Press.

National Research Council. 1999. *Our Common Journey: A Transition Toward Sustainability*. National Academy Press, Washington, D.C.

Perry, C.J., M. Rock and D. Seckler. 1997. Water as an Economic Good: A Solution, or a Problem. Research Report 14. International Irrigation Management Institute, Colombo, Sri Lanka: pp. 16.

Pingali, P.L. and R.V. Gerpacio. 1997. Toward Reduced Pesticide Use for Cereal Crops in Asia. Working Paper 9704. International Maize and Wheat Improvement Center, Mexico, D.F.: pp. 22.

Pinstrup-Andersen, P., R. Pandya-Lorch, and M.W. Rosegrant. 1999. The World Food Prospects: Critical Issues for the early Twenty-First Century. International Food Policy Research Institute, Washington, D.C.: pp. 32.

Robertson, G.P. and R.R. Harwood. 2001. Sustainable agriculture. In *Encyclopedia of Biodiversity*, Vol I., S. A. Levin, (Ed.) Academic Press, New York: pp. 99-108.

Ruttan, V.W. 1996. Meeting the food needs of the world. Staff Paper. Presented at the World Food Prize Symposium, Des Moines, IA, October 18, 1996. In: V.W. Ruttan, (Ed.) International Agricultural Research: Four Papers, Department of Applied Economics, University of Minnesota, St. Paul, MI: pp. 98-104.

Scherr, S. J., and S. Yadav. 1996. Land Degradation in the Developing World: Implications for Food, Agriculture, and the Environment to 2020. International Food Policy Research Institute, Washington, D.C.: pp. 35.

Seckler, D., and U. Amarasinghe. 2000. Water supply and demand, 1995-2025. In International Water Institute Annual Report, 1999-2000: Water for Food, Nature and Rural Livelihoods, International Water Management Institute, Colombo, Sri Lanka: pp. 9-17.

Swaminathan, M.S. 2001. *Century of Hope: Harmony with Nature and Freedom from Hunger*. East West Books, Chennai, India: pp. 154.

Part Four

Poverty and Equity

22 Global Food Security, Environmental Sustainability and Poverty Alleviation: Complementary or Contradictory Goals?

*William B. Lacy, Laura R. Lacy
and David O. Hansen**

CONTENTS

INTRODUCTION

One of the most significant challenges facing humanity during the 21st century will be how to pursue three key goals simultaneously: global food security, environmental

* The first two authors are at the University of California, Davis, and the third author is at The Ohio State University.

1-5667-0594-0/02/$0.00+$1.50
© 2002 by CRC Press LLC

sustainability and poverty alleviation. These goals, which have been described as the critical triangle (Vosti and Reardon, 1997), are not necessarily or always complementary. Thus, achieving them simultaneously cannot be taken for granted, particularly in the short term. Hundreds of millions of people labor to produce food from already depleted soils, degraded hillsides, tropical rain forests and dry areas that are threatened by desertification. Their efforts further harm the environment, thereby worsening their poverty. This contributes to a vicious cycle and jeopardizes their precarious food security. The three goals of the critical triangle are inextricably linked and successful pursuit of each will require policies, institutions and technologies that make them more compatible.

DIMENSIONS OF FOOD SECURITY

The first and continuing challenge is food security, or how to produce and ensure access to enough food to feed the still-growing population. Food security is a necessary, although not sufficient, condition for the well-being of a society. Recent archeological evidence demonstrates that ancient civilizations rose and fell based on their ability to maintain a secure, stable food supply. Conceptually, the problem of food security has been with us at least since Malthus. In the previous century, the frightening specter of population demand's outstripping the food supply arose at least once each decade.

Food security has been defined in many ways. For example, the Food and Agriculture Organization's World Food Security Compact (1985) states that the "ultimate objective of food security is to ensure all people, at all times, are in a position to produce or procure the basic food they need and that it should be an integral objective of economic and social plans."

For the purpose of this chapter, food security is viewed broadly to include at least three important dimensions: availability, adequacy and accessibility (Lacy and Busch, 1986). Some view availability simply in terms of sufficient production. Availability should encompass the concept of food sufficiency to sustain human life for the entire population in the short, as well as the long term. Availability also implies that food production and supply are dependable in the face of possible production shortages due to general causes such as climatic changes, natural disasters and civil disturbances. In addition, availability concerns go beyond the immediate feeding of the population to include issues of natural resource preservation, regeneration and sustainability for future generations. Consequently, embedded in this concept is concern for the long-term ecological balance of natural systems. Food security requires that concerted attention be given to conservation and enhancement of the natural resource base for food production.

The dimension of adequacy refers to differing nutritional needs of various segments of the population. It can be conceptualized in terms of balanced diets and having a variety of foods throughout the year. Also, an adequate food supply must include concern for the long-term health effects of continuous, and largely unmonitored, changes in the types and supply of food available to a population.

A third essential element in food security is accessibility. It encompasses not only transportation and marketing, but also the means by which food is acquired.

Producing an adequate food supply is not enough. Consumers must be in a position to purchase or obtain the necessary food. For people in poverty, markets are not an effective means for distributing food. Food security for the poor requires a careful examination of prevailing societal values and commitment to providing all its members with fair access to the food supply.

WORLD FOOD SUPPLIES

Despite frequent concerns about famine and starvation, the performance of our food and agricultural system has been rather phenomenal over the past 200 years. During this period, there has been a sixfold increase in the world's population. At the same time, global agricultural production has generally kept pace. Falling real grain prices during most of the 20th century are evidence of this remarkable success.

Factors contributing to this increased food production have changed over time. During the 19th century, increased output was achieved primarily by expanding the land area under production. This additional land was mainly located in newly settled areas of the Americas and Australia. During the 20th century, new mechanical, chemical and biological technologies produced a science-based agriculture that led to dramatic increases in yields in certain parts of the world and to substantial increases in food production. For example, from 1960 to 1990: (1) global cereal production doubled; (2) per capita food availability increased by 37%; (3) per capita calories available per day increased by 35% and (4) real food prices declined by 50%. Even in countries like India, where severe famine was predicted, food grain production increased from 50 million tons in 1947, at the time of independence, to 200 million tons in 1998–99. Agricultural production in the decade ending in 1991 increased by 125% and India's per capita food production rose steadily during the later part of the 20th century despite significant increases in population during that period (McCalla, 2001).

These increases in global food production were not, however, common to all regions in the world. Significant regional differences have occurred. In sub-Saharan Africa, for example, per capita food availability decreased between 1960 and 1990, with a 1% annual decline of annual grain output during that period. Droughts during the mid-1980s have resulted in approximately one fifth of Africa's people being sustained by imported grain.

ENVIRONMENTAL DEGRADATION AND POVERTY

At the same time, significant environmental degradation is occurring worldwide, in part due to the use by farmers of inappropriate agricultural practices. Two billion hectares of land, an area the size of North America, have experienced severe environmental damage in the past 50 years, with 5 to 10 million hectares worldwide becoming unproductive every year because of severe degradation. In sub-Saharan Africa, natural resource degradation is advancing at a startling rate, particularly in the form of desertification in dry land areas, soil erosion and deforestation on hillsides, biodiversity losses, increased siltation and flooding and loss of soil fertility in many cropped areas. Some estimates suggest that land degradation affects two thirds of the total cropland of Africa and one third of the pastureland. Much of this degradation is irreversible, or can be reversed only at very high cost.

Food security and agriculture also depend on genetic diversity and water resources. However, it is estimated by some scientists that 40% of the world's species could be extinct within 25 years. Furthermore, increased intensification of agriculture, which has resulted from expanded use of irrigation, has been a major factor in increasing production in certain areas. Nonetheless, these productivity increases may be difficult to sustain because of increased competition for water and salinization and waterlogging of irrigated soil that result in yield stagnation or land's being removed from production.

Equally important to considerations of environmental sustainability for food security is the extent and depth of poverty in both the developed and developing world. Globally, 1.3 billion people, nearly a quarter of the world's population, live in absolute poverty. They earn the equivalent of only US$1 per day per person or less, and must use this meager income to meet their food, shelter and other needs. Not surprisingly, hunger, malnutrition and associated diseases are widespread. More than 840 million people lack access to sufficient food to lead healthy, productive lives. Every second person in Africa and sub-Saharan Africa is absolutely poor.

In most of the developing world, poverty is a rural phenomenon. Approximately 70% of the poor live in rural settings and a majority of these individuals are involved in agriculture. In many sub-Saharan African and Asian countries, over three quarters of the poor live in rural areas. Latin America's higher urbanization rates have led to a greater prevalence of urban poverty, but, even in this region, the majority of the poor are rural. Literally millions of small subsistence farmers live in poverty. Even when the rural-based poor are not engaged in their own agricultural activities, they rely on nonfarm employment and income that are in one way or another linked to agriculture (McCalla, 2001; Pinstrup-Andersen and Pandya-Lorch, 1995).

These conditions dictate that policy makers around the world and in particular in developing countries, are faced with a need to simultaneously meet three inter-related and challenging goals — global food security, environmental sustainability and poverty alleviation. Agriculture and food production must continue to expand to keep up with rapidly increasing populations. At the same time, this growth must not jeopardize the underlying natural resource base but instead should enhance its sustainability. This process must also be equitable if it is to help alleviate poverty and food insecurity.

POVERTY ALLEVIATION: A COMPLEMENTARY GOAL

Poverty alleviation is an essential component of any successful strategy to achieve food security and environmental sustainability. Poverty undermines development and enhancement of the environment, threatens a steady and reliable food supply, destabilizes communities and regions and ruins lives. Poverty-alleviation strategies, policies and activities need to be undertaken with a clear understanding of (a) the characteristics of the poor, (b) the causes of their poverty, (c) where they are located and (d) their movement into and out of poverty. This approach requires a sound understanding of the multifaceted nature of poverty. Being impoverished may include inadequate access to food, housing, clothing, education, meaningful work and health care, as well as a general diminution of the quality of life. Alleviating poverty is a highly complex process that requires multiple approaches and policies. It varies by location and culture.

While opportunities for progress on these three goals depend considerably on specific social, economic and agro-ecological circumstances, much more remains to be learned about how these three critical and interrelated goals are linked and about the factors that condition these relationships. Links between poverty and the environment are often more complex than previously described. For example, many farmers are poor because they do not own farmland. They depend on the commons (open-access land such as rain forest) for their livelihood. Barring access to the commons will reduce environmental degradation but hurt the poor. Another example of how complex these relationships are can be seen in the use of agricultural chemicals and intensive grazing systems for animals. Alleviation of poverty through these methods may not prevent degradation and may even increase it, because richer farmers tend to use more agricultural chemicals than do poorer ones and rich landholders often hold a greater proportion of wealth in cattle, thereby creating pressure to turn forested land and hillsides into pasture.

Many factors affect the relationship among these three goals. Policies, technologies, institutions, population, agro-ecology and climate change can all modify the links among environmental sustainability, food security and poverty alleviation by affecting the choices of rural households and communities and the context in which these choices are made.

The interrelationship between natural-resource management, agriculture and poverty alleviation is illustrated by new research projects on gender, poverty and water management recently initiated in six countries by the International Water Management Institute (Cleaver, 1998). This gender, poverty and water initiative explores how irrigation development, improvement and reform can result in gendered poverty alleviation in rural areas of the developing world. The central assumptions are that water and irrigated land are major assets with which poor women and men can improve their well-being and that agencies can alleviate poverty more effectively by targeting their support to the poor. Such inclusive intervention methods primarily strengthen the rights to water and irrigated land of poor people. Poor cultivators who obtain access to water and irrigated land tend to make highly productive use of these resources under most conditions. Consequently, poverty alleviation through improved resource rights of the poor can also be a viable path to agricultural growth and environmental sustainability.

AGRICULTURAL GROWTH AND EQUITY

Several leading authorities have argued that agricultural growth is key to meeting the challenges of food security, environmental sustainability and poverty alleviation. Hazell (1999) has proposed that a high degree of complementarity among these three goals is more likely when agricultural development and food security are: a) broadly based and involve small and medium-sized farms, b) market-driven, c) participatory and decentralized and d) driven by technological changes that enhance productivity but do not degrade the natural resource base. Food security pursued in this manner can reduce real food prices while increasing farm incomes. It is employment-intensive and increases the effective demand for nonfood goods and services, particularly in small towns and market centers. By reducing poverty and promoting economic diversification in rural areas, this strategy can also relieve livelihood demands on the natural resource base.

During the 1970s and 1980s, policy makers and development experts learned that agricultural development could be used to both reduce poverty and increase food security while contributing to economic growth under certain circumstances. Hazell (1999) characterized "lessons learned" as six equity modifiers for agricultural growth. First, agricultural growth needs to promote broad-based agricultural development through a *focus on small producers*. Few economies of scale for agricultural production exist in developing countries. Therefore, targeting family farms was attractive on both equity and efficiency grounds.

Indeed, Manning, in his book *Food's Frontier: The Next Green Revolution* (2000), has documented through accounts from Ethiopia, Zimbabwe, Uganda, India, China, Chile, Brazil, Mexico and Peru that improvements in the food, environment and poverty triangle seem most likely to be achieved in the developing world when alternative methods and philosophies based on indigenous knowledge and native crops, as well as on cutting-edge technology are all considered. His case studies and stories indicate that in these places, information and knowledge often do not flow from top to bottom, but rather originate in and reverberate through every part of the system.

Alex McCalla (2001) has pointed out that 90% of the world's food production is consumed in the country where it is produced. As a consequence, to be effective, most food production increases need to occur within the countries experiencing population increases. Thus, in the next 25 years, most of the food needed to meet increased demand must be produced in tropical and subtropical farming systems, where rapid population growth will occur.

This will be difficult, because these farming systems are complex, highly heterogeneous, fragile, generally low in productivity and dominated by small-scale, resource-constrained farmers. To support the necessary improvements, priority must be given in publicly funded research and extension to the issues of small and medium-sized farms in these locations, building on indigenous knowledge and adopting heterogeneous approaches.

Hazell's other five "lessons learned" about agricultural growth are summarized below. The second lesson learned about utilizing agricultural development to reduce poverty is to *undertake land reforms* where necessary. This was particularly important where productive land was too narrowly concentrated among large farms. Successful approaches could include securing farmers' property rights and privatizing common property resources or, where this was not desirable, strengthening community management systems. A third lesson is to *invest in human capital* through such means as rural education, clean water, health, family planning and nutrition programs to improve the productivity of poor people and increase their opportunities for gainful employment. Fourth, the agricultural extension and education system as well as credit programs assisting small businesses need to be organized to *reach rural women*, because women play a key role in farming and auxiliary activities. Fifth is the need to *involve all rural stakeholders*, not just the rich and powerful, as participants in setting priorities for public investments that they expect to benefit from or finance. Finally, Hazell observes that it is important to actively *encourage the rural nonfarm economy*. This economy is not only an important source of income and employment in rural areas, especially for the poor, but also benefits from powerful income and employment multipliers when agriculture grows. Free zones, which have been established in many central American and Caribbean nations, are good examples of how this type of employment opportunity can impact agriculture, rural poverty and related natural-resource management.

Although past patterns of agricultural growth and development have sometimes degraded the environment and exacerbated poverty and food insecurity among rural people, this is not an inevitable outcome of agricultural growth. Instead, these negative effects are usually the product of (a) inappropriate economic incentives for managing modern inputs in intensive agricultural systems, (b) insufficient investment in many heavily populated less-developed areas, (c) inadequate social and poverty programs and (d) political systems that are biased against rural people. With appropriate government policies and investments, institutional development and agricultural research, agricultural development can provide a triple-win situation by contributing to poverty alleviation, food security and improved natural-resource management and environmental sustainability.

Pinstrup-Andersen and Pandya-Lorch (1995) have argued forcefully and persuasively that agricultural growth is the key to poverty alleviation in low-income developing countries. They note that very few countries have experienced rapid economic growth without agricultural growth either preceding or accompanying it. Furthermore, economic growth is strongly linked to poverty reduction. In most low-income countries, agricultural growth is a catalyst for broad-based economic growth and development. Finally, poverty is the most serious threat to the environment in developing countries. Because they lack the means to intensify their agriculture appropriately, the poor are often forced to overuse or misuse the natural resource base to meet their basic needs.

AGRICULTURAL RESEARCH AND TECHNOLOGY

Pinstrup-Andersen and Pandya-Lorch (1995) further maintain that agricultural research and technological improvements are crucial to increased agricultural productivity and financial returns to farmers and farm labor, thereby reducing poverty, meeting future food needs and protecting the environment. Accelerated investment in agricultural research is particularly urgent for low-income developing countries, partly because they will not achieve reasonable growth and poverty alleviation without increases in agricultural productivity and partly because comparatively little research is currently undertaken in any of these countries.

Ironically, many poor countries that depend the most on productivity increases, grossly underinvest in agricultural research. Per capita agricultural research expenditures in low-income countries are one tenth of those in high-income countries, despite the fact that agriculture accounts for a much larger share of average income in these low-income countries. For example, in sub-Saharan Africa, which desperately needs increases in agricultural productivity, there are fewer than 50 agricultural researchers per million individuals employed in agriculture. This is in sharp contrast to industrialized countries, where there are over 2,400 agricultural researchers per million economically active persons in agriculture.

One new research area in particular, agricultural biotechnology, is seen as presenting opportunities for reducing poverty, food insecurity, child malnutrition and natural resource degradation. Pinstrup-Andersen and Cohen (2000) observe that developing countries are faced with many problems and constraints that biotechnology may actually help to resolve. They acknowledge that agricultural biotechnology is not a silver bullet to achieving food security and, to date, has not focused on these broader environmental and societal issues. However, when biotechnology is used in conjunction with traditional knowledge and conventional agricultural research methods, it may be a powerful tool for addressing food, environment and poverty issues. Consequently, policies must expand and guide traditional research and technology development, as well as the new sciences, to solve problems of importance to poor people.

At the same time, adequate attention and policy development must be given to the wide range of environmental, biosafety, social and value concerns associated with these new technologies (Lacy 2000a). Manning (2000) concluded that the "prime directive for those who would help the world's poor ought to be 'first do no harm.' " Research should focus on crop and animal production relevant to small farmers and poor consumers in developing countries, such as bananas, cassava, yams and sweet potatoes, rice, millet and certain livestock. Failure to significantly expand agricultural research in and for developing countries and to invest in agricultural development will make poverty eradication and alleviation a more elusive goal (Pinstrup-Andersen and Cohen, 2000).

Peter Senker (2000) reminds us that major multinational corporations do not pursue the objective of alleviating world poverty because they are essentially market-creating and -satisfying organizations. They neglect the poor basically because they do not offer attractive markets. Companies are motivated to seek profits from rich markets in the developed world rather than moved by a drive to feed the hungry. Given the current emphasis on using these new biotechnologies to develop proprietary products, there is a strong potential that genetic engineering will further increase poor farmers' dependence on the corporate sector for seeds and agricultural inputs and on associated chemical herbicides and fertilizers.

Strong public-sector investment in biotechnology, therefore, has an important role to play. Pinstrup-Andersen and Pandya-Lorch (1995) observe that much of the research needed to reduce poverty is of a public goods nature. The benefits of such research are not easily captured by individual farmers or firms but extend to society as a whole and, as a consequence, are unlikely to be undertaken by the private sector. Thus, the research institutions of these developing countries should receive substantial public investments and be further supported by the international research community.

INFRASTRUCTURE AND POLICY

Research and technology alone will not drive agricultural growth, environmental sustainability and poverty alleviation. The interaction between technology and policy is critical and the beneficial effects of the research will occur only if government policies are appropriate. Distortion of input and output markets and of asset ownership as well as other institutional and market conditions that adversely affect the poor must be minimized or eliminated. Access by the poor to productive resources such as land and capital needs to be enhanced. Zeller and Sharma (1998) note, for example, that microfinance institutions designed to finance the poor, such as the Grameen Bank in Bangladesh, offer services proven to alleviate poverty that the marketplace is not willing to provide on its own. Sharma (2000) also observes that the impact of microfinance on poverty alleviation continues to be substantial, but is conditional on access to other complementary inputs such as seeds, irrigation water or market access.

Further, rural infrastructure and economic, legal and governmental institutions must be strengthened. An insightful study by Fan, Hazell and Thorat (1998) examined government spending, growth and poverty in rural India. The study showed that government spending on productivity-enhancing investments, such as agricultural research and development, irrigation, rural infrastructure including roads and electricity and rural development initiatives targeted directly to the rural poor, have all contributed to reductions in rural poverty and most have contributed to growth in agricultural productivity. They conclude that, to reduce rural poverty, the Indian government should give priority to increasing its spending on rural roads and on agricultural research and extension, investments that have both a large impact on poverty and the greatest impact on agricultural productivity growth.

In addition, human resources must be improved through standard investments in education, health care, nutrition and sanitary environments. The director of the Harvard Center for International Development, Jeffery Sachs, has proposed that malignant poverty may be primarily the product of wretched public health (Birch, 2000). He and his colleagues have joined health advocates to lobby the world's pharmaceutical companies to develop vaccines against tuberculosis, AIDS and malaria, some of the world's biggest killers.

Finally, the policy environment must be conducive to analyzing the complex factors affecting poverty and must be supportive of actions to alleviate poverty and implement sustainable management of natural resources.

Although the focus of this chapter has been on the interrelationship among food, environment and poverty, a similar, more general relationship has recently been proposed between enhancing economic growth and productivity and fighting poverty. In May 2001, U.S. Treasury Secretary Paul O'Neill criticized the World Bank for digressing from its core purpose, which he defined as raising productivity and increasing income in developing countries. The World Bank responded that significant attention to the distribution of income to benefit the poor was a proper priority for the Bank.

Anthony Lanyi, director for economic policy at the IRIS Center of the University of Maryland, argued in response (2001) that these are not mutually exclusive, but rather complementary goals, as proposed also in this chapter. Lanyi observed that conventional wisdom has held that the best way to help the poor is to raise overall income, typically through big infrastructure and industrial projects. However, these strategies often have proved ineffective for both growth and poverty reduction. Instead, Lanyi argues that, in reality, the best way to pursue both of these goals is through reform of economic, legal and governmental institutions.

Specifically, these two goals are more likely to be achieved where contracts and property rights are legally established and enforced fairly; government is publicly accountable for the fair and equitable delivery of public services such as education and health; governmental corruption is legally restrained and carefully scrutinized; financial institutions are properly capitalized and supervised; and government, through legal measures, prevents predatory behavior within the private sector while leaving as much of the economic activity as possible to that sector.

While it is generally obvious how these policies would enhance growth, it may be less apparent how they would reduce poverty. However, studies demonstrate that faulty government, legal and financial institutions tend to disenfranchise the poor by restricting their entry into business, appropriating their property and weakening the quality of their health and educational services. Lanyi concludes that we should not pit the two goals of growth and poverty reduction against each other, but should focus instead on the policies and key incentives that produce both enhanced growth and dramatically more equitable distribution of its benefits.

COMMUNITY EMPOWERMENT

Any discussion of the interrelationship of food, environment and poverty, and of strategies for reducing or alleviating poverty, must address the role of community. Communities are the basic building blocks and foundations of our society, making critical contributions to the quality of food systems, environment, education, health, economy and overall well-being. Unfortunately, communities are experiencing rapid transformations that may significantly erode their capacity to remain viable and sustainable both domestically and internationally. Local communities and their economies have become increasingly enmeshed in a global economic system characterized by extreme mobility of capital and by the use of places as little more than production sites. Furthermore, this globalization has decreased the importance of community as a social unit, particularly in the developed world.

The issues of empowering communities in this context are critical to alleviating poverty and ensuring food security, as well as to meeting other important social goals. Without trying to address the full array of processes and structures that empower communities, we have argued elsewhere (Lacy 2000b) that four are particularly key: public work, science, food systems and democracy. The ways in which we (a) view and structure our work in terms of its public purpose; (b) generate and disseminate new knowledge through science and technology; (c) produce, distribute and consume food and (d) make political decisions, contribute to our sense of self and community and, thereby, are essential to shaping the viability of our food system, our environment and our communities.

One component of democracy, civic spiritedness, which encourages responsibility, respect and equality among community members over individual rights, must be nurtured in particular if democracy is to work and communities are to thrive. Liberty Hyde Bailey, dean of the College of Agriculture at Cornell University at the turn of the last century, said it eloquently:

> "Every movement that tends to weaken local responsibility and initiative is a distinct menace to the people .... Our present greatest need is the development of what may be called the community sense, the idea of a community, as a whole, working together towards one work(1996: 43, 51)."

CONCLUSION

The key goals of global food security, environmental sustainability and poverty alleviation are not necessarily or always complementary. However, they should be pursued simultaneously with policies, technologies, infrastructures and economic, legal and governmental institutions that make the three more compatible. At the same time, policies designed to foster food security or growth combined with improved natural-resource management will not always and everywhere alleviate poverty, particularly if a narrowly defined food security objective is pursued.

When food security includes all three dimensions — availability, adequacy and accessibility — poverty alleviation is more likely. Food security should be viewed as a social goal existing within a broader set of societal goals for national and global food and agricultural systems. It is inextricability linked to such goals as equity, equality of opportunity, justice, stewardship and community security and sustainability. Furthermore, eradicating poverty must always be a central pillar of development goals and of efforts to achieve food security and environmental sustainability. Moreover, these efforts must not be pursued merely within the community, but also by the community as a whole on behalf of the community. Mahatma Gandhi said it well when he stated, "We must be the change we wish to see in the world."

ACKNOWLEDGMENT

The authors gratefully acknowledge the assistance of Gina Anderson and Wendy Kercher, Office of University Outreach and International Programs, UC Davis, in the preparation of this paper.

REFERENCES

Bailey, L.H. 1996. *The State and the Farmer*. St. Paul, MN: Minnesota Extension Service, University of Minnesota. First published in 1908.
Birch, D. 2000. Ailing People, Ailing Economies." *Baltimore Sun*, November 12.
Cleaver, F. 1998. Special issue: Choice, complexity and change: Gendered livelihoods and the management of water. *Agriculture and Human Values*, 15:4.

Fan, S., P. Hazell and S. Thorat. 1998. Government spending, growth and poverty: An analysis of inter-linkages in rural India. IFPRI Environmental and Protection Division Discussion Paper 33:1-28. International Food Policy Research Institute. Washington, D.C.

FAO. 1985. *Report of the Tenth Session of the Committee on World Food Security.* Food and Agriculture Organization of the United Nations. Rome, Italy.

Hazell, P. 1999. Agricultural growth, poverty alleviation and environmental sustainability: Having it all. International Food Policy Research Institute 2020 Brief 59:1-5. International Food Policy Research Institute. Washington, D.C.

Lacy, W. 2000a. Agricultural biotechnology, socio-economic issues and the fourth criterion. In *Encyclopedia of Ethical, Legal and Policy Issues in Biotechnology,* T. Murray and M. Mehlman, Eds. 76-89. John Wiley. New York.

Lacy, W. 2000b. Empowering communities through public work, science and local food systems: Revisiting democracy and globalization. *Rural Sociology,* 65:1, 3-26.

Lacy, W. and L. Busch. 1986. Food security in the United States: Myth or reality. In *New Dimensions in Rural Policy: Building Upon Our Heritage,* 222-223. Subcommittee on Agriculture and Transportation, Joint Economic Committee, Congress of the United States. U.S. Government Printing Office. Washington, D.C.

Lanyi, A. 2001. Enhancing growth and fighting poverty: Is there a contradiction? www.iris.umd.edu.

Manning, R. 2000. *Foods Frontier: The Next Green Revolution.* North Point Press. New York.

McCalla, A. 2001. Challenges to world agriculture in the 21st century. *Agricultural and Resource Economics Update – University of California, Davis,* 4:(3), 1, 2, 10, 11.

Pinstrup-Andersen, P. and M.J. Cohen. 2000. Agricultural biotechnology: Risks and opportunities for developing country food security. *International Journal of Biotechnology,* 2: 1-3, 145-163.

Pinstrup-Andersen, P. and R. Pandya-Lorch. 1995. Agricultural growth is the key to poverty alleviation in low-income developing countries. *International Food Policy Research Institute 2020 Vision Brief,* 15:1-4.

Senker, P. 2001. Biotech and inequality. Center for International Development at Harvard University. www.cid.harvard.edu/cidbiotech/comments/comments66.html. 3/6/01/.

Vosti, S. and T. Reardon. 1997. Sustainability, growth and poverty alleviation: A policy and agro-ecological perspective. *International Food Policy Research Institute – Food Policy Statement,* 25:1-4.

Zeller, L. and M. Sharma. 1998. Rural finance and poverty alleviation. *International Food Policy Research Institute — Food Policy Statement,* 27: 1, 2.

23 Poverty and Inequality: A Life Chances Perspective

Norman Uphoff

CONTENTS

INTRODUCTION

One of India's most significant accomplishments over the last five decades has been to change the world's image of its poverty. At independence, India was widely regarded as a poor country with almost all of its people living in poverty, albeit overlaid by a very thin but rich upper crust of maharajahs and commercial moguls. The Nizam of Hyderabad and other anachronistic remnants of the precolonial era particularly intrigued many people in other countries, who considered such huge personal accumulations of gold and jewels to be real wealth. These extravagant riches and their accompanying regal lifestyles made even less tolerable the contrasting miserable, constrained conditions of most Indians.

For India to be known and respected now for its large middle class, which is practically as large as the population of the United States, represents progress. However, the number of Indians who subsist below the poverty line, nearly 400 million in 1992 (World Bank, 2000), is greater than India's total population was at independence. It is also progress that the rest of the world has come to understand better that the wealth of nations — to use Adam Smith's term — lies not in precious metals and stones but in the capacities of its people to produce, to invent and to organize.

It is a matter of debate whether inequality is greater in a country when the vast majority are poor and a few people are very wealthy, with a small middle class, or when there is a considerably larger middle class that has distanced itself from the poor, with the number of rich persons, few of them opulently wealthy, increased by several multiples. It would seem that the answer is more a matter of values than arithmetic. The latter situation has a lower Gini coefficient (developed by Italian statistician Gini to provide a mathematical expression of the degree of concentration of wealth or income), but it makes more visible to the poor on a continuous and very obvious basis how different are their lives and their life chances from those of the rest of their society.

INDICATORS OF INEQUALITY
AND ITS JUSTIFICATION

It is worthwhile to know what the Gini coefficients are, as well as other such measures of inequality (discussed in Chapter 29), so that we can assess differences between countries and, over time, in specific and reasonably comparable terms. Having myself undertaken to compare differences in income inequality among 16 Asian countries some years ago as part of an evaluation of their rural development strategies and accomplishments (Uphoff and Esman, 1974), I have come to favor the ratio of income that is accruing to the highest and lowest quintiles of the population. Wealth, the stock of income-producing assets, is probably a better indicator than income, but measurement problems are daunting enough to make such figures even less reliable than income data.

This measure of inequality considers the income of the top 20% as a multiple of that which goes to the bottom 20%. This is a comprehensible and meaningful indicator of equality or, conversely, inequality. It is concrete, not abstract, and moreover, it highlights the most extreme differentials, reflecting the nonlinearity that goes with living very far below or very far above the poverty line.

Our analysis of the Asian experience, which was supported by the U.S Agency for International Development (USAID) in the days when it was more concerned with the substance and strategy of development than now, was undertaken to make some objective comparisons among countries that ranged geographically, politically and economically from China and Japan in East Asia to Turkey and Yugoslavia on the western edge of Asia. The analysis was done in the positivist spirit of the time, emphasizing measurement. While income distribution was not a central focus of our research project, it was something that had to be considered. At that time, there was still a strong argument in the literature justifying inequality as good for promoting economic growth.

We found, on the contrary, that those countries with the best records in rural development across a wide range of measures (agricultural, nutritional, educational, public health, etc.) also had definitely more equal distributions of income. Unfortunately, however, the relationships between different development measures and income distribution were too many and too complex for us to attribute which was cause and which effect. Because our research was not intended to illuminate inequality, we reported our findings and left them for readers to consider (Uphoff and Esman, 1983: 292-294).*

What could be concluded from our data, however, was that having a more equal income distribution was not unfavorable for rural development. Analyses based on a simplistic understanding of the Harrod-Domar model of economic growth had argued that unequal distribution of income should promote economic growth and greater employment because — it seemed logical — (a) richer people would save more than poor people, (b) a greater volume of domestic savings would increase the supply of resources available for investment and (c) accelerated capital formation would raise gross domestic product and resulting incomes, a virtuous cycle feeding back into greater savings. Thus, income inequality, even that reflecting widespread poverty, was regarded as good for development. It would contribute to more savings, investment and growth of GNP.

By the mid-1960s, however, at the height of development thinking that equated development with GNP growth, a few economists were already pointing out that this logical construction was not empirically supported by evidence.** Although these empirical challenges appeared in leading journals at the time, they were ignored. Why? They went against the prevailing paradigm, which seemed so logical.*** Moreover, they went against predominant economic interests. If development was regarded as depending almost entirely on capital as the scarcest and thus as the most valuable factor of production, this justified the owners of capital receiving the largest share of the benefits from development.

---

* It might have been expected that per capita income levels would explain both higher performance on measures such as rate of agricultural production growth, adult literacy, life expectancy and caloric intakes per capita and greater income equality, all measured in the early 1970s. In fact, the eight country cases classified as having more kinds and greater degrees, of local organization involved in rural development functions, the independent variable in this analysis, had per capita incomes almost three times greater in 1973 than the average for the "less organized" cases — US$352 vs. $119. Just 20 years earlier, per capita income levels for the two sets of countries had been practically the same, $74 and $78, respectively. There were good grounds for inferring that the functioning and performance of local governments, cooperatives, farmer associations, etc., had contributed something to the general advancement in agricultural, economic and human resource terms.

** In an extensive review of the economic literature that documented the dominance of these savings and investment theories, Hahn and Matthews (1964) noted that these theories failed to account adequately for the phenomena under consideration. When Hamberg (1969) reviewed a number of cross-country studies that used UN, OECD and other data sets, he found that the correlations between gross domestic investment and gross domestic income growth seldom reached even .33, which would not account for more than 10% of variation and were anyway not statistically significant. Various other studies, summarized by Owens and Shaw (1972), showed that persons with higher incomes did not necessarily save higher proportions of their income, and greater savings did not necessarily lead to more investment. In addition, even though investment was correlated with national income growth, the correlation was not a strong one, as returns to investment varied widely across countries. In my own research on Ghana's economic development in the 1950s and 1960s, I found that gross domestic capital formation reached 21% in 1965, yet Ghana's economic growth the next year was negative. Per capita growth had already been declining since 1961 despite GCDF rates over 15%. Even a modest incremental capital-to-output ratio (ICOR) of 5:1 should have given growth of 3% if capital is indeed a determinant of economic growth.

*** On the power of paradigms to restrict valid new ideas, see Krugman (1995) on how and why spatial relations have been largely excluded from development economics analysis.

About this time (1967), President Julius Nyerere of Tanzania presented in The Arusha Declaration a conceptual, not just empirical, challenge to the prevalent view. If poor countries have little capital and an abundance of labor, he asked, why not use whatever capital is available to make the most abundant resource, labor, more productive — rather than use labor, often wastefully and certainly with poor remuneration, to make the resource they had least of, capital, particularly foreigners' capital, more productive? Why should the poor seek to fight their war against poverty with the weapons of the rich? Nyerere asked pointedly. This was dismissed as ideology rather a legitimate question.

There were some stirrings within the economics discipline during the 1960s that questioned the dominant capital-favoring paradigm.* But it took another 20 years before the case for more equitable paths to development gained acceptance, though still not dominance. The proponents of meeting basic human needs in the 1970s justified this more on grounds of equity and fairness than as a way to raise productivity.

What finally seems to have gained the most ground for equitable distribution of income and wealth was the success of the East Asian "Tigers" — Japan, South Korea, Taiwan and Singapore. By the 1980s, this success was too apparent to overlook, as was their more equal distribution of income.** In these countries, policies ranging from land reform and universal basic education to public housing, and primary health care had contributed to political-economic systems that sought to contradict the Biblical admonition: "The poor you shall have always with you." These countries considered poverty to be unacceptable, and a drag on their economies. There was also evidence accumulating, such as that from Berry and Cline (1979), that more equal distributions of land contributed to aggregate agricultural production as well as economic growth. Certainly, East Asian land reforms, including those in China, were important impetuses for economic growth in a number of ways.

In recent years, the economic performance of some of the Tigers has flagged. Some might want to attribute this to their relative income equality, because capital formation in Japan and South Korea was only 1.1% and 1.6% during the 1990s. But this was the period in which income distribution in these countries became

---

* Singer (1966) suggested that the problem of development was not the creation of wealth but the creation of the capacity to produce wealth and this favored investment in human resources. Kuznets (1966) showed that incomes were almost always more equal in the more developed countries, though this did not resolve the question of what was cause and what effect. A cross-national analysis by Leibenstein (1965) of sources of productivity found that efficient allocation of the conventional factors of production could not account for more than 30% of variation in productivity, attributing the rest to human, cultural, organizational and other sources. Lewis (1965) concluded that efforts to accelerate the pace of development by devoting (or diverting) a larger share of increased output into savings and capital investment, rather than to consumption, was self-defeating. This was partly because it could give rise to political unrest, but also because output cannot be sustainably increased unless consumption also increases. Capital formation, in Lewis's, view was more of an intervening or even resulting variable than an independent one. On the fallacious dichotomy between consumption and investment expenditure, see also Morgan (1969).

** Current data from the World Development Report 2000/2001, which unfortunately does not include data on income distribution for Singapore and Taiwan, show income ratios (top 20%:bottom 20%) to be 3.4 in Japan and 5.2 in South Korea. For comparison, the ratios in China and U.S. are 7.9 and 8.9, respectively. In the four countries of Scandinavia, on the other hand, the ratio is 3.6, less than half these latter figures. India's ratio is 5.7, as discussed further below.

less equal, with fortunes made (and later lost) in real estate and corporate and other dealings. As economic behavior occurs "at the margin" rather than being based on averages, it is difficult to draw definite causal inferences with such complex relationships.

I had the good fortune to have W. Arthur Lewis as a teacher of development economics and I acquired from him a skepticism about capital formation as the cause of economic growth. He considered it to be, in general, a consequence of growth, not being persuaded of the validity of neoclassical economics' assumptions and preferring to think along the lines of more classical economic theory. He did not regard market prices as an infallible equalizer of values, whereby $100 worth of deodorants would be equal to $100 worth of productive land. I share Sir Arthur's reservations, though the current theory and practice of economics is quite happy to equate everything by market prices, even when it is acknowledged that these prices reflect very unequal distributions of income that distort the forces of demand and supply. The price system, except under unattainable conditions, is better able to maximize profits than to maximize human welfare. It also fails to reflect adequately the needs and interests of future generations. But this is not the time or place for a fundamental debate on economics.*

In any case, there is enough evidence now accumulated and analyzed to assure us that relative income equality is not a necessary drag on growth and we should know that there is some significant evidence showing positive effects from relative equality. Reducing inequality can thus be seen as a spur to economic growth, reducing poverty by that complex path rather than by a direct process of income redistribution or transfer.

A LIFE CHANCES VIEW OF POVERTY

As a social scientist who works on development, rather than an economist who tries to explain all those things that can be denominated in terms of money, I would suggest the following perspective on poverty and inequality. If I were doing today the kind of analysis I undertook with Cornell colleagues 25 years ago, I would still want to look at income distribution data and to compare statistics such as top 20%: bottom 20% ratio to assess magnitudes and trends.

If there are data detailed and extensive enough to use more refined indicators such as the Foster-Greer-Thorbecke measure of inequality (Foster et al., 1984), this would be desirable because it maps disparities in income distribution in more precise and meaningful ways. I would also want to have some of the kind of qualitative (one might better say phenomenological) assessments of poverty that were done for the World Bank by Deepa Narayan and her associates to show the human face of poverty for its 2000-2001 World Development Report on poverty (Narayan, 2000; and Narayan et al., 2000).

But increasingly, the most meaningful measure of poverty, in my view, is one not found in the literature. Assessing the lives of people — their present living

---

* For critiques of contemporary economic theory and theorizing offering well-considered alternative formulations, see Leibenstein (1976), Daly (1990) and Ormerod (1998).

standards and conditions — is important, but I think poverty is most significant in terms of what it does to people's life chances — their opportunities to get educated, to have food, shelter and clothing that meet basic needs (and more), to move not just a little way up the ladder of income distribution but to be able to make some significant jumps and — most important — to give their children greater opportunities.

A life-chances indicator would tell us what is most significant and oppressive about poverty: its stratification of society into relatively static as well as separate groups. This should be of concern to almost everyone, not just those persons who bear the brunt of poverty. To be sure, not everyone loses equally from a social arrangement of group stratification, but the losers, who are more than just the poor, greatly outnumber the gainers from inequality. The poor consume less because they produce fewer goods and services that are consequently not available to the rest of society.

What is the probability that people who are born into poverty will, in the course of their lives end up, reasonably stably, above the poverty line? Or put another way, what is the probability that people born into families in that lowest 20%of households would eventually head or co-head households in the next higher 20%, so that their children will have definitely "moved up the ladder," even if not out of poverty. One would like to know this for persons born in the next higher quintile as well.* Perhaps the worse thing about poverty is the inescapability it creates from the problems, constraints and insults that are imposed upon the poor, documented by Narayan and her collaborators in the recent World Bank studies cited above. These deprivations and humiliations are of concern not just within a single generation, but even more, from generation to generation.

Not everyone within a country or community can be above average — by definition, a fifth of people must always be in the lowest quintile. So whenever incomes or standards of living are compared, some inequality is unavoidable, though this can be a greater or lesser degree. If one can move from a zero-sum to a positive-sum framework for thinking about wealth and poverty, of course, it is easier to address this problem, both analytically and psychologically.

It is an important question practically and ethically whether persons in the lowest bracket are those who have the least physical or mental capacity — or whether they are persons who have, through no action or failing of their own, been deprived of effective opportunities to develop their productive capacities to the fullest and to attain concurrent status and security. The latter situation represents a loss not only for people who are so constrained by economic, social, cultural or political circumstances, but also for the whole society. Its aggregate loss may be even greater than that for the poor.

---

* In fact, there is often considerable transient movement into and out of a given quintile year to year as incomes rise or fall. "Hard core" poverty occurs where persons remain within the bottom category for their entire lifetime. In many countries, people in the fourth quintile from the top qualify as poor because of their deprivation of income, goods and services considered in absolute terms. In particularly impoverished countries, people even in the third (middle) quintile may also be classified as poor. The fifth (botton) quintile is invariably poor, and some or all of them may be considered the poorest of the poor, so this is the focus of discussion here.

All in society remain somewhat poorer when others' productive potential is unfulfilled. Not only are there fewer goods and services to be enjoyed, but there are fewer contributions to the life and culture of a country, fewer songs and poems, fewer self-respecting friends to enrich social relations, fewer persons of talent and integrity to hold political office, etc. Poverty reduction is thus not something to be done just to benefit the poor. It is good for everyone except for those persons who derive their wealth from extractive relations that are zero-sum, or worse, negative-sum.*

If there is one core process that underlies development, it is that of creating positive-sum relationships, such as through the production of value-added, creation of consumer surplus, economies of scale from market integration and trade or through broader friendship networks (Uphoff and Ilchman, 1972: 75-121; Uphoff, 1996: 284-289). Economic relations that are only zero-sum contribute little to development, even though they may add to GNP as conventionally measured. Ironically, some negative-sum transactions, such as waste disposal and pollution abatement, also add to GNP. What truly accelerates development are positive-sum effects.

Such an understanding makes issues of poverty and inequality more central to development theory, policy and practice. Poverty and inequality are not just a blemish on the development record of a country, nor are they just unfinished business to be taken care of once development has progressed fairly far. Where poverty is of the locked-in variety, with stagnant life chances for the poor, it reflects a pattern of development that is not basically driven by positive-sum dynamics. It is a stunted form of development.

Thus, some strong practical as well as ethical reasons for "attacking poverty" exist, to use the subtitle of the World Development Report 2000/2001. The conditions of life for the poor can be improved in various ways, directly through assistance, or indirectly but more sustainably, by enhancing people's productivity. The latter can be accomplished: (a) by upgrading the factor endowments of the poor (their human as well as physical resources), (b) by ensuring them greater access to opportunities through general or specific processes of market integration that enable them to employ their factors more productively, (c) by enhancing bargaining power (where it is weak) to get more return for factors of production or goods and services, usually through organization or (d) by innovative initiatives of entrepreneurship and leadership that alter structures of economic, social and political production in more productive directions.** Returns to factors of production are affected more by bargaining power than by intrinsic value, because the market by itself offers no means to appraise the latter.

This dynamic view of poverty and inequality should be of interest to both individuals — especially those within categories of the poor — and to society as a

---

* Many economic relationships, such as certain share-cropping arrangements or petty trading sustained by obligations of indebtedness, are negative-sum for society in that they impoverish a large number (or class) of people by holding down the productive potential of large numbers, while only a few gain from these exploitative relationships. The gains of the latter are less than the aggregate imputable loss, but, because the latter never get considered, the social cost to society of such relationships is invisible.

** These generic processes contributing to attaining higher levels of productivity were addressed analytically in Uphoff and Ilchman (1972), esp. pp. 82–86.

whole. Living in poverty has myriad degradations and debilitations, well documented in Narayan (2000), but being locked into this status, with its attendant diminutions of life quality, makes a bad situation worse. The prospect that one's children will, through no fault of their own, have no better chances of living a more productive and fulfilled life, adds greatly to the psychological burden of poverty. (This was not adequately addressed, in my view, in the surveys that Narayan and associates drew on for their two volumes.) From a societal point of view, to the extent that more people and more talent are locked into poverty, their contributions to GNP, but also to cultural creations and to political and social life, are diminished.

Life chances can be measured fairly precisely at any point in time, at least retrospectively, by tracking intergenerational mobility in economic and social terms through interviews with persons according to some appropriate simple classification or scaling of economic and social status (class). The implications for policy are that steps should be taken and investments made that most surely increase the probability that people can move to a higher rank, level or category in the future and particularly that their children will be able to live stably in a higher one.

RELATIVE VS. ABSOLUTE MEASURES

This approach to assessing and attacking poverty leads into some sticky analytical and evaluative terrain. It also argues against my preferred measure of inequality — comparing as a ratio the income going to the top 20% and that to the bottom 20% of households (or individuals). Such a measure is zero-sum in that it uses a fixed proportion. The ratio can improve, i.e., move lower, but it can only approach, never reaching, zero; 2:1 or 3:1 ratios would represent a great victory in reducing inequality and alleviating poverty, when the ratio can exceed 25:1 as in Brazil or El Salvador.

To assess progress in improving life chances, one would use appropriate poverty line measures, secondarily looking at movement between quintiles of distributed income. There could be considerable poverty alleviation if all households simply moved up in income level without any change in rank-ordering (seen from an analysis of which households are in which quintiles). However, the creation of greater opportunities for achievement and mobility based on merit will not have been achieved, because one of the few things we know with some certainty in the social sciences is that there is, in intergenerational terms, invariably some regression toward the mean in terms of intelligence and other talents. A "rising tide that lifts all boats" should be welcomed as an unprecedented policy achievement, but it would not represent a full-fledged victory in the war against poverty. While more individuals would be better off, society as a whole would not have gained as much as it could by opening up more opportunities for leadership and responsibility based on talent and innovativeness.

Whether persons are in poverty can thus be viewed in either absolute terms (poverty lines) or relative terms (ratios). Having a high degree of equality in a situation where everyone is poor in terms of their possibilities for consumption and living a good life is hardly satisfactory. For this reason, we are concerned with both poverty and inequality together, even though they can be and should be analyzed separately.*

There is always some tension between the absolute and relative concepts of poverty. Poverty lines get conceived and drawn as something absolute, producing certain numbers of persons below them who thus belong in the category of the poor. Even such lines are, however, relative to some conception of human needs or social acceptability and the data on which such calculations are based are themselves often very debatable, the products of sampling and surveys that can be contested.** So one should not regard the numbers as being true or real in any absolute sense. Rather, they are constructs, worth knowing and of special value when they are tracked over time or compared across regions or social groups, using the same standards for derivation.

IMPLICATIONS OF SOCIAL MOBILITY FOR SOCIETAL EFFICIENCY AND EQUITY

For assessing life chances, one needs to ascertain how much socioeconomic mobility there is in a society through surveys and observations that do not rely so much on measurement as on simple categorizations such as job classifications or possession of certain kinds of assets, which are not very ambiguous. Comparing that status (category) of persons with that for their parents can be reasonably objective even if recall must be used, because the things being recalled are simple and discrete. (The information solicited in surveys that enable analysts to report or estimate current incomes is more subject to error). One can put aside the fact that there will always be some persons below average, even way below average; the important question in this kind of analysis is whether they are always the same persons, or persons from the same families.

This life chances approach to understanding and evaluating poverty and inequality can be justified by efficiency as well as equity concerns. As noted above, one of the few things about human beings known with reasonable certainty is that intelligence, or at least potential intelligence, as well as other talents, are distributed quite evenly across all populations, all races, both genders, etc. In a country with a high degree of access to positions of higher income, status and authority based on merit, the offspring of the families in the upper quintile, biologically speaking, have some greater chance of being in that quintile in the next generation simply based on natural talent. This will be augmented by various acquired, as distinguished from innate, characteristics.

---

* This dual concern is the focus of a new interdisciplinary program at Cornell on poverty and inequality. It is directed by Prof. Ravi Kanbur, previously the World Bank's chief economist for Africa and then the head of its task force for preparation of the World Development Report 2000/2001. Inequality is seen as both a cause and consequence of poverty.
** Anyone who places much confidence in such data should read the evaluation of such data in rural Nepal that has been done by Gabriel Campbell et al. (1979). They found 50 to 200% discrepancies between the results of surveys carried out conventionally and the reality that could be ascertained through in-depth anthropological methods. Income data were indeed some of the most difficult to obtain accurately.

But this is only a chance, not a certainty. If all of those persons in the top quintile come from parents who themselves have had that status, the country's economic, political and other institutions are being directed by persons who have less than the greatest natural innate capability. They may have certain advantages of education and social connections that make them effective in such positions and this is not to be neglected. But the very highest intelligence and other talents will not be among their endowments.

The law of regression toward the mean means that most children of the most privileged group in a society will be less capable than their parents were and will deserve to end up in a lower quintile than they were born into. The converse implication of this law — that persons of highest intelligence can and will be born into any and all social categories — means that, on the basis of merit, there should be many persons, indeed a majority in any generation, in the top category who were born into lower quintiles and, on the basis of their talent, were able to rise up the socioeconomic ladder.

In fact, it is unlikely that any social policy aiming to end poverty and inequality can ever succeed fully. The chances of the first really becoming last are negligible, even though it can be fruitful to think about the implications of this (Chambers, 1997). The advantages of being brought up in an advantaged family with social contacts, psychological confidence, role models, etc., cannot be redistributed except by heinous measures that are destructive for everyone in society, as seen from the Khmer Rouge experience in Cambodia, which tried to expunge all past privileges by force.

What is possible, however, is to have an active policy of investment in developing human capabilities including universal, high quality education and health care, with effective programs of prenatal maternal as well as childhood nutrition. A progressive inheritance tax that levels the economic playing field between generations could finance a good part of this, offering at the same time the social utility of its becoming easier for persons with talent, imagination, energy and social skills to rise, their way not blocked by less capable persons who had extrinsic inherited advantages.

In India, there is a special problem that few people are willing to talk about. Even after 50 years, there is still strong residual discrimination against persons born into scheduled-caste or scheduled-tribe families. There are some exceptions, as some of these households have been able to climb up some rungs on the socioeconomic ladder. But the continuing effect of a caste system several thousand years old is one of the most glaring sources of poverty and inequality in India. A life-chances approach to evaluating poverty is particularly relevant where we know that there are certain sociocultural impediments to upward mobility.

ISSUES FOR INDIA TODAY

The good news is that income distribution in India appears to have become more equal over the past 35 years. When calculating the ratio of incomes in India going to the top 20% and the bottom 20%, we found two sets of figures; one from 1964–65 (National Sample Survey) analyzed by Pranab Bardhan, and the other from 1967–68 (National Council of Applied Economic Research) analyzed by K.R. Ranadive. These data sets produced quite different ratios, 6.0:1 and 10.9:1, which we averaged to consider 8.5:1 as a representative figure for India (Uphoff and Esman 1974: 147). The most current figures on income distribution in India (World Bank 2000: Table 5) give a ratio of 5.7:1, as a result of 46.1% of income going to the top 20%, while 8.1% of income goes to the bottom 20%. This suggests that India has made some progress in reducing inequality compared with earlier NCAER data, though not with regard to NSS surveys.

But what vision and strategy of development will the Indian government and its citizens pursue? Will it be purely incremental, being satisfied to have moved annually some number of individuals or households above the poverty line? Will there be longitudinal tracking to know how this number compares with those who have, in this same time period, fallen below the line? Will we know what kinds of persons are moving out of poverty and what kinds are sinking into it? Aggregate numbers that balance these two groups out, perhaps with little net change, are not very informative. A life-chances conception of poverty will focus on such data and on what can be done to create "one-way tickets" out of poverty because that is what reducing poverty is taken to mean.

As suggested above, poverty should be seen as bad for everyone, not just for the poor. Looking for ways to help people get themselves out of poverty — note that I did not say ways to get people out of poverty — would focus on the obstacles for different categories of persons defined as being among the poor. Often, these will derive from socioeconomic and sometimes political relationships that are extractive and exploitative, i.e., negative-sum, where the gains of the few are, in total, fewer than the losses of the many. If improvements in life chances are the measure and criterion of success, these relationships become unavoidable focuses of concern, whereas, with conventional poverty or inequality measures, any net incremental changes are interpreted as positive and there is no need to address structural impediments or resistances.

What will most improve life chances of the poor in India? Education and health care are the two most obvious measures, having the advantage of being positive-sum and not requiring anyone else to lose thereby, except, perhaps, those who have been exploiting cheap labor. Having a more educated population is good for the large majority in a country and having better health has positive payoffs by reducing disease that can harm those who are better off. Programs for fair hiring and promotion are more difficult to install because they involve some reallocation of opportunities, from less qualified to more qualified. But they are not impossible to promote as a kind of fair employment practices system that would benefit employers because they are supported in hiring and promoting on the basis of merit, which should improve the efficiency and profitability of enterprises.

THE SPECIAL ISSUE OF LAND DISTRIBUTION AND ACCESS

A controversial but sound policy would be to pursue a kind of land reform or redistribution that is different from the classical "land to the tiller" program. I call this universal access to land. It would not try to give every household in the agricultural sector a holding large enough to produce a subsistence income, as has been the usual policy objective when such redistribution has been contemplated. In many places, there is not enough arable land to set up every household wanting to practice agriculture with a so-called economic unit. This constraint has been a sufficient argument to get land distribution kept off the development agenda for the past several decades.*

But the image of agriculture that underlies — and is used to discredit — the classical form of land reform is an outmoded one. In most countries, including India, an increasing share of rural incomes is derived from nonfarm and nonagricultural sources. In part, this represents a high degree of desperation, as poor rural households find that they must turn to other sources of income to meet their basic needs. But it can also represent modernization and diversification of a rural economy which is no longer solely dependent on agricultural and own-enterprise activities for output and employment.

Two lines of argument support this suggestion, one emphasizing agricultural productivity and the other human productivity. First, as arable land becomes relatively scarcer with population growth, and demand for production continues to rise for the same reason, higher productivity per unit of land becomes critical for further development. In almost all situations, smaller holdings are more productive per hectare than larger ones because smaller ones are more intensively farmed, while larger ones are farmed more extensively (Berry and Cline 1979). Mechanized production that substitutes machines for labor raises profits more than it raises production. Only where mechanization increases intensification, as with plowing that permits cultivation of an extra crop, does it increase output. It is true that larger units of production produce higher incomes, but not because of higher output per unit of land. Most of the gains are due to economies of size rather than to technical economies of scale. Gains are based on advantages of bargaining power rather than on real gains in efficiency.

Second, there can be very real gains in welfare that contribute to the productivity obtainable from providing poor households with even small holdings. These units may be considered "subeconomic" by analysts if one expects households to get all their income from agricultural and own-farm pursuits, but they can add to the health, productivity and security (bargaining power) that can help households begin moving up out of poverty.

In India, research by Kumar (1977) found that, other things being equal, that is, for the same level of household income, children's nutritional status was higher if the household owned some — even a small piece — of land.** This could be easily explained. If a household had an opportunity to produce even a small share of the food that it needed, it had more control over its food supply and would not be as vulnerable to hunger periods. The land did not even need to be high quality, because

---

* I am pleased that it has been resurfacing recently in discussions of development strategy, e.g., Binswanger and Deininger (1997).
** She also found that nutritional status was higher — for any given level of household income — the larger the share of this that was contributed by the mother. This is not counterintuitive once the relationship is pointed out: the more a mother contributes to income, the more influence she can have over how income is used and she   ....:.:~~ +~ insist that a larger share be devoted to nourishing small children.

good management of the soil could improve it sufficiently for growing vegetables and fruits and maybe some staple crop. If a household had only a small plot, it was worthwhile to invest labor in raising its productivity.

Research in Indonesia by Hart (1978) showed that, other things being equal, including controlling for level of education and thus for the inferred level of human capital, households that owned even a small amount of land had higher returns for their labor, i.e., they had higher net wages per hour. If a family had even one-quarter acre (a tenth of a hectare) on which to grow some food to meet its subsistence needs, its workers could hold out for more than the very lowest wages being offered. Those completely landless workers who had no land to fall back on had to accept whatever work was available. These were often jobs to which, being desperate, they had to travel several hours in both directions. The hourly returns for such employment were thus pitifully low. Poor families with even some small amount of land were considered to be more desirable "clients" by the more powerful "patrons" in the village, those persons who had larger landholdings, so these poor families were better able to find employment locally and received better wages for their labor. Also, they were more likely to get benefits like gleaning rights on larger farmers' fields after harvest.*

The issue of land access should be put on the development agenda, even for a country like India, where formal land reform efforts have been mostly a failure (Herring, 1982) and where person:land ratios are, in many parts of the country, quite high. Two generations of population growth and resulting subdivision of land have accomplished at the upper ends of the land tenure system part of what land reforms intended: the breakup of very huge landholdings. But there has been concentration in lower-upper and upper-middle echelons, and the number of landless has continued to grow.

Exclusion from access to land in rural areas, coupled with poor or inaccessible schools, no or nonfunctioning medical facilities and social discrimination, means that several hundred million Indians are now — or will in the next generation be — denied the kinds of life chances that ought to be a human right. Such life chances are essential for the progress of an economy that is prosperous and dynamic in the modern world. The absence of life chances will slow an economy due to the inertia of millions of persons who have been marginalized and made dispensable by the economic system. They nevertheless need to meet their survival needs and can adversely affect the economy by becoming, in small or large numbers, strongly negative social and political forces.

---

* This analysis empirically contradicted the claims of "the new household economics" that were in vogue at the time. They claimed that returns to labor are, in general, efficiently and equitably determined because they reflect productivity as embodied in human resources, particularly as improved through education. In this study of rural realities that are surely not very dissimilar from those in rural India, it could be shown econometrically that the returns to education were less than those to owning even a small amount of land. People got more for their labor from owning land, or having secure usufruct rights, than from having small increments in education, mostly because, by not owning land, they were so vulnerable that they could not, in a labor-abundant market, capture the value of their labor productivity. With some land, they could afford to get a larger share of the value they added to production. This research was elaborated and published in Hart (1986).

This perspective on poverty and inequality has various measurement and normative aspects that can be addressed with more or less elegance, but it also has very practical and political implications that need to be addressed with some sense of urgency. A danger of preoccupation with the measurement aspects of poverty and inequality, especially if divorced from normative considerations, is that analysis will have nothing to contribute to the redress of practical needs and political pressures.

REFERENCES

Berry, R. A. and W. Cline. 1979. *Agrarian Structure and Productivity in Developing Countries*. Johns Hopkins University Press. Baltimore, MD.

Binswanger, H. and K. Deininger. 1997. Explaining Agricultural and Agrarian Policies in Developing Countries. Policy Research Working Paper 1764. World Bank. Washington, D.C.

Campbell, J.G., R. Shrestha and L. Stone. 1979. The Use and Misuse of Social Science in Nepal. Centre for Nepal and Asian Studies, Tribhuvan University. Kathmandu.

Chambers, Robert. 1997. *Whose Reality Counts? Putting the First Last*. Intermediate Technology Publications. London.

Daly, Herman. 1990. *Steady-State Economics*. Island Press. Washington, D.C.

Foster, J., J. Greer and E. Thorbecke. 1984. A class of decomposable poverty measures. *Econometrica*, 52: 761-766.

Hahn, F.H. and R.C.O. Matthews. 1964. The theory of economic growth. *Economic Journal*, December, 779-902.

Hamberg, David. 1966. Savings and economic growth. *Economic Development and Cultural Change*, 17:4, 760-782.

Hart, Gillian. 1978. Labor Allocation Strategies of Rural Javanese Households. Ph.D thesis, Department of Agricultural Economics, Cornell University. Ithaca, NY.

Hart, Gillian. 1986. *Power, Land and Livelihoods: Processes of Change in Rural Java*. University of California Press. Berkeley.

Herring, Ronald J. 1983. *Land to the Tiller: The Political Economy of Agrarian Reform in South Asia*. Yale University Press. New Haven, CT.

Krugman, Paul. 1995. *Development, Geography and Economic Theory*. Massachusetts Institute of Technology Press. Cambridge, MA.

Kumar, Shubh. 1977. Composition of Economic Constraints on Child Nutrition: Impact of Maternal Incomes and Employment in Low Income Households. Ph.D. thesis, Division of Nutritional Sciences, Cornell University. Ithaca, NY.

Kuznets, Simon. 1966. *Modern Economic Growth: Rate, Structure and Spread*. Yale University Press. New Haven, CT.

Leibenstein, H. 1966. Allocation efficiency vs. 'x-efficiency.' *American Economic Review*, June, 392-415.

Leibenstein, H. 1976. *Beyond Economic Man: A New Foundation for Microeconomics*. Harvard University Press. Cambridge, MA.

Lewis, W. Arthur. 1965. A review of economic development. *American Economic Review*, May, 1-16.

Morgan, T. 1969. Investment vs. economic growth. *Economic Development and Cultural Change*, 17:3, 392-414.

Narayan, D., with R. Patel, K. Schafft, A. Rademacher and S. Koch-Schulte. 2000. *Voices of the Poor: Can Anyone Hear Us?* Oxford University Press. New York.

Narayan, D., R. Chambers, M.K. Shah and P. Petesch. 2000. *Voices of the Poor: Crying Out for Change*. Oxford University Press. New York.

Ormerod, Paul. 1998. *Butterfly Economics: A New General Theory of Social and Economic Behavior*. Pantheon. New York.

Singer, Hans. 1966. The notion of human investment. *Review of Social Economy*, March, 1-14.

Uphoff, Norman. 1996. *Learning from Gal Oya: Possibilities for Participatory Development and Post-Newtonian Social Science*. Intermediate Technology Publications. London.

Uphoff, Norman and Milton J. Esman. 1974. Local Organization for Rural Development: Analysis of Asian Experience. Rural Development Committee, Cornell University, Ithaca, NY. Report of comparative research project for the Asia Bureau, USAID.

Uphoff, Norman and Milton J. Esman. 1983. Comparative Analysis of Asian Experience with Local Organization and Rural Development. In *Rural Development and Local Organization in Asia*, Volume III: South East Asia, Ed. Norman Uphoff, 263-399. Macmillan. New Delhi, India.

Uphoff, Norman and Warren F. Ilchman. 1972. *The Political Economy of Development*. University of California Press. Berkeley.

World Bank. 2000. *World Development Report 2000-2001: Attacking Poverty*. Oxford University Press. New York.

24 Microfinance, Poverty Alleviation and Improving Food Security: Implications for India*

Richard L. Meyer

CONTENTS

INTRODUCTION

Analysts are becoming increasingly aware that microfinance can play multiple roles in reducing poverty and improving food security for poor people. This chapter discusses these roles and applies them to India. It begins by summarizing the changes in perceptions about poverty reduction that have occurred during the past couple of decades. Then there is a brief discussion of the relationship between finance and food security. The following section considers microfinance as a win–win proposi-

* I acknowledge with appreciation the information and comments received on an earlier draft from Dale W Adams, Hema Bansal, Nimal Fernando, P.B. Ghate, Brigitte Klein, Geetha Nagarajan, V. Puhazhendhi, Shubhankar Sengupta, Girija Srinivasan, N.Srinivasan, Mather Titus and Norman Uphoff. However, the conclusions and any remaining errors are my responsibility alone.

tion in the provision of financial services. This is followed by a discussion of microfinance in India, noting important strengths and weaknesses of current policies and programs. The concluding section outlines ways in which microfinance could be strengthened to improve its contribution to poverty alleviation and food security in India.

CHANGING PERCEPTIONS OF POVERTY AND FINANCE

Historically, poverty has been viewed mostly as a problem of the poor earning too little income, consequently consuming too little to attain a socially acceptable standard of living and possessing too few assets to protect themselves against unforeseen problems. Poverty-alleviation strategies, therefore, included employment creation, skills development and, occasionally, redistribution of assets from the rich to the poor.

Technological change for small farmers has been a part of most rural poverty programs. Improving access to financial services, especially credit, has also been viewed as an important weapon in the arsenal to fight rural poverty. As shown in Table 24.1, granting production loans to small farmers was viewed as a means to augment food production under the now discredited "directed credit" approach to

TABLE 24.1
Changing Perceptions of Poverty and Finance

Poverty	Finance	Expected Results of Finance
Narrow view:	Single role:	Production and investment:
Income/consumption	Small production loans for food	Virtuous circle of investment,
Assets	production	production, income,
		consumption, savings and
		investment
Broad view:	Multiple roles:	Multiple results:
Income/consumption	Loans for wider uses, leasing,	Virtuous circle of investment,
Assets	savings, insurance,	production, income,
Vulnerability	payment/money transfer and	consumption, savings and
Health	financial intermediation	investment
Education		Consumption smoothing (food
		security)
Voicelessness		Capacity to bear risk
Powerlessness		Empowerment
Food insecurity		Education
		Health
		Nutrition
		Contraceptive use and other
		social impacts

small-farmer development pursued by many donors and governments in developing countries during the 1960s and 1970s.* These loans were expected to contribute to a virtuous cycle: credit would increase production and raise incomes, permit greater consumption and savings and lead to further investment. The borrowing farm households would gain and so would society because of greater food supplies. In this traditional view, finance was largely limited to the role of augmenting production through loans to producers, often at concessional interest rates.

During the past two decades, analysts concluded that this traditional view of poverty was too narrow and simplistic. A recent example of the expanded view of poverty is found in the World Bank's *World Development Report 2000/2001*. It notes that, not only do the poor lack income, they lack adequate food, shelter, education and health. They are vulnerable to ill health, economic dislocation and natural disasters. They are often exposed to ill treatment by the state and are powerless to influence decisions that affect their lives.

Paralleling this change in perceptions about poverty has been an evolution in understanding the role of finance in development. As noted in Table 24.1, financial services are recognized now as playing multiple roles in development so that improved access can have a far greater and more comprehensive impact on poor households than previously assumed. In addition to the virtuous production and investment cycle, financial services can smooth consumption and improve food security. Moreover, supplying financial services to women may be an especially important way to empower them to play more active economic and social roles in society. As the microfinance industry matures, many microfinance institutions (MFIs) are redesigning their financial products and services so they make a stronger contribution to these broader poverty impacts.

FINANCIAL SERVICES AND FOOD SECURITY

Critics of the directed credit approach frequently argue that an overemphasis on lending distracted attention from the fact that poor households need — and increasingly demand — a variety of financial services including savings and insurance. A recent statement of these arguments, emphasizing how financial services affect household food security, is found in a monograph from the International Food Policy Research Institute (Zeller et al., 1997).

The authors discuss three pathways or channels through which financial services affect food security. The first is through the familiar poverty-reducing path of improved income generation. The effects are expected to be twofold. First, there is the traditional argument that loans can temporarily enhance a household's productive human and physical capital. Second, savings and credit services can increase a household's risk-bearing potential, leading to the adoption of more risky but potentially more profitable income-generating activities. The profitability and mix of

---

* The flaws in the "supply-led" approach to agricultural credit, which dominated thinking in the height of the Green Revolution, ultimately contributed to its demise. These flaws are summarized in studies such as Adams et al. (1984), Meyer and Larson (1997) and Meyer and Nagarajan (2000).

productive activities may change, leading to increased income that contributes to the virtuous production and investment cycle.

In the second pathway, finance contributes to poverty reduction by decreasing the rural household's cost of self-insurance. Improved access to credit, savings and insurance services can induce changes in household assets and liabilities. For example, the holding of "precautionary savings" in the form of nonremunerative physical assets, such as cash, jewelry, staple foods and livestock, may decline. The emergency sale of productive assets at low prices may decrease and the storage of crops for later sale at higher prices may rise. The importance of more expensive informal financial services may decline. Reductions in the cost of stabilizing consumption will release resources to finance more consumption and investment.

The third pathway, consumption credit, represents the greatest divergence from the narrow production- and investment-oriented view of finance. Households attempt to smooth consumption over time by adjusting their disposable income. In the event of adverse shocks such as bad weather, accidents and illness, rural households use traditional consumption-smoothing measures such as the emergency sale of assets, depletion of stocks and inventories and grants and loans from family and the informal sector. Formal credit, savings and insurance services may help households smooth consumption so they use fewer traditional methods, which are often inefficient and bind households into unproductive social relationships that discourage savings and wealth accumulation.*

Financial policies will be more beneficial for the poor in developing countries if they pursue all three pathways rather than expanding only production credit. Poor households may use loans in immediately productive ways as envisioned in the narrow view so that incomes and food supplies rise, but they may also use loans to finance education or health expenses or to smooth consumption. Savings and insurance services must be emphasized and savings programs for the poor recognizing that liquidity and transaction-cost considerations may be more important than interest rates should be designed. Financial institutions that supply multiple financial services have a better chance of alleviating poverty along its multiple dimensions than those that focus exclusively on loans.

MICROFINANCE: A WIN–WIN PROPOSITION

Microfinance refers to the provision of financial services, usually in the form of small financial transactions, to people who usually fall outside the reach of formal finance. They tend to be the poorest members of all societies. Commercial banks usually ignore them to avoid high transaction costs incurred in servicing small loans and savings deposits. Moreover, most of the poor do not possess assets normally demanded as collateral, and they are perceived as being too risky to be granted loans.

The microfinance sector experienced considerable growth during the 1990s. The World Bank reported that 206 institutions surveyed in September 1995 held about

---

* Townsend (1995) summarized the literature that reveals how traditional risk-sharing methods in developing countries are incomplete and inefficient and why formal markets for credit and insurance services may improve welfare.

US$7 billion in outstanding loans made to more than 13 million individuals and groups (Paxton, 1996). This was an admittedly incomplete inventory and the number of microfinance institutions and the volume of lending and savings mobilization have grown since then.

Bangladesh is one of the pioneer microfinance countries. It was estimated that, by the end of 2000, more than 1,000 MFIs were operating in the country. As of December 2000, 585 MFIs reported loans outstanding to almost 8 million borrowers with a total amount of over US$400 million (Credit and Development Forum, 2001). Most of these MFIs serve only rural areas. The growth in their lending has more than offset the fall in traditional agricultural lending of the commercial and agricultural development banks. The MFIs have reached the scale where they may have an important influence on the country's rural poverty (Meyer and Nagarajan, 2000).

The microfinance industry consists of nongovernmental organizations (NGOs), village banks, credit unions, specialized banks for the poor and commercial banks. It is difficult to generalize about such a heterogeneous group, but an important segment of the industry, especially in Latin America, operates on the so-called win–win proposition: when the poor can obtain financial services otherwise unavailable to them and benefit from these services, they are willing and able to pay high interest rates and fees that permit the MFIs to be sustainable (Morduch, 2000). Therefore, the MFIs that apply good banking principles are also expected to be those that alleviate the most poverty.

Although the industry is beginning to offer broader financial services, it is still dominated by MFIs that specialize in lending. They target the poor but, unlike the traditional small-farmer lenders, they do not impose strict restrictions on the use of loan funds. They acknowledge the fungibility of money and recognize that borrowers will use funds to earn the highest economic return or meet their greatest needs, especially emergencies, consumption smoothing and medical expenses. Therefore, the MFIs educate clients to be prudent and to expect that they will have to repay their loans regardless of how the funds are spent.

Incentives such as interest rebates and automatic access to new larger loans encourage clients to repay on time. Many MFIs use some form of joint-liability, group-lending procedure so that group members screen out those who are less likely to repay and apply peer pressure on those delinquent in payments. The most successful MFIs recover most loans and experience loss rates of only 1 or 2%, a record far superior to most financial institutions under the directed agricultural credit paradigm.

MFIs are evaluated using three objectives. The first is outreach, to reach a large number of poor clients. The second is long-term sustainability, so the MFI can continue to provide financial services after any initial government or donor start-up funds have been exhausted. The third is impact on the clients served, improving incomes sustainably and alleviating poverty.

There are complementarities among these objectives. For example, MFIs that serve a large number of clients may achieve economies of scale that contribute to their sustainability. But there may also be trade-offs. If MFIs try to serve very poor clients, i.e., improve their depth of outreach and impact on the poor, average loans and savings deposits will be small and costs will be high, so sustainability may be difficult to achieve (Conning, 1999). This has prompted some analysts (e.g., Hulme and Mosley, 1996) to fear mission drift because MFIs that strive for sustainability may avoid serving poorer clients.

The objective of institutional sustainability is one of the most fundamental changes in the paradigm shift from directed agricultural credit to market-oriented microfinance. While this objective is difficult to achieve, a few "flagship" institutions are highly successful. For example, the *unit desa* system of Bank Rakyat Indonesia (BRI) serves several millions of rural clients and has been so successful that each year the equivalent of millions of dollars in profits and surplus funds are transferred to bank headquarters to finance urban operations (Parhusip and Seibel, 2000). BancoSol in Bolivia is an example of an NGO that successfully converted itself into a specialized bank for the poor and currently manages a portfolio of over US$75 million.

Nevertheless, less than 1% of all MFIs have reached the ability to cover all costs and mobilize funds on a commercial basis. That is one reason that some MFIs are beginning to mobilize voluntary savings aggressively rather than rely exclusively on donors or government funds. Some are experimenting with leasing, insurance and other financial services to attract more clients and increase revenues. By offering more services desired by the poor, MFIs will also contribute more to poverty alleviation and food security.

MICROFINANCE IN INDIA

Unlike neighboring Bangladesh, India has not been a leading country in microfinance despite massive rural poverty. Until reforms were recently introduced, it was a prime example of a country that aggressively pursued the directed-credit strategy for rural and agricultural finance. Credit policies were designed to cater to the rural population, a major voting block for political parties (Meyer and Nagarajan, 2000). Poverty alleviation has been a major political appeal since the late 1970s and the expansion of formal finance to serve the poor has been perceived as an important strategy to achieve it. The government has intervened in the banking sector with policies for setting up bank branches in rural areas, mandatory lending quotas and below-market interest rate loans for the priority sector, waivers of loan principal or interest (referred to as loan melas) and recapitalization and refinancing of loss-making financial institutions.

The bank branching policies contributed to the expansion of commercial banks in rural areas and loans to the rural population. The average population covered by each bank branch declined from 65,000 in 1969 to 15,000 in 1998. Agricultural cooperatives and regional rural banks were also created to help improve rural access to financial services. But directed credit, loan waivers, subsidies and bailing out nonperforming institutions weakened the financial system and contributed to a breakdown in loan repayment discipline. By 1994, 196 regional rural banks had accumulated arrears of R. 13 billion (about US$5 billion) (Mosley, 1996). Overdue loans on some categories of rural loans were as high as 94% in 1997 (NABARD, 1997). The weak financial sector has not performed financial intermediation satisfactorily, nor has it contributed to efficient rural development (Vyas, 2001).

In 1978, the government launched the Integrated Rural Development Program (IRDP) intended to alleviate poverty. It provided loans to the rural poor through the banking system at subsidized rates. In addition, a cash subsidy was paid to borrowers equal to 25% of the total cost for projects financed for small farmers, 33% for projects for agricultural laborers and 50% for lower-caste persons. These subsidies were disbursed when the loans were disbursed. Loans made by commercial banks were subject to a nominal interest rate ceiling of 12% per year and were made for a maximum of 3 years. Serious questions about its impact exist, however, and loan recovery had fallen to only 31% by 2000 (Vyas, et al., 2001). It was finally discontinued in 2000 and replaced by the Golden Jubilee Rural Self-employment Programme (SGSY). Subsidies are now deposited with the lending bank and released only after the client repays the loan. The interest rate should not exceed the bank's prime lending rate.

In 1982, the government created the National Bank for Agricultural and Rural Development (NABARD) as an apex bank to provide credit for agricultural and rural development. Besides its role in refinancing loans made by financial institutions in rural areas, NABARD has emerged as a major institution to support institutional development, to regulate and supervise rural financial institutions and to develop and implement programs for channeling credit, often at subsidized interest rates, into agricultural and rural activities.

In 1992, the government began to move away from directed credit, to liberalize the financial sector and to strengthen it by reorienting banks and other financial institutions toward a market-based financial system by increasing competition and improving the quality of services. Microfinance programs began on a large scale in the early 1990s and they are considered essential for the provision of working capital and financing nonfarm activities for the rural poor.

Three major microfinance approaches have emerged in the country. First, some MFIs that specialize in serving the poor have emerged. They make retail loans directly to the poor and wholesale loans to NGOs that specialize in reaching the poor. Second, several apex organizations have been created that wholesale funds to NGOs and nonbank institutions that lend to the poor. These apex organizations also support institutional development through training and technical assistance. Third, the government has undertaken a massive program through NABARD to create and link self-help (SHG) groups to banking institutions. A spinoff of this approach is that some banks have developed their own linkage programs with SHGs.

NABARD began to support microfinance in February 1992 with a pilot project to test the SHG bank linkage approach, set up to cover 500 SHGs (Wright, 2000).* The intent was to utilize the large existing banking network rather than create special MFIs. An SHG is a small homogeneous group of rural poor coming together to save small amounts regularly and mutually contribute to a common fund to be lent to individual members per group decisions (Nanda, no date). Often some organization,

---

* The Asian and Pacific Regional Credit Association (APRACA), with financial support from the German technical assistance agency GTZ, began a program in Asia in the late 1980s to promote SHGs as financial intermediaries (Kropp et al., 1989). Indonesia was the first country to begin large-scale field activities in the project.

usually an NGO, forms the group and links it with a bank as part of a broader package of activities implemented by the NGO in the village. Group formation may take 6 months to a year and representatives selected by the group members are responsible for management. The NGOs can simply be facilitators in linking the groups with financial institutions or can act as financial intermediaries themselves.

Banks can lend to the NGOs or directly to the SHGs; increasingly they have chosen to lend to the SHGs using NGOs as facilitators. A few experiments are occurring with banks hiring promotional agents to form groups. The banks are now authorized to count SHG loans against their required lending to priority sectors. Most loans are scheduled to be repaid over a 2- or 3-year period. NABARD refinances up to 100% of the loans made by banks to the groups at the interest rate of 6.5% (recently raised to 7%). On June 1, 1999, the rate the banks charge the NGOs or the SHGs, the rates that NGOs charge the SHGs and the rates the SHGs charge their members were completely deregulated (NABARD, 2000).* Still, many banks are reported to be skeptical of the SHG approach because of their past poor experience with lending in rural areas.

NABARD reported that, as of March 31, 2001, more than 260,000 SHGs involving over 4 million poor families had been linked to banks. In excess of 70% were concentrated in five of the wealthier states in the south, with more than 85% of the groups composed of women. Over 750 NGOs and 14,000 branches of 318 banks were associated with the program. Banks serving some 213,000 SHGs received refinancing from NABARD in a cumulative amount of Rs. 4.8 billion (US$100 million). Almost 87% of the loan portfolio was financed by NABARD, but that percentage is expected to fall in the future as many banks have excess liquidity, a fact that calls into question the rationale for the refinance facility.** NABARD has set a target of reaching a million groups covering 100 million rural poor by 2008 (Nanda, 1998).***

NABARD also provides grants for strengthening NGOs and operates training programs for bankers, NGOs and SHGs.**** It led a task force that in October 1999 recommended a series of measures to strengthen microfinance, including a regulatory framework for MFIs, and equity and start-up capital and capacity building funds for

---

* NABARD Annual Report 1999-2000, reported in http://www.nabard.org/annr2000/chap6.htm, April 29, 2001.
** To date, most of the banks that have granted loans are commercial state banks and Regional Rural Banks. The share of loans made by District Cooperative Central Banks has been increasing (Klein, 2001).
*** Another increasingly important source of refinance for MFIs is the Small Industries Development Bank of India (SIDBI), which started a microcredit lending scheme in 1994 that evolved into a Foundation for Micro Credit in 1998. It provides loans to well-managed MFIs for on-lending to poor groups and individuals, with an emphasis on women taking up micro-industrial activities. The MFIs can lend to smaller MFIs and NGOs or directly to SHGs and individuals. The minimum MFI loan is Rs.1 million (about US$24,000) and the maximum loan to a single borrower or SHG member must not exceed Rs. 25,000 (approximately US$600). The nominal annual interest rate on the MFIs is 11% and they are to charge market rates. As of March 2001, the total cumulative disbursement to 169 MFIs exceeded Rs. 810 million (US$17 million). It also supports training and capacity building of NGOs.
**** The NABARD annual report for 2000–2001 indicates that the cumulative grant support through March 2001 totaled Rs. 462,000 (US$20,000) provided to 198 NGOs that supported almost 37,000 SHGs. The German aid agency GTZ is providing financial assistance for capacity building.

institutions engaged in microfinance. Much remains to be done to implement these recommendations.

The emphasis of microfinance in India has been to expand outreach and· disbursements rather than to create sustainable institutions or improve impact on the clients. NABARD claims that the SHG linkage program has made a great impact on participating members, but few studies are available that evaluate how it operates in the field or its contribution to poverty alleviation and food security. A case study of NGO–bank linkages in one district in the state of Gujarat in 1997–98 found that, at that time, many NGOs and banks were not aware of the program (Bansal, 1998). That situation has probably changed in recent years due to the massive expansion of SHGs sponsored by NABARD and other organizations. The NGOs studied were engaged in a variety of village-level development activities. Some had organized savings groups, but there was little borrowing reported from banks. Some SHGs had started to lend their own funds and reported high recovery. Little information was provided on the sustainability of the operations.

Two NABARD studies attempted to evaluate impacts on the banks and households that participate in the SHG linkage program. In 1997, Puhazhendhi (2000) studied SHGs in Tamil Nadu, where almost 80% of the 427 SHGs in the state were linked to just eight banks. Some banks had begun to organize groups on their own but most were organized by NGOs. The study focused on four NGOs responsible for over 90% of the groups. Seventy SHGs were sampled: 80% were women's groups; average group size was 19; and two thirds of the groups were 3 to 4 years old. About half of the group members were reported to be landless laborers with seasonal employment so they were considered to be among the poorest of the poor.

The total funds managed by the SHGs were composed of savings, donations from NGOs and bank loans. Annual savings per group member rose from about Rs. 500 (US$12) in the groups of 1 to 2 years of age, to over Rs. 1,000 (US$24) for groups 4 years of age or older. The average size of loan granted per member rose over time from nearly Rs 1,000 in the first year to almost Rs 4,000 (US$100) when groups were 4 years old or older; therefore, the multiple of loan size to savings grew from 1.64 to 3.75. Length of loans grew from 2 to 4 months to 8 to 12 months over the same period, so members borrowed fewer loans per year as groups aged. Interest rates ranged from 3 to 5% per month initially but fell to 2 to 3% per month after about 3 years.

A study of the banks showed that, for 1996–97, the average loan granted to an SHG was small at about Rs. 9,700 (US$230), or about Rs. 500 (US$12) per member for a group of 19 members. Almost two thirds of the loans fell into the range of Rs. 5,000 to 10,000 (US$120-240). The repayment period was 3 years, but over 70% of the SHGs repaid early and received second and third loans of larger sizes. Interest rates charged by banks ranged between 12 and 14% annually, so SHGs earned a significant spread between their cost of funds and the rates charged to their members.*

Two important benefits were reported for the banks that made SHG loans compared with other types of loans. First, the recovery rate in 1996–97 for SHG

---

* No information was provided on the rates of interest that the SHGs pay on member savings.

loans was reported to be over 90% compared with a range of 37 to 68% for agricultural loans generally and 31 to 43% for IRDP loans. Second, a study of one commercial bank revealed that the transaction costs of making the first SHG loan were only 60% as high as for making IRDP or general loans, while the costs for the second SHG loan were only 43% as expensive. No data were reported on average loan sizes, but if loan sizes were roughly comparable, the cost per rupee of making SHG loans would have been much lower and the probability of recovery much higher. However, the banks had made little progress in streamlining operations and SHG loans were treated much like other types of loans.

The economic and social impacts on SHG members were estimated. The average annual net family income was estimated at almost Rs. 4,440 (US$105), roughly double pre-linkage income of about Rs. 2,000 (US$47). Members who were agricultural laborers were able to undertake income-generating activities that increased family income. Income growth was correlated with the quality of group performance measured on a multi-variable scale. Groups that performed better had a higher growth of member income. However, no information was provided about how income growth was measured, nor was income growth reported from any comparative control group; therefore, the role of the SHGs in causing these income changes is open to question. Several positive social impacts were also noted, including greater consumption of wheat, rice and vegetables and a better capacity to stock food for the lean season.

A more comprehensive impact evaluation was conducted by Puhazhendi and Satyasai (2000) of 223 SHGs sampled in 11 states. Two to three members were interviewed from each household for a total of 560 members. Roughly a third of the SHGs were drawn from each of the three models: a) groups developed by banks, b) groups with NGOs as only facilitators and c) groups with NGOs as financial intermediaries. Impact was measured by comparing the members' pre-group situation (apparently established by member recall) compared with the post-linkage situation of 1999. Poverty lines were established using state government standards for monthly consumption levels to evaluate whether member households moved out of poverty during the time of membership.

The characteristics of the groups were similar to those found in the Tamil Nadu study. Average group size was 16 members and agricultural laborers represented the largest number of members, followed by small farmers (2.5 to 5.0 acres), then marginal farmers (fewer than 2.5 acres). Almost 35% reported to be engaged in a mix of farm and nonfarm activities and 20% reported only nonfarm activities. The groups formed by banks tended to be somewhat smaller than the other two types of groups, but even so, they saved significantly larger amounts and received larger loans. The explanation may be that bank-organized groups are encouraged to emphasize financial services rather than other developmental activities, or perhaps persons more interested in obtaining financial services chose bank-promoted groups. When banks organize the groups and are able to monitor them more closely, they may be willing to lend more than to groups associated with NGOs.

The total size of the loan portfolios grew with the age of the groups and the share of income-generating to nonincome-generating loans rose over time. However, because of the fungibility of money, it is impossible to know for certain how loans were actually used. These data may simply reflect reporting bias if income-generating loans are considered by the banks and NGOs to be more desirable. Annual interest rates tended to fall in the range of 12 to 24% and the term of most loans was 6 to 12 months. The repayment rate for loans received from all sources rose from about 84 to 94%, with the most dramatic increase noted for bank loans.

This evaluation concluded that the SHG linkage program had significant economic and social impacts on members. For example, member households were reported to experience a more than 70% increase in assets, a more than tripling of annual savings and almost a doubling of annual borrowing. Average net household income reportedly rose by a third compared with pre-SHG levels and the greatest increase was observed for the groups with NGOs as facilitators. Perhaps the assistance provided by NGOs in the form of services other than finance contributed to this difference.

The proportion of members below the poverty line before joining the SHG (42%) fell to half that level at the time of the survey. The proportion of members that rose out of poverty was higher if they engaged in off-farm activities, had smaller families and had higher levels of income before joining. Estimated household monthly consumption levels rose by 24%. Total food expenditures rose, but, following Engel's law, more slowly than did other categories such as expenditures for health and clothing. The social impacts included improved feelings of confidence and self-worth, a reduction in social problems such as wife beating and better access to improved health and sanitary services. Surprisingly, fewer than a quarter of the members reported receiving training and that proportion was higher for both NGO models than the bank model.

Although promising, these results must be interpreted with caution. The evaluation did not address possible problems of self-selection bias, measurement errors in using recall data and the lack of a control group to help determine whether the changes reported for the members should be attributed to the SHGs rather than to other factors.

Other studies in India have focused on alternative microfinance models and the operational performance and sustainability of MFIs. One study concluded that few Indian MFIs had achieved great success in their microfinance operations. Quiñones (1997) evaluated ten NGOs considered among the best in outreach and sustainability. However, few could cover their operating costs and none could operate completely free of subsidies.

One of the largest MFIs in the study was SEWA, the Self-Employed Women's Association, which began in 1972 as a trade union.* It currently has about 320,000 members with a trend toward attracting more rural members in recent years. SEWA organized a cooperative bank for members in 1974, which makes loans, mobilizes savings and utilizes mobile vans and field agents for the daily collection of loan payments and savings deposits. It emphasizes savings and considers making a loan only after observing a member's saving behavior. About 70% of the bank's members are now urban. It also provides financial services to savings and credit groups in

---

* The information reported here was obtained from the SEWA Web site at the end of April 2001. For more on SEWA, see Rose (1992) and Chen and Snodgrass (2001).

Gujarat state. As of March 1999, there were 298 groups with over 8,500 members that had collected over Rs. 2.5 million (US$60,000) in savings and had Rs. 1.1 million (US$26,000) in loans outstanding (Chen and Snodgrass, 2001). The bank currently reports 130,000 depositors and shows profits, but Quiñones found that, in 1997, it covered operational but not financial costs. In addition, SEWA organizes rural women into savings groups and teaches them how to manage these funds. SEWA has also initiated various types of life, accident, health and other insurance products for its members to reduce their vulnerability and increase their ability to withstand negative shocks. Several of these products are offered in conjunction with established insurance companies.

One of the most rapidly growing MFIs included in the Quiñones study was SHARE (Society for Helping, Awakening Rural Poverty through Education) that provides credit and savings services in over 500 villages in the state of Andhra Pradesh. During the year 2000, its active borrowers grew from 29,000 to 59,000 while the number of savers climbed from 37,000 to 72,000. By September 2001, the number of savers reached 104,000 and the borrowers totaled almost 70,000. It had about US$5.8 million in outstanding loans at the end of September 2001. Although it aggressively mobilizes savings, the savings balance was just over US$2.9 million, reflecting the organization's continuous dependence on outside resources*

SHARE follows the Grameen Bank methodology of group lending in which borrowers, mostly rural women, are organized into five-person groups, then seven groups are organized into centers. The centers meet weekly to discuss matters related to loan approval, disbursement and repayment structure. Loans are granted on a graduated scale starting with a maximum of Rs. 4,000 (US$95) for the first loan to a maximum of Rs. 8,000 (US$190) in the fifth year. Additional seasonal loans, which range from Rs. 3,000 to Rs. 6,000 (US$70-140) are available in the 5th year. Clients are eligible for a housing loan of Rs. 12,000 (US$285) after the second year. All loans are to be repaid in 50 equal installments. Clients are required to deposit Rs. 5 (US$0.12) per week in a compulsory savings program and 5% of the loan amount is retained in a group fund. The members were generally poorer than nonmembers in the villages, suggesting that it has good depth of outreach (Sharma et al., 2000)**

Recent results reported by an organization that has rated some of the best MFIs in South Asia revealed a pattern of financial weakness. The results were heavily influenced by India because of the 53 MFIs rated, 44 MFIs were in India, four in Nepal and three in Bangladesh (M-CRIL, 2001). They had been operating for an average of 5.9 years and employed different lending methodologies with the largest

---

* These data are reported in the newsletter *Credit for the Poor,* April 2001 and October 2001. The dependency on external funds is reflected in the information included in the December 2000 issue of the newsletter. It reported that SHARE was going to receive a soft loan as part of US$1.2 million raised by the Grameen Foundation-USA for expanding Grameen-type programs in India. It also reported on a meeting held with Steven C. Rockefeller to discuss mobilizing foreign funds to expand microlending rapidly in India, with 70% of the on-lending funds to come from commercial banks in India at market interest rates.

** A preliminary report of an impact study conducted of SHARE clients was summarized in *Credit for the Poor,* Vol. 30, April 2001. It reported that 38% of the SHARE clients moved out of poverty after 3 or 4 years in the program. Another 22% reported no change in status, however. No details were presented on the methodology used in the study in order to evaluate the robustness of the results.

number, 31, using self-help groups. Ten employed the Grameen model and ten used individual lending.

Total outreach and coverage was almost 700,000 persons, of which the SHGs served about 50%. About 300,000 were borrowers, but the total volume of lending was only about US$23.5 million, less than one third of the portfolio of BancoSol in Bolivia and much smaller than just one of the large MFIs in Bangladesh. Savings mobilization, including amounts generated by SHGs but not deposited with MFIs, totaled only US$12.3 million, in spite of the supposed emphasis on savings and self-help. Moreover, nonbank finance companies and other MFIs face constraints in being denied authorization to mobilize voluntary savings. Obligatory savings are permitted because they can be interpreted as ownership shares and can only be withdrawn when the savers' loans have been repaid. The total savings represented 34% of the amount lent to clients, reflecting the great dependency on nonmember sources of funds for lending.

Low levels of productivity were also evident as the SHG programs averaged only 50 clients per staff member compared with 94 clients for Grameen programs and more than 100 for the individual lenders. This was due to the heavy staff input required during the start-up phase of some SHGs and the social rather than business orientation of many SHGs in India. The sample average of 15% of the portfolio at risk (defined as 60 days overdue) was a cause for concern and the SHGs averaged a much larger 29% of portfolio at risk. The average operating cost ratio of 23% exceeded the average portfolio yield of 17%, so this means that most MFIs are not able to cover their costs. The average return on assets was a negative 4.8% showing how far they are from commercial viability.* The report concluded that the microfinance sector in South Asia has, generally speaking, a long way to go before it can achieve any form of commercial viability and can become a dynamic and sustainable component of the region's poverty reduction efforts.

A new Indian MFI, BASIX, is taking a different approach to microfinance. It was organized in 1996 as a group of financial services and technical assistance companies, one of which is a nonbanking finance company.** Recently, it established a local area bank with a license from the Reserve Bank of India to offer full-fledged banking services. It is pursuing a more commercially oriented strategy than most MFIs, but faces challenges from the subsidized credit provided by the formal financial sector. Rather than specifically targeting the poor, BASIX believes that making larger loans to the nonpoor in rural areas will create employment for the poor. By the middle of 2000, it had over 12,000 borrowers with more than US$2.5 million in loans outstanding with an average loan size of over US$200.

BASIX operates in the states of Andhra Pradesh and Karnataka and offers a variety of group and individual loan products granted through self-help groups, intermediaries such as trader organizations and agroprocessing firms and NGOs. About half the loans are for agricultural activities. Interest rates range from 15 to

---

* The report noted that many of these performance measures are similar to the average results shown for the MFIs included in *The MicroBanking Bulletin*, but the peer group of large Asian MFIs reported in that source have more positive results than were shown for these 51 MFIs.
** Bhartiya Samruddhi Finance Limited.

24% and vary in term from 1 to 5 years. Most loans require monthly payment schedules. Almost 90% of the assets are financed from commercial sources of funds.* This, plus rapid growth and competition, has prevented the organization from making rapid progress in achieving operational and financial self-sufficiency. Interest rates on some products have been reduced due to competition. There is pressure to improve efficiency by increasing the client load per loan officer and to improve on-time loan recovery. Commercial banks, subsidized NGOs and government programs are increasingly attracting self-help groups by providing loans at cheaper rates (Nagarajan, 2000).

Although most Indian MFIs seek cheap foreign or government funds, in part to cover start-up and institution building costs, there is no serious shortage of loanable funds for MFIs that meet the standards established by apex institutions. A total of four apex institutions, including NABARD, now provide funds to MFIs. Yet, in spite of such support, the country has not yet succeeded in achieving large outreach or in creating vibrant and sustainable MFIs. Most are far from reaching the stage where they can operate on a commercial basis. No single "flagship" institution has emerged to demonstrate how to achieve large-scale outreach and sustainability. Microfinance is not yet making much of an aggregate impact in the country and is far from reaching its potential in contributing to poverty reduction and food security.**

IMPROVING MICROFINANCE IN INDIA

India is attempting to expand microfinance and it is logical that the country would look to its large bank network as the primary way to supply microfinance services. That approach, however, continues the country's long tradition of a top-down, non-market strategy of mandates, quotas and refinance funding to expand access to financial services for priority sectors.

The microfinance strategy also continues the strong bias toward emphasizing targets for achievements, i.e., outreach, rather than stressing financial efficiency and self-sustainability. Many policymakers are encumbered by the outdated view that the poor cannot save and need subsidized loans so lending is often emphasized relative to savings mobilization and other financial services. Recent government poverty initiatives threaten to undermine commercial microfinance by providing more subsidies for the poor. Although some liberalization has occurred, the financial system is still highly constrained by governmental regulations. State-owned or -controlled banks and cooperatives thwart the development of nonbank financial institutions and other types of MFIs that attempt to operate on a commercial basis (Mahajan, 2000). The following changes should be considered in India's microfinance strategy to achieve three major objectives: increase outreach and coverage, both total numbers of poor served as well as the depth of poverty of the clients;

enhance the sustainability of MFIs and reduce dependency on subsidies and improve impact on clients.

ADJUST INTEREST RATES AND THE USE OF SUBSIDIES

The interest rates charged on most microloans are too low to cover lending costs and risks, as shown by the M-CRIL analysis (2001) of MFIs and by experience in other countries. Low interest rates undermine financial institutions, destroy institutional sustainability, discourage MFIs from trying to serve the poorest of the poor and constrain the emergence of market-oriented MFIs and the innovations that arise from increased competition. Many banks do not need NABARD funding and might be induced to invest some of their excess liquidity in microfinance if the rates of return were more attractive.

The remaining regulatory and social barriers to charging cost-covering interest rates need to be removed and MFIs should utilize the flexibility they already have to set realistic rates. It is interesting that rural people often set high rates for loans made from savings they mobilize themselves, as found in the study of Tamil Nadu SHGs. They understand the need to compensate savers and to ration credit use through proper pricing.* In the long run, interest rates will fall as MFIs expand, become more efficient and reach economies of scale. Market forces will eventually determine the level of rates appropriate for the poorest segment of the market.

Governments have an obvious interest in providing subsidies to aid the poor. But alternatives other than financial institutions should be used as channels for such subsidies to avoid confusing the poor about the difference between grants and loans. Subsidies for MFIs would be better used to cover start-up costs rather than being passed on directly to clients in the SHGs. Financial discipline within the SHGs could be damaged if this subsidy issue is not handled with care (Sheokand, No date).

BROADEN THE SCOPE OF FINANCIAL SERVICES

The poor need and demand financial services other than just loans. The poor, as well as the rich, value savings, insurance and other financial services. Loans are useful for those with good investment alternatives, but secure savings services are valued by everyone regardless of their investment opportunities. Flexible access to voluntary savings is especially important to the poor so they can smooth household consumption in emergencies. With its large network of regulated financial institutions, India should be a leader rather than a laggard among low-income countries in savings mobilization and in offering secure savings outlets for the poor. Postal offices also offer a vast untapped network to mobilize savings. Unfortunately, government policies continue to emphasize lending and treat obligatory savings largely as a way to screen potential borrowers and reduce lending risks, rather than viewing savings instruments as important in themselves.

---

* In a program to introduce participatory irrigation management through farmer organizations in Sri Lanka, when farmers set up their own savings and credit scheme, they decided on a loan interest rate of 16% per month. This sounds usurious, but farmers had paid about 25% per month to informal money-lenders and wanted to build up their capital quickly so they could displace the latter (Uphoff, 1996).

In group-based financial services, the logical tendency is for all members to be offered the same product: loans of the same size and maturity and a fixed requirement for obligatory savings. There are advantages for MFIs in offering only a few highly standardized products in terms of lower costs and simplicity in internal control. But the Bangladesh experience has shown that standardization can contribute to client dissatisfaction and high rates of dropout (Meyer, 2002). Moreover, peer pressure to repay loans seems to wane after several loan cycles. These observations support the argument favoring more flexibility in the products offered to the poor (Wright, 2000). Some MFIs in Bangladesh and elsewhere have begun to add individual loan products and have completely discontinued making group loans. This leads to a fundamental question: should group-based financial services be viewed as just an intermediate step in the long-term process of developing sustainable individual financial services for the poor?

Evaluate the Long-Term Strengths and Weaknesses of SHGs

The APRACA-GTZ financial linkage concept* was based on the idea that indigenous self-help groups could be strengthened by becoming formally linked with formal-sector financial institutions, perhaps with the assistance of NGOs (Kropp et al., 1989). The Indian approach is more oriented toward creating and linking self-help groups of a uniform size rather than formalizing existing indigenous groups. It represents a hybrid model with characteristics borrowed from models that link indigenous groups to banks, from models that create self-governed village banks and from models that create joint-liability borrowing groups.

The concept of providing financial services through groups offers some advantages, one of the most important being the potential to reduce transaction costs for MFIs as noted in the Tamil Nadu study. Once SHGs are formed, they can serve as efficient mechanisms through which to provide training and other services. Undoubtedly, many NGOs find the linkage model attractive for this reason. NGOs have their own objectives, however, which may conflict with those of banks.** Some NGOs may object to standardized loans that treat all borrowers the same and may even ally with borrowers against the banks when they face difficulties in making loan payments.

Group-based microfinance systems have demonstrated major limitations. As noted above, the effectiveness of peer pressure as a contract-enforcement mechanism for group lending may decline after several loan cycles. All member-owned institu-

---

* See footnote 4.
** Reid (no date) describes a case where the objectives and perceptions of an NGO diverged from those of financial institutions. The case describes the evolution in the relationships between the NGO Youth Charitable Organisation (YCO) in Andhra Pradesh and the banks. At first, YCO was enthusiastic when its women's groups were successfully linked with local banks. With the assistance of NABARD, the groups mobilized savings, deposited them with banks and received loans. The relationship soured, however, when the groups became disenchanted with what they believed were excessively bureaucratic bank regulations, lack of access to their savings and the small interest margin the NGO was permitted to earn relative to the margins permitted for the banks. Eventually, YCO bypassed the financial system and borrowed from the National Women's Fund of the government. YCO is now promoting a mutually aided cooperative society with the hope it can eventually become a bank.

tions face serious governance challenges as the poor performance of many Indian cooperatives amply demonstrates. Local elites may eventually dominate the groups and monopolize access to loans at the expense of weaker members. The poorest may be systematically excluded from groups because wealthier members fear that they will not fulfill their loan contracts. Women take on huge additional burdens to receive financial services through SHGs and the potential negative aspects do not seem to be as well appreciated in India as they are in Bangladesh (e.g., Kabeer, 2001).

Reaching the poorest, who are most food insecure, may require more complex programs such as the combined food aid and credit approach being tested in Bangladesh (Hashemi, 2001). Groups face inherent instabilities because financial interests of group members may diverge over time. The most successful members may decide to drop out and seek larger individual loans, which presumably would be the preference of banks. It will be a huge and expensive task for the banks and NGOs to create, strengthen and effectively monitor and supervise the performance of hundreds of thousands of small groups to minimize potential governance and other problems and preserve the financial integrity of the SHGs. Huge costs are involved in reaching the targeted numbers of SHGs and no system is yet in place to cover these costs (Sheokand). Perhaps these resources would be better utilized if they were directly invested to meet a long-term goal of sustainable individual lending.

Other unresolved issues include who should bear the cost of NGOs and banks in forming and nurturing groups, the role and sustainability of federations of SHGs, the ability of SHGs to effectively engage in self-regulation and promotion and the potential for other subsidized enterprise development and poverty alleviation activities to undermine the self-help philosophy introduced in the linkage model.* Apex institutions and federations of SHGs are expanding, but their sustainability is uncertain and their future role in promoting and strengthening the system is unclear.

Strengthen Alternative Institutional Forms

The SHG linkage strategy is complex and represents huge principal-agent challenges. To succeed, the staff of NABARD, the banks, the NGOs and the members of the SHGs themselves must all perform their respective tasks effectively if SHG members are to receive efficient financial services. This complexity offers many opportunities for failures, inefficiencies and unproductive rent-seeking behavior. The financial layering involved means that the costs of several institutions must be covered for the entire system to be financially sustainable. Considering the enormity of the task and limited resources, the target of 1 million SHGs by 2008 seems unrealistic if they are to achieve efficiency and sustainability. Moreover, banks and other types of formal financial institutions have not been the most innovative MFIs in most countries. Generally NGOs, foundations and nonbank financial institutions have been more dynamic and banks have become interested only after profitable models have been created, tested and proven. This pattern seems to be evident regarding SHG linkages.

---

* Fukaya et al. (2001) report on a study made of MFIs in which self-regulation was proposed by many of the participants in response to the threats and opportunities represented by the explosion of organizations and resources devoted to microfinance in the country.

Competition between rival models and approaches has proven to be beneficial for the entire microfinance industry. The examples of SEWA, SHARE and BASIX suggest this will likely be the case in India as well, but the current strategy is heavily biased in favor of formal banks. Competitive conditions need to be improved and a more level playing field created so that multiple forms of financial institutions can emerge and seek solutions to the challenge of increasing access, achieving sustainability and enhancing impact. Regulations need to be changed so that nonbank financial institutions can more easily mobilize savings for rural and microlending (Mahajan, 2000).

Natural disasters are frequent in India and MFIs face challenges in servicing affected clients. An appropriate role for NABARD would be to monitor the sector and assure that all market participants implement prudent policies so they can survive these challenges. NABARD might explore the development of a safety-net mechanism in the form of a lender of last resort to support MFIs most affected by disasters.

EXPAND THE ANALYSIS OF MICROFINANCE PROBLEMS AND PERFORMANCE

Many countries have benefited from an active program of analysis of microfinance problems and performance. In these countries, baseline data are collected from clients so that their progress over time can be more accurately monitored; marketing studies are conducted to evaluate how MFIs can offer more attractive products and services. Investments are made in good management information systems so that MFI managers have the information necessary to make timely decisions; a few carefully designed impact studies are done so that the effects and limitations of expanded financial services for the poor can be better understood. Although fiercely competitive, some MFIs have found ways to exchange ideas about common problems and the identity of their delinquent borrowers.

The relatively small amount of information on microcredit operations that is available publicly suggests that India is underinvesting in microfinance research and analysis. More information is needed about what is actually happening on the ground so that policymakers can assess how the microfinance industry is evolving, both in terms of performance and problems. Good analysis of the level of interest rates required for MFI sustainability is needed, as well as of the problems and constraints as observed by all agents in the system, client perceptions about changes needed in products and services and the dynamics of SHG operations.

CONCLUSION

Microfinance can contribute to poverty alleviation and food security. It does this through supplying loans, savings and other financial services that enhance investment, reduce the cost of self-insurance and contribute to consumption smoothing. India has expanded microfinance, but it has not yet developed a strong system capable of serving massive numbers of poor in a sustainable fashion. Undoubtedly, the legacy of directed credit with its top-down approach to lending and the prevalence of highly subsidized state and national poverty projects and programs retard the development of true market-oriented rural microfinance. The policy of supporting SHG linkages with banks has merit in a country with a large bank network, but it should not be the only model encouraged. Additional efforts are needed to create and nurture competitive MFIs willing to experiment with other models.

Policymakers face a dilemma in channeling microfinance funds through the rural banking system. On the one hand, it represents a fairly quick way to expand outreach. On the other hand, the country faces the serious challenge of repairing the damage done to the rural financial system by years of political involvement. The current system of unviable and proliferated rural cooperatives and banks needs to be restructured and rationalized. This will involve closing unprofitable branches, merging and privatizing some institutions and building institutional capacity to provide quality services to rural clients. Considering the poor performance of much of the rural financial system, providing it with access to the SHG linkage refinance window may complicate rather than simplify the reform process.

REFERENCES

Adams, D.W, D.H. Graham and J.D. Von Pischke. 1984. *Undermining Rural Development with Cheap Credit.* Westview Press, Boulder, Colorado.

Bansal, H. 1998. Self-help groups — NGOs–bank linkage program in India: A case study. Unpublished paper. Department of Banking and Business Finance, Faculty of Commerce, M. S. University of Baroda, India.

CDF Statistics. 2001. Vol. 11, December 2000. Credit and Development Forum, Dhaka, Bangladesh.

Chen, M. and D. Snodgrass. 2001. Managing resources, activities and risk in urban India: The impact of SEWA Bank. Management Systems International, Washington, D.C.

Conning, J. 1999. Outreach, sustainability and leverage in monitored and peer-monitored lending. *Journal of Development Economics* 60: 51-77.

Credit for the Poor. Various issues. Newsletter of Cashpoor, Inc., Negri Sembilan, Malaysia.

Fukaya, K. et al. 2001. Microfinance Regulation in India. Sa-Dhan, New Delhi, India.

Hashemi, S. 2001. Linking microfinance and safety-net programs to include the poorest: The case of IGVGD in Bangladesh. Unpublished paper. World Bank, Washington, D.C.

Hulme, D. and P. Mosley. 1996. *Finance against Poverty.* Volume 1. Routledge, London.

Kabeer, N. 2001. Conflicts over credit: Re-evaluating the empowerment potential of loans to women in rural Bangladesh. *World Development* 29: 63-84.

Klein, B. 2001. GTZ-NABARD linkage banking project. Unpublished paper. GTZ, New Delhi, India.

Kropp, E., M.T. Marx, B. Pramod, B.R. Quiñones and H.D. Seibel. 1989. *Linking Self-Help Groups and Banks in Developing Countries.* GTZ, Eschborn, Germany.

Mahajan, V. 2000. A framework for building a sustainable rural financial system for India. BASIX, Hyderabad, India.

M-CRIL. 2001. *The M-CRIL Report, 2000.* Micro-Credit Ratings and Guarantees India Ltd., Gurgaon, India.

Meyer, R. L. 2002. The demand for flexible microfinance products: Lessons from Bangladesh, *Journal of International Development* Vol. 14, no. 3, April 2002 (in press).

Meyer, R.L. and D.W. Larson. 1997. Issues in providing agricultural services in developing countries. In: L.G. Tweeten and D.G. McClelland (Eds.) *Promoting Third-World Development and Food Security,* Praeger, Westport, CT: 119-151.

Meyer, R.L. and G. Nagarajan. 2000. *Rural Financial Markets in Asia: Policies, Paradigms and Performance*. Oxford University Press and Asian Development Bank, Hong Kong.

MicroBanking Bulletin, 2000. Issue No. 5, September. Calmeadow, Washington, D.C.

Morduch, J. 2000. The microfinance schism. *World Development* 28: 617-629.

Mosley, P. 1996. India: The regional rural banks. In: David Hulme and Paul Mosley (Eds.) *Finance against Poverty*, Volume 2, Routledge, London.

NABARD. 1997. Annual Report. National Bank for Agriculture and Rural Development, Mumbai, India.

Nagarajan, G. 2000. Bulletin case study: BASIX, India. *The MicroBanking Bulletin*, No. 5, Calmeadow, Washington, D.C.

Nanda, Y.C. No date. Microfinance: An important tool for empowerment of poor women. Unpublished paper. National Bank for Agriculture and Rural Development, Mumbai, India.

Parhusip, Uben and H. Dieter Seibel. 2000. Microfinance in Indonesia: Experiments in Linkage and Policy Reform. In: J. Remenyi and M. Quiñones, Jr. (Eds.) *Microfinance and Poverty Alleviation: Case Studies from Asia and the Pacific*, Pinter, New York: 153-179.

Paxton, J. 1996. A Worldwide Inventory of Microfinance Institutions: Sustainable Banking with the Poor. World Bank, Washington, D.C.

Puhazhendhi, V. 2000. *Evaluation of Self Help Groups in Tamil Nadu*. National Bank for Agriculture and Rural Development, Mumbai, India.

Puhazhendhi, V. and K. J. S. Satyasai. 2000. *Microfinance for Rural People: An Impact Evaluation*. National Bank for Agriculture and Rural Development, Mumbai, India.

Quiñones, B. 1997. Evaluation of the Linkage Banking Programme in India. Unpublished paper. Asia Pacific Rural and Agricultural Credit Association, Bangkok, Thailand.

Reid, K. J. No date. Financial relationships between community-based organizations and the formal banking sector: YCO's perspective. Youth Charitable Organization, Andhra Pradesh, India.

Rose, Kalima. 1992. *Where Women Are Leaders: The SEWA Movement in India*. Zed Books, London.

Sharma, M., S.V. Rangacharyulu, K.H. Rao and S. Reddy. 2000. Synthesis report for the Case Study of SHARE, India. Working paper. International Food Policy Research Institute, Washington, D.C.

Sheokand, S.M. No date. Reorienting Banking with the Poor: The SHG-Bank Linkage Way. Paper presented at the Bankers Institute for Rural Development, Workshop on Kick-starting Micro-Finance, A Challenge for the Indian Banks. National Bank for Agriculture and Rural Development, Mumbai, India.

Sinha, S., J. Samuel and B. Quiñones, Jr. 2000. Microfinance in India: Adjusting to economic liberalization. In: J. Remenyi and M. Quiñones, Jr.(Eds.) *Microfinance and Poverty Alleviation: Case Studies from Asia and the Pacific*, Pinter, New York: 84-106.

Townsend, R.M. 1995. Consumption insurance: An evaluation of risk-bearing systems in low-income economies. *Journal of Economic Perspectives*, 9: 83-102.

Uphoff, N. 1996. *Learning from Gal Oya: Possibilities for Participatory Development and Post-Newtonian Social Science*. Intermediate Technology Publications, London.

Vyas, V.S. et al. 2001. Expert committee report on rural credit. National Bank for Agriculture and Rural Development, Mumbai, India.

World Bank. 2000. *World Development Report 2000/2001: Attacking Poverty*. World Bank, Washington, D.C.

Wright, G.A. N. 2000. *Microfinance Systems: Designing Quality Financial Services for the Poor*. Zed Books, London.

Zeller, M., G. Schrieder, J. Von Braun and F. Heidhues. 1997. *Rural Finance for Food Security for the Poor: Implications for Research and Policy.* International Food Policy Research Institute, Washington, D.C.

25 Poverty and Gender in Indian Food Security: Assessing Measures of Inequity

Paul Robbins

CONTENTS

INTRODUCTION

The magnitude of the Green Revolution's agricultural intensification in India is not to be denied. The gains in productivity throughout the country that have followed on technological innovation and extension are remarkable and represent a fundamental change to Indian food security and the nation's relative position in global commodity markets. Its detractors point to instabilities that such changes have formed in the social and political fabric of the countryside (Shiva 1991), but even these critical accounts emphasize the degree to which the revolution has been widespread and comprehensive.

The extent to which this revolution has transformed the lives, health and labor of rural women in India remains considerably more obscure. Has the revolution for women been similarly widespread and comprehensive? What has it done to household practices, equity and the balance of gender power? This chapter looks at the impacts of high-input agrofood systems on Indian women and households, paying specific attention to questions of equity in food production and consumption. It addresses the following three questions:

1. Has the quest for food security led to an increased share of food and nutrients for women?
2. Has the Green Revolution led to agricultural employment opportunities for women and are their wages on parity with those of men?
3. Has the intensification and extensification of agricultural development adversely affected land use and land cover and improved the conditions of women's labor in the household and family?

These questions are addressed in turn, using secondary data, as well as a field case study from Rajasthan, to explore the effect of the Green Revolution on gender equity and to assess the role of gender in the changing agricultural landscape of India. It concludes that, despite the massive technical assistance deployed in solving food scarcity problems, better distribution of benefits remains woefully unaccomplished. Moreover, many of the metrics used in the evaluation of the problem are inadequate to the task and serious thought must be given to developing new categories and measurements for examining women's lives in the context of Indian agricultural development.

THE GENDERED AGRARIAN HOUSEHOLD

Several prior factors affecting women's labor and household power should be noted at the outset. First, gendered labor in most of India involves work in both the productive and reproductive spheres of the household. Women are increasingly engaged in wage-earning activities. These forms of work vary, ranging from agricultural labor to road construction and tanning of leather. Agricultural labor in some places is often disproportionately female. In paddy cultivation in Jharkand/Bihar, for example, it is reported that of the 80 days of human labor required per acre, female workers supply 65 (Sharan and Dayal, 1993). At the same time, women are also laboring to reproduce the household through the procurement of wood for fuel and other minor forest products, the acquisition of fodder, the herding of animals, as well as child care, cooking and cleaning, and home construction and maintenance.

Second, despite the high proportion of women's labor outside the home, control over key decision-making in production, inputs and the allocation of labor commonly rests in the hands of men. A patrilineal and patrilocal pattern of social organization dominates most of India and young women typically marry into household economies where they become the least powerful and least well-established participants. This, along with the male-oriented inheritance structure, makes women's position in agricultural labor and the distribution of surpluses extremely precarious (Liddle and Joshi, 1986).

Having said this, it is also important to keep in mind the vast regional, caste and class differences among the various positions of women in agricultural production. High-caste rajput women in Uttar Pradesh, for example, traditionally live in the exclusion of purdah throughout much of their adult lives, so that public labor is less common in this community (Minturn, 1993). Low-caste meghwal women in Rajasthan, on the other hand, are prominent in the regional labor force and have a larger voice in household decision-making than their upper-caste neighbors (Robbins, 1998). An assessment of gender equity in food security viewed on a national scale is, therefore, pretty much a convenient analytical fiction.

Even so, aggregate trends and questions should be explored. The problems of food security and gender equity on a national scale can, at a minimum, determine whether women are getting more food, better wages and a reduced labor burden as a result of new agricultural practices and technologies. The first of these questions — the intra-household allocation of food — is one of the most direct and fundamental measures of the impact of the Green Revolution on women.

INTRA-HOUSEHOLD ALLOCATION OF FOOD

The most direct measure of the gendered impact of increased food grain production is the intra-household distribution of Green Revolution surpluses to women and girls. Are women getting enough to eat and are the larger harvests finding their way to meeting women's survival needs relative to those of men?

Data for answering this question are limited. Basic measures of health, female infant mortality and caloric intake stand as rough indices for the overall problem of nutritional equity. Though not perhaps the best direct measures of the problem, all of these indicators suggest persistent unevenness in the intra-household distribution of food and nutrients.

Asian demographics and specifically those of India continue to show an antifemale bias in nutrient intakes and fewer surviving females in most age cohorts. Gender-skewed mortality, morbidity and malnutrition are still in evidence at both the household and national levels. Table 25.1 shows infant and child mortality figures at a national scale. Notably, in the first 12 months of life, girl children in India stand a higher chance of survival across all but the poorest quintile of families. In the next 5 years, however, female mortality rates overtake those of males, in all but the richest quintile of households.

Other indicators suggest a crisis of nutrition for women throughout their lives; 52% of women in India are anemic (defined as hemoglobin levels <11 grams/deciliter) while 16.7 are moderately or severely so (hemoglobin levels <10 grams/deciliter) (Government of India, 2000). Overall, these results suggest a systematic and disproportionate undersupply of nutrients to girl children, especially in the most marginal households.

Recent countervailing data, however, does indicate some positive findings for women's nutrition, suggesting that females sometimes enjoy better nutrition than males when measured in terms of individual caloric intake relative to nutritional standards (Millman and DeRose, 1996). The actual mean caloric intake of women and girl children relative to generalized demands varies significantly from state to state. Table 25.2, based upon a study by India's National Nutrition Monitoring Bureau (NNMB 1980, cited in DeRose et al., 1998), shows high levels of variance in caloric intake relative to requirements. A pattern of relative equity prevails, however, and in many cases, female caloric intake relative to requirements is higher than that of males.

TABLE 25.1
Income and Nutrition Effects on Mortality in India by Gender

Indicator	Population Average	Poorest Quintile (M/F)	Second Quintile (M/F)	Middle Quintile (M/F)	Fourth Quintile (M/F)	Richest Quintile (M/F)
Deaths under 12 months (/1000 births)	86	108/110	109/102	94/84	69/62	45/42
Deaths under 5 years (/1000 births)	118	146/163	147/158	118/120	86/87	55/54

Source: Gwatkin et al. (2000)

TABLE 25.2
Caloric Intake as Proportion of Requirement

Age Range	Kerala (M/F)	Tamil Nadu (M/F)	Maharastra (M/F)	Gujarat (M/F)	West Bengal (M/F)	Uttar Pradesh (M/F)
13–16	0.48/0.65	0.69/0.76	0.67/0.73	0.70/0.82	0.69/0.63	0.81/0.63
16–18	0.53/0.60	0.66/0.84	0.63/0.73	0.69/0.74	0.62/0.69	0.74/0.82
Adult (moderate activity)	0.61/0.52	0.82/0.88	0.79/0.82	0.79/0.78	0.71/0.66	0.77/0.85

Nutrition experts, however, have criticized using individual caloric intakes as an indicator of nutrition because they fail to measure the intake of proteins, micronutrients and other key elements of quality diets. Indian data suggest that, even where women and girls consume calorie amounts similar to men's, they are often denied milk, fruits and other important dietary staples that other family members receive (Das Gupta, 1987, 1995). As a result, recent advances in nutrition policy have suggested that a broader concept of nutritional security be adopted, one that embraces both food quantity and the overall quality of nutrition as "food, health and care" (International Conference on Nutrition, 1992). Such approaches use anthropological assessments of food equity to explore and explain nutritional discrepancies between men and women. These approaches, however, only further underline the disproportionate nutritional deprivation of girls and women in India (Messer, 1997).

Some examinations of women's general nutritional neglect (despite increasing food surpluses) in the country stress male-based "investment" strategies of rural households that are common in patrilineal, patrilocal societies like that of North India. These explain deprivation in terms of "benign neglect" that is "less intentional than routine" (Messer, 1997: 1677). Such approaches note that parents follow cultural norms for food distribution that set normative standards for feeding and health care in local cultural terms. In this case, women are understood as following institution-alized norms, eating last and meeting expectations of self-sacrifice and frugality (Cassidy, 1980; Scheper-Hughes, 1992).

More critical readings of nutritional inequity suggest a more calculated and grim view of the problem, suggesting an intentional strategy on the part of families to control and manipulate family size and composition by supplying differential nutri-ents to girl children (Das Gupta 1995). Here men and older women intentionally deprive younger female household members of the quality and quantity of food needed, whether these are girl children or recently wed daughters-in-law who cus-tomarily enjoy fewer rights in patrilocal households.

Ethnographic accounts of female nutrition deprivation take a position somewhere between these two conclusions. While emphasizing that female micronutrient dep-rivation is "intentional" on the part of family decision-makers, they conclude that the complex cultural standards and household economics set the stage for such inequity (Messer 1997). As wage earning becomes an increasing determinant of individual power in rural households, moreover, an increasing share of nutrients will flow toward males who labor outside the household.

In rural states like Rajasthan, where male agricultural employment is roughly four times that of female employment (Government of India, 2000), such discrep-ancies are common and increasing. While, from a normative point of view, the deprivation of women is an appalling outcome of household dynamics, more atten-tion must be paid to the structure of household labor if such imbalances are to be addressed. More food in the household, it is clear, does not necessarily translate to more food for women. Specifically, the increase of cash orientation in household economics puts women in an increasingly precarious position relative to the axes of power and control of resources in the home. Nutrition programs alone may do little to rectify the problem.

In sum, despite increases in agricultural surpluses and increasing food supplies in India at the aggregate level, women in India continue to suffer from dispropor-tionate lack of access to nutrients and food. This deprivation is a result of traditional cultural norms but may also be linked to increasing cash-based labor relations of households following the Green Revolution, which disempower women even as they appear to benefit the household as a whole. Thus, the next and perhaps more fundamental question becomes: are women increasing their wage earnings in the wake of the quest for Indian food security?

AGRICULTURAL WAGE EQUITY

From such findings, it is suggested that increasing participation of women in the food production chain and its related wage activities leads to an overall increase in household food availability, especially for women (Soysa, 1987) and to increased power and autonomy for women in the household (Holmboe-Ottesen et al., 1988).

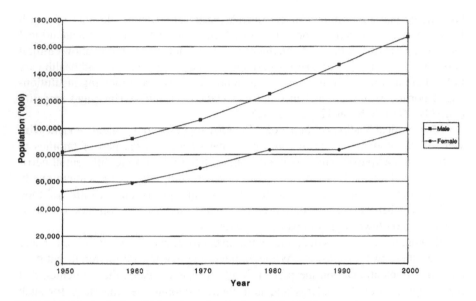

FIGURE 25.1 Agricultural labor force by gender (FAO, 2000).

On the other side of the question, increased productive wage activity tends to increase the total work burden of women, given that household reproduction remains the responsibility of women despite increases in their wage labor time. Still, in a country where the female literacy rate is 42% overall and only 74% among women in urban areas, agricultural labor represents one of the most viable income generation paths for women. Figure 25.1 shows the agricultural labor force in India and the increasing participation of women in agricultural labor.

Leaving aside the crucial concern of reproductive labor, another indicator of the benefits of the food production revolution in India would be the distribution of wages to female agricultural workers in India. Here again, the indicators are not clear.

On initial investigation, the prospects for increasing wages for women's labor in agriculture appear promising. The period between 1966 and 1988 is marked by a steady increase in female agricultural wage workers' earnings. Daily real wage rates rose from an average of 2 rupees to 4. Though these wages remain lower than those of male wages (by 1 or 2 rupees), the gap shows a slow overall decline over the past 30 years. There is, however, significant regional differentiation in both the measure of women's wages and its difference from that of male workers. Singh reports that the wage rate is low in many southern Indian states like Tamilnadu (Rs. 1.72) but quite high in northern Indian states like Punjab (Rs4.77). Moreover, the differential between female and male wage rates also varies dramatically. In poorer states like Bihar and Orissa, with relatively less-well-developed agricultural infrastructure and intensity, women workers earn between 85 and 87% of their male counterparts' wage, while, in advanced agricultural states like Maharastra, they earn as little as 67% of the male daily wage (Singh, 1996).

Observers disagree about the overall trajectory of changes in this area. Some economists suggest that an overall decline in the wage gap is the result of new agricultural technology that has improved the relative bargaining power of female labor (Rao, 1989). Others conclude, however, that the wage gap has not appreciably narrowed and that economic growth has had minimal effect on women's labor status (Agarwal, 1985; Jain and Banerjee, 1985). The degree to which these changes in the overall wage rate of women workers and the male–female differential have been affected by Green Revolution intensification of agriculture in the subcontinent is thus uncertain, although recent models point to some explanatory trends.

A time-series cross-state model of the problem by Singh (1996) examined both the impact of new technologies on wage differentiation — including differences between mechanical technologies (like tractors) and biological advances (new seed stock) — and of labor supply, especially the limited mobility of women and the prospects for alternative employment. His study concluded that differential wage elasticities with respect to increases in yield have caused very uneven increases in overall female wage level. Moreover, despite the increased opportunities for women, the crowding of the female labor force in agriculture has tended to depress women's wages overall. Further, the study concluded that other factors, including changing views of women and labor laws, have had some favorable impact on the wage gap.

Some more detailed case study work, however, suggests that a combination of agricultural intensification and the implementation of the new economic policy (NEP) has resulted in an overall increase in women's labor with disproportionately low compensation and return. Increasing exploitation of female labor power is especially acute in traditionally impoverished states like Bihar (Jharkand), where wages are generally low and labor burdens are extremely high. Here, the expansion of women's wage agricultural work in the wake of the Green Revolution has not meant any offset in reproductive responsibilities, which are portrayed as "natural" duties in daily life. In both surplus-creating cultivator households as well as marginal and landless homes, the high and rising burden of labor for women is, in turn, passed along to girls, who increasingly drop out of school as a result (Saihjee 1996).

Thus, the progress made in real wage increases for women in India as a result of agricultural modernization and intensification is offset somewhat by the hidden costs of domestic reproduction. Even as women are gaining power and autonomy through wage employment, the increasing yields and profits of industrialized agriculture are partly realized by the increased labor burden of women. These conclusions further suggest the importance of new indicators for the costs and benefits of agricultural change. Moreover, they suggest a more fundamental question: if the reproductive labor burden remains high (or growing) for women and girls, what are the environmental conditions upon which that labor depends and what has the been impact of the Green Revolution on the exploitation of that labor?

THE CRISIS OF REPRODUCTION: GENDERED LAND-USE PROBLEMS

The Green Revolution in India has meant massive changes in land use and land cover since the time of independence. Specific land-use changes over the past 30 years include: 1) an increase in the proportion of land cropped more than once, 2) a decrease in the extent of cultivable "waste" lands and 3) an increase in the extent of closed or reserved forest areas. All of these changes are the direct and deliberate result of efforts to modernize agricultural management. The increase in double cropping has been achieved by tremendous outlays of capital for the construction of irrigation systems and through direct subsidies to producers for the procurement of inputs. The increase in yields that resulted from this process of intensification is discussed more extensively in other chapters in this volume. Simultaneously, the extensification of agricultural lands into previously underutilized or waste lands allowed cropping in areas where none had occurred before. Conservation efforts simultaneously enclosed many village forests for the protection of tree stocks and, more recently, of biodiversity and wild fauna. While this last change is not a direct result of agricultural intensification, conservation can be viewed as a parallel environmental development practice during the same period.

Beyond their immediate environmental effects, these changes have a direct bearing on the reproductive duties of women in rural households. Fallow lands provide key inputs for household reproduction including animal browse, thatch and medicinal plants. So-called waste lands are often dominated by village forests or pasturage. Formal forest lands provide fuel-wood sources as well as construction materials and a number of important inputs into household maintenance. Thus, the loss of dry season fallow to increased cropping, the decline of village wastes and the enclosure of forest lands have all served to reduce the resource base for household reproduction; rural women travel farther for fuel, fodder and construction materials, and parts of the annual resource calendar are marked by scarcity.

LAND-USE CHANGE AND GENDER EQUITY CASE: RAJASTHAN

The case of Rajasthan is instructive. While the state has in no way undergone the sweeping Green Revolution changes of neighboring Uttar Pradesh, Punjab and Haryana, its recent increase in agricultural productivity has been rapid, with concomitant changes in land use and rural household economics that well represent the question of gender and household reproduction more generally. The state is largely arid and semi-arid with rainfall as low as 200 mm in the far west and as high as 900 mm in the north and east. Like most of the subcontinent, traditional agricultural production depends on the southwest monsoon and single cropping of subsistence and market crops, especially bajra (pearl millet) and wheat. Like other marginal and underdeveloped states in India, Rajasthan's literacy rate is extremely low (30%), especially that of women (16%). The largest employment categories for rural women are those of "cultivator" and "agricultural laborer," and males outnumber females in all but the oldest age rank cohorts (Government of India, 2000). Thus, Rajasthan is in many ways typical of the traditionally underdeveloped agrarian states of India, with a significant proportion of the laboring agricultural poor made up of women and with traditional agricultural production predominating.

But, in recent years, Green Revolution inputs have also swept across the region into even the most remote areas. Where only 2.6 million hectares of land in Rajasthan were under irrigation in 1976, 5.5 million were irrigated in 1999. The area under high-yielding varieties (HYVs) has also climbed; 255,000 hectares of land were under improved varieties of bajra (pearl millet) in 1976, while nearly 1.4 million

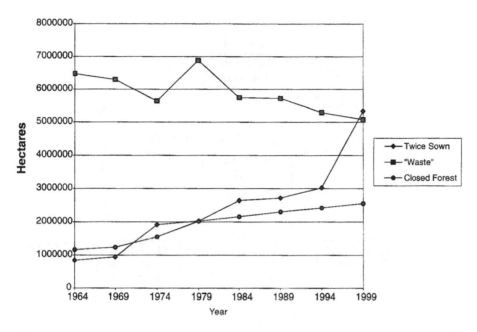

FIGURE 25.2 Land use change in Rajasthan.

hectares of HYV millet were grown in 1999. HYV wheat production climbed by more than 250% between 1976 and 1999 to a total of 2 million hectares. Though not part of the first wave of Green Revolution advancements of the 1970s, Rajasthan has undergone tremendous technological acceleration in recent years with intensive high input systems spreading to even the most marginal villages in the state.

The land-use changes that accompanied this growth are shown in Figure 25.2 and these too fit the profile of other Indian states. Rapid growth in land cultivated twice or more per year is accompanied by a rise in enclosed reserve forest land and a fall in the extent of "cultivable waste" land. This last category includes a vast range of actual kinds and qualities of land cover, but, most prominently, it includes village scrub and pastureland and local, non-enclosed forest.

These land-use changes, though an inevitable part of the process of intensification and extensification, allied with the quest for food security in the region, further overburden rural women in Rajasthan who are necessarily involved in most aspects of the reproduction of the household. As fallow lands are taken out of circulation, wastes are put under the plow and forests are enclosed, resource bottlenecks begin to appear in women's labor, causing difficulties and crises in women's work.

RESOURCE BOTTLENECKS IN HOUSEHOLD REPRODUCTION

Women's reproductive labor outside of agricultural production in Rajasthan depends on the uses of inputs from scrub forest and fallow land, including the browsing of livestock on forest species and the harvesting of fodder and thatch grass species from land in fallow. Many wild species, including *Acacia nilotica*, *A. catechu* and

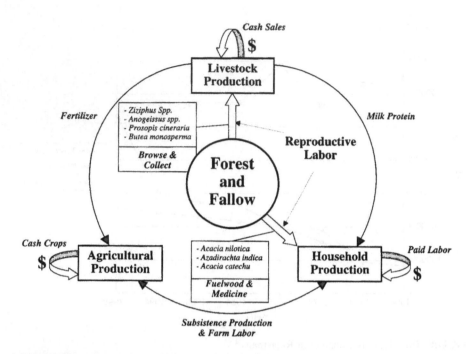

FIGURE 25.3 Women's use of forest and fallow in household reproduction.

Azadirachta indica provide fuel wood and medicine for households. Figure 25.3 shows the role of reproductive labor in production, identifying the inputs into livestock and household activities that indirectly support agricultural production, including and especially Green Revolution agriculture. These species and landscapes are managed almost exclusively by women, and the intensification of agriculture has in no way lowered demand for these key inputs. Indeed, the land cover changes attendant in intensification have arguably made them scarcer.

Figure 25.4 shows schematically how land cover changes have created bottle-necks in household reproduction and have increased the labor demands on women. Each resource calendar is based on the mean land coverages described for 1955 and 1994 in the misl bandobast (land settlement) and jamabandi (land coverage) records for a sample of 29 randomly sampled survey villages in western Rajasthan. The availability of resources for household reproduction, as described above, is mapped over the year based on the results of a 100-meter transect-intercept measurement of key species in a sample of 38 resource lands conducted twice in 1994. These data were supplemented with environmental history accounts, revealed in discussion with older men and women (see Robbins, 1998). The vertical extent of each resource area represents the coverage of that area in hectares and the shading denotes the quality of resource availability based on species coverage and local knowledge of land users.

The 1955 arrangement shows a system in which losses of productive farmland to cultivation during the traditional monsoonal growing season (July to October) are offset by the availability of alternative resource lands, especially village forests

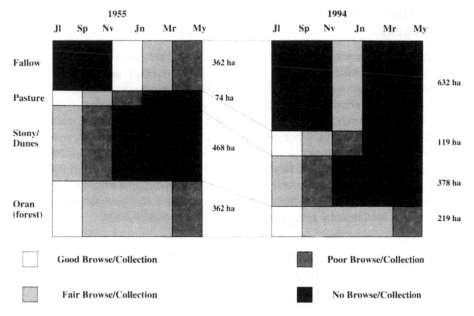

FIGURE 25.4 Changes in annual resource-use calendar for women's labor.

(orans) and otherwise marginal stony lands, pasturage and sand dunes fixed with grass and scrubby vegetation. When these lands become less productive in the later winter and dry season, fallow lands become available for use and harvesting and traditional forests continue to supply some materials for forage and fuel. Women's labor, therefore, is a spatial solution to the problem of scarcity, where demands for reproductive resources are met in differing areas at differing times of year.

By 1994, however, the coverage of these lands had changed dramatically. Forest lands and dunes had been lost to enclosure and cropping. The prime oran (forest) lands in the region, though acting as a village resource, fall largely under the official classification of cultivable waste and so were enclosed for cropping. The slight expansion of official village pasturage over the period did little to offset the loss of resources during the rainy season. Moreover, as the intensification of agriculture becomes ubiquitous in the region through the cropping of the land twice (or sometimes three times) annually, the dry season fodder and fuel reserves traditionally harvested on fallow lands between March and May have become scarcer as well.

The effects of these changes fall as a heavy labor burden upon women. Women in the survey villages all report that their daily fuel-wood and fodder-gathering activities have become more time consuming and difficult as these resources become fewer and farther away. So too, the employment of young girls in these activities is reported to be on the rise, especially during the dry season, when resources are most scarce. Whether these changes have resulted in the removal of children from schools or in changes in women's health or autonomy is not known. Nevertheless, the unintended consequence of extensification and intensification has been a resource squeeze in the reproductive realm of the household and the resulting burden has fallen disproportionately on women.

LAND COVER, HOUSEHOLD REPRODUCTION AND LANDSCAPE SCALE ANALYSIS

The pattern of resource scarcity for women shown here manifests itself in other forms throughout India. The fuel-wood and fodder crises of the region, though exacerbated by growing population size, are rooted in the unintended consequences of agricultural development. The increasingly capitalized agricultural system of Rajasthan is, as in other states, leveraged on the support of important resources in the form of women's unpaid labor and harvested wild species. In short, women and men are not sharing equally the costs and benefits of the emergent institutional arrangements of an intensified landscape. This pattern becomes apparent only when viewed at the landscape scale and in terms of geographic distribution of resources over space and time. Like nutrition and wage equity, a more realistic measure of inequity is a fundamental prerequisite to its discovery and solution.

CONCLUSION: MEASURING AND RECTIFYING INEQUITY

The intensification and extensification of agricultural production in India in the wake of the Green Revolution has proven to be a mixed blessing for rural women. Certainly women's wages have risen in the last 30 years as a result of agricultural expansion. Also, more calories are available and flowing into the household for consumption by all parties. At the same time, however, women continue to receive unequal shares of both food and nutrients and capital or wages relative to their male counterparts. Moreover, the expansion of agriculture has come at the expense of women's unpaid labor in an increasingly unstable natural-resource environment.

Practical solutions to these problems do exist. Fodder deficits created by agricultural change might be offset by systems of fodder banking. Investments in pasture and woodlot development might provide new resources for women's labor. More research and development might go toward protecting and cultivating the wild species such as wild fodder grasses and wood trees that indirectly sustain the production of high-yielding cereal varieties.

But to determine, locate and map the location of areas for these kinds of interventions requires that we supplement our traditional assessment metrics — including calories, wages, output levels — with new methods and tools for seeing and measuring inequality. Ethnographic approaches to nutritional equity, for example, have pointed to new measures of food security, as explained above. Does caloric intake provide a meaningful measurement of food security for individuals or vulnerable groups such as women in India? If not, what other forms of measure might be used to explore the effects of the Green Revolution on the lives of people in the subcontinent?

Similarly, measures of women's labor will of necessity require more serious attention to the costs associated with the reproductive sphere of the household. Can wage disparities fully capture the complexity of systems of an unequal labor burden? A better measure of differential impact would be paid and unpaid labor hours, for example. In national and state-level accounting, girls' enrollment in schools might also provide a better and more accurate measure of in-house labor demands.

Finally, the relationship between agricultural change and other land use and land cover changes must be measured with more care. What are the landscape-level effects of agricultural change? These effects need to be understood, moreover, not only in terms of their environmental implications (biodiversity, genetic resources, etc.) but also in terms of their costs to the stability of agricultural production overall, through the avenue of women's household reproductive labor.

Thus, it is the very categories of analysis that come into question when we explore gendered impacts on food equity or inequality in the wake of the Green Revolution. To meet the promises of food security in the region, the manner in which we measure and explain the effects of agricultural change, therefore, will be as important as the technical means by which it is achieved.

REFERENCES

Agarwal, B. 1985. Agricultural Modernization and Third World Women: Pointers from the Literature and an Empirical Analysis. Geneva: International Labor Organization.

Cassidy, C. 1980. Benign neglect and toddler malnutrition. In *Social and Biological Predictors of Nutritional Status, Physical Growth and Neurological Development*, L.S. Greene and F.E. Johnston, Eds. 109-133. Academic Press. New York.

Das Gupta, M. 1987. Selective discrimination against female children in rural Punjab: *Indian Population Development Review*, 13(1): 77-100.

Das Gupta, M. 1995. Life course perspectives on women's autonomy and health outcomes: *American Anthropologist*, 97:3, 481-491.

DeRose, L., E. Messer and S. Millman. 1998. *Who's Hungry? And How Do We Know?* United Nations University Press. New York.

Government of Rajasthan.2000. *Basic Statistics, Rajasthan*. Directorate of Economics and Statistics. Jaipur, India.

Gwatkin, D.R., S. Rustein, K. Johnson, R. Pande and A. Wagstaff. 2000. Socio-economic differences in health, nutrition and population in India. HNP/Poverty Thematic Group paper. World Bank. Washington, D.C.

Holmboe-Ottesen, G., O. Mascarenhas, et al. 1988. Women's role in food production and nutrition: Implications for their quality of life. Food and Nutrition Bulletin, 10:3, 1-12.

International Conference on Nutrition. 1992. World Declaration and Plan of Action for Nutrition. FAO/WHO. Rome, Italy.

International Institute of Population Sciences. 2000. National Family Health Survey (NFHS-2). International Institute of Population Sciences. Mumbai, India.

Jain, D. and N. Banerjee. 1985. *Tyranny of the Household: Investigative Essays on Women's Work*. Shakti. New Delhi, India.

Liddle, J. and R. Joshi. 1986. *Daughters of Independence: Gender, Caste and Class in India*. Rutgers University Press. New Brunswick, NJ.

Messer, E. 1997. Intra-household allocation of food and health care: current findings and understandings — Introduction. *Social Science and Medicine*, 44:11, 1675-1684.

Millman, S.F. and L. DeRose. 1996. Food deprivation: report to the United Nations University. In *Who's Hungry? And How Do We Know?* L. DeRose, L.E. Messer and S. Millman. Eds. United Nations University. Tokyo.

Minturn, L. 1993. *Sita's Daughters: Coming Out of Purdah.* Oxford University Press. New York.

National Nutrition Monitoring Bureau. 1980. Report for the Year 1979. National Institute of Nutrition. Hyderabad, India.

Rao, C.H.H. 1989. Technological change in Indian agriculture: Emerging trends and perspectives. *Indian Journal of Agricultural Economics*, 44:4, 385-98.

Robbins, P. 1998. Authority and environment: institutional landscapes in Rajasthan, India. *Annals of the Association of American Geographers*, 88:3, 410-435.

Saihjee, A. 1996. Politics of economic adjustment: The changing nature of women's work in Jharkand, India. *Asia Pacific Viewpoint*, 37:2, 165-180.

Scheper-Hughes, N. 1992. *Death Without Weeping: The Violence of Everyday Life in Brazil.* University of California Press. Berkeley.

Sharon, R. and H. Dayal. 1993. Deprivation of female farm laborers in Jharkand region of Bihar. *Social Change*, 23:4, 95-99.

Shiva, Vandana 1991. *The Violence of the Green Revolution.* Third World Network. Penand, Malaysia.

Singh, R. D. 1996. Female agricultural workers' wages, male-female wage differentials and agricultural growth in a developing country: India. *Economic Development and Cultural Change*, 45:1, 89-123.

Soysa, Priyane. 1987. Women and nutrition. *World Review of Nutrition and Dietetics*, G.H. Bourne, Ed. Karger. 52: 1-70. Basel, Switzerland.

Part Five

Policy Issues

26 Priorities for Policy Reform in Indian Agriculture

Peter Hazell

CONTENTS

INTRODUCTION

India has made impressive gains in agricultural growth, food security and rural poverty reduction since the food crisis years of the mid-1960s. Food-grain production has approximately doubled since that time and India is now self sufficient in cereals, producing about 180 million tons of cereals each year, more than enough to meet current market demand. Although many Indians still do not have an adequate diet, the per capita availability of cereals has improved and the incidence of rural poverty has fallen from about two thirds of the rural population to one third today. But these favorable trends are now stalling and there is urgent need for new approaches if agriculture is to contribute to future national economic growth, employment creation and poverty reduction.

This chapter addresses three connected sets of issues. The first concerns the recent cutbacks in public investment in agriculture that threaten future productivity growth and poverty reduction in the rural sector. Second, past patterns of agricultural growth have been environmentally destructive and there is need to redress this problem on a national scale to sustain future productivity growth in agriculture. Third, there are new and favorable opportunities for market-driven growth in the agricultural sector with trade liberalization and increasing diversification of the national diet. If these opportunities are properly managed, they could make significant contributions to further reductions in rural poverty.

PUBLIC SPENDING ON AGRICULTURE

The Indian government's "development" expenditures on agriculture, irrigation, transportation, power and rural development grew at an average annual rate of 15.1% during the 1970s, by 5.1% in the 1980s and by 1.3% in the early 1990s (Fan et al., 1999). Despite an increase in private investment in the early 1990s, there is little evidence to suggest that it is substituting for public investment, either in its level or composition. Given the strong links between government investment in the rural sector and agricultural growth and poverty reduction overall (Fan et al., 1999), there is a real danger that future growth and poverty reduction will now also slow. Moreover, to make matters worse, the government is wasting a good deal of its resources by paying too much and charging farmers too little for basic services in agriculture (power, fertilizers, water, credit, etc.) that could be financed privately or provided more efficiently. This approach has a number of serious consequences:

- It places a huge and growing financial burden on the government. Subsidies currently consume more than half of total government spending on agriculture (World Bank, 1999). While many of these subsidies played a useful role in launching the Green Revolution in the late 1960s and helped ensure that small farmers and not just large farmers gained access to new technologies, today they are largely unproductive and detract from the public resources that are available for investment in future agricultural growth. Gulati (2000) has estimated that the subsidies for power and fertilizer alone now cost the government about $6 billion per year, equivalent to about 2% of national gross domestic product (GDP). Water, credit and the food distribution system also call for large subsidies. If these are added in, it seems likely that at least $10 billion is spent each year on unnecessary subsidies. Thus, enormous opportunities exist for doing much more with the resources that are already being allocated to agriculture.
- A highly subsidized agricultural support system fosters inefficiency in the supply of key inputs and services. In India, the fertilizer industry receives the lion's share of the subsidy, and production costs for urea from a majority of firms in the industry are well in excess of the world price (Gulati, 2000). The public supply systems for power and irrigation water are also notoriously inefficient, a direct result of their having no need or incentive to perform better when they are almost fully financed by government rather than by their clients.
- Because farmers pay too little for the inputs and services they receive, they have little incentive to use them carefully, which leads to overuse and waste. This is costly to the country and, in some cases, it has high environmental costs, e.g., fertilizer and pesticide runoff into waterways and waterlogging of irrigated lands.

India thus does not need large-scale foreign aid for agriculture. Rather, the resources that are already available need to be used to greater effect. India already spends relatively more on promoting agriculture (public investment as a percentage of agricultural GDP) than do most other Asian countries (World Bank, 1999). The challenge is not to spend more, but rather to get better value for the money that is spent.

AGRICULTURE AND THE ENVIRONMENT

The Green Revolution played a key role in achieving national food security and in reducing rural poverty. By raising yields, it also avoided having to increase the total cultivated area significantly, thereby helping to preserve remaining forest areas and avoiding crop expansion into environmentally fragile areas (e.g., hillsides and dry-lands). Even so, there is no question that the Green Revolution was environmentally damaging in many of the areas in which it occurred. The problems created or worsened include:

- Salinization of some of the best irrigated lands
- Fertilizer and pesticide contamination of waterways
- Pesticide poisoning
- Falling water tables

The problems began in the 1970s and appear to be getting worse. There is also mounting evidence that yield growth in many of the intensively farmed areas has now peaked and, in some cases, is even declining (Rosegrant and Hazell, 2000). Even where there is no absolute yield decline, there is diminishing factor produc-tivity. Growing voices are arguing that Indian farmers should revert back to the low external-input farming technologies of pre-Green Revolution days (Shiva, 1999). This would have disastrous impacts on yields and food supplies and would destroy the environment on an even larger scale because of the need to rapidly expand the planted area.

Realistic prospects exist for making modern technologies more environmentally benign and reversing resource-degradation problems on a national scale (Pingali and Rosegrant, 2000). But it will take significant and determined action by the govern-ment. Needed actions include:

- Development and dissemination of technologies and natural-resource management practices that are more environmentally sound than those currently used on many farmers' fields. Some technologies that already exist include precision farming, crop diversification, ecological approaches to pest management, pest-resistant varieties and improved water management practices. The challenge is to get these technologies adopted more widely on farmers' fields. Managed properly, some of these technologies can even increase yields while they reduce environmental damage. Further agricultural research is needed to create additional tech-nology options for farmers, which should include interdisciplinary work on pest control, soil management and crop diversification. Modern biology should also be used to develop improved crop varieties even better suited to the stresses of intensive farming but with reduced dependence on chemicals, e.g., varieties that are more resistant to pest, disease, drought and saline stresses. Agricultural research and extension systems will need to give much higher priority to sustainability problems than they have in the past.

- Reform of policies that create inappropriate incentives for farmers in their choices of technology and natural resource management practices. As mentioned above, current large subsidies for water, power, fertilizer and pesticides make these inputs too cheap and encourage excessive, even wasteful use, with dire environmental consequences. Pricing these inputs at their true cost would save the government much money while also improving their management. This would reduce environmental degradation and, in the case of scarce inputs such as water, lead to important efficiency gains. Improvements in land tenancy rules would also improve the incentives for many smaller farmers to take a longer-term view in their choice of technologies and management practices. Strengthening community rights and control over common property resources like grazing areas, woodlots and water resources could also improve incentives for the more careful and sustainable use of these natural resources. Additionally, the farm credit system needs to offer more medium- and long-term loans for investment in the conservation and improvement of natural resources, especially for smallholders and women farmers.
- Reform of public institutions that manage water to improve the timing and amounts of water that are delivered relative to farmers' needs and to get better maintenance of irrigation and drainage structures. When farmers have little control over the flow of water through their fields, they have reduced capacity to prevent waterlogging or salinization of their land and to use water more efficiently. Forestry departments also need to work more closely with local communities, devolving responsibilities where possible, to improve incentives for the sustainable management of public forest and grazing areas.
- Assistance to farmers in diversifying their cropping patterns to relieve the stress resulting from intensive monoculture. Investments in marketing and information infrastructure, trade liberalization, more flexible irrigation systems and so forth, can increase opportunities for farmers to diversify. Unfortunately, the kinds of diversification that the market wants are not always consistent with the kinds of on-farm diversification that are needed for sound crop rotations.
- Resolution of widespread "externality" problems that arise when all or part of the consequences of environmental degradation are borne by people who have not caused the problem, e.g., pollution of waterways or siltation of dam reservoirs due to soil erosion in upstream watershed areas. Possible solutions include taxes on polluters and degraders, regulation of resource use, empowerment of local organizations and appropriate changes in property rights. Effective enforcement of rules and regulations is much more difficult than writing new laws, so attention needs to be given to ensuring implementation.

FROM FOOD SECURITY
TO MARKET-DRIVEN GROWTH

India presently has a good supply of food grain (about 60 million tons of cereals are held in stock at the present time). Future agricultural growth will be constrained if the country does not move beyond its past concerns with national food self sufficiency to better exploit its comparative advantages. The overall food-grains balance should be monitored, but this seems unlikely to be a major problem, at least within the next several decades (Bhalla et al., 1999). Food security has become primarily a distribution problem that requires other solutions than simply growing more food grains. It requires more focused and targeted efforts to raise the incomes of the poor, most of whom are rural and live in rain-fed areas, which are often backward areas with limited agricultural potential or infrastructure and market access (Fan et al., 2000).

New growth opportunities for agriculture are arising from a number of sources:

- Changes in the national diet are occurring with the accelerated national economic growth achieved in recent years. With the rising affluence of the middle classes, domestic demand for livestock products (especially milk and milk products), fruits, vegetables, flowers and vegetable oils has shot up. This creates new growth opportunities for farmers to diversify (even specialize) in higher-value products, especially those farmers who have ready access to markets, information and inputs.
- Ongoing policy reforms are slowly opening up export markets for Indian farmers. This, together with the removal of restrictions on interstate trade within India, should enable more farmers to specialize in those crops in which they have comparative advantage and can best compete in the market. These opportunities should further improve if the next round of world trade negotiations sponsored by the World Trade Organization succeed in freeing up more agricultural markets in countries around the world.
- There are also good opportunities for generating greater value added in agroprocessing, particularly if agroindustry is liberated from current protective policies and can became more competitive with imports (World Bank, 1999; Gulati and Kelley, 2000). Oil seed processing, for example, is highly protected at present, making domestic vegetable oils noncompetitive with imports. Producers of vegetable oil seeds can compete as growers, but they are penalized when competing in the international vegetables oils market because their products have to be processed by a highly inefficient domestic industry (Gulati and Kelley, 2000).

Agricultural growth of these types can make important contributions to increasing rural Indian incomes. But, like the Green Revolution, such growth is likely to leave many poorer regions and poor people behind. Farmers will prosper most in those regions that can best compete in the market. Competitiveness will require investments in rural infrastructure and technology (roads, transport, electricity, improved varieties, disease control, etc.) and improvements in marketing and distribution systems for higher-value perishable foods (refrigeration, communications, food processing and storage, food safety regulations, etc.). If poorer farmers and regions are to benefit from these new opportunities, policy makers will have to assist them rather than leaving everything to market forces alone.

Helping small-scale producers capture part of these growing markets will require that agricultural research systems give greater attention to the problems of smaller farms and not just large ones. The private sector seems likely to play a greater role in undertaking the research needed for many higher-value products. But private research firms will be more attracted to the needs of larger farms than of small ones and to regions with good infrastructure and market access. Public research institutions will need to play a key role in ensuring that small farmers and more remote regions do not get left out.

Smallhold farmers will also need to be organized more effectively for efficient marketing and input supply. While smallholders are typically more efficient producers of many labor-intensive livestock and horticultural products, they are at a major disadvantage in the marketplace because they have poor information and marketing contacts and their smaller volumes traded (both inputs and outputs) lead to less-favorable prices than larger-scale farmers receive. Contracting arrangements with wholesalers and retailers has proven useful in some contexts, but, for the mass of smallhold farmers in India, some kind of cooperative marketing institutions probably offer a more realistic option, even recognizing the many faults of previously government-sponsored co-ops. Operation Flood is a good example of what can be done with good leadership, use of modern technologies and commitment to serving farmer and consumer interests rather than those of intermediaries (Doornbos and Nair, 1990; Kurien 1997). This program supports dairy cooperatives for the collection, treatment and marketing of milk, produced by many millions of small-scale producers, including landless laborers and women. Many of the smallholders produce only 1 or 2 liters per day. In 1996, Operation Flood involved 9.3 million farmers, yet still accounted for only 22% of all marketed milk in India (Candler and Kumar, 1998). The Government assists the program through technical support (e.g., research and extension, veterinary services and the regulation of milk quality), but otherwise the program is run by the cooperatives themselves with no direct financial support from government.

Spreading the benefits of new growth opportunities to less-favored areas will also require focused policies and investments. These will need to include greater investment in research, infrastructure and human capital to improve the ability of less-favored areas and producers to compete in the market place. Policy makers have been reluctant to do this in the past, preferring to rely on the "trickle down" benefits from investments in high-potential areas, i.e., increased employment there, migration opportunities and cheaper food. But this approach has proven insufficient to resolve the problems of many less-favored areas. Although people migrate to better areas and urban jobs, rural population is nevertheless increasing. Population densities are still increasing in many less-favored areas and seem likely to do so for at least a few more decades. Without adequate investments in basic infrastructure, technology and human development, less-favored areas will lose out even further as agricultural markets become more liberalized and competitive. They will become victims, not beneficiaries, of market liberalization and globalization, with worsening poverty and environmental degradation.

Does investing in less-favored areas have to mean less growth per dollar of investment than investing that money in high-potential areas? Few would dispute the possibility of achieving bigger direct reductions in poverty by investing in less-favored areas, but are there significant tradeoffs with long-term economic growth and poverty reduction? Will present investments in less-favored areas reduce the long-term prospects of the poor and the country? Recent IFPRI research on India says no. In fact, many investments in less-favored areas offer a win–win strategy for India, giving both more growth and less poverty (Fan et al., 2000). This is true also for investments in research and development, though not necessarily in the most difficult agroclimatic zones.

CONCLUSIONS

India has made impressive gains in agricultural growth, food security and rural poverty reduction since the crisis years of the 1960s. Agricultural growth continues to be critical for addressing the livelihood needs of large numbers of rural people, including most of the country's poor. But future growth will need to be different from the past. It will be less driven by growth in food-grain production and more by new growth opportunities for higher-value livestock, horticultural and agroforestry products for the domestic market, by increased value-added opportunities in agro-industry and by export opportunities. Moreover, if future agricultural growth is to benefit the poor, it must be more focused on rain-fed areas than in the past, including many of the less-favored and backward regions that gained relatively little from the Green Revolution.

Future agricultural growth will also need to be more environmentally benign and sustainable than in the past, with greater attention to the problems of intensive farming areas. This will require policy reforms to change incentives in favor of more sustainable technologies and natural-resource management practices, as well as appropriate types of agricultural research.

Meeting these challenges will require serious policy and institutional reforms, including the phasing out of input subsidies, trade liberalization (with removal of trade protection for agroindustry), reform of public institutions serving agriculture and increases in productive investment in agriculture and the rural sector. (This also means spending less, not more, on agricultural subsidies.) India has been flirting with some of these changes for over a decade, but progress has been impeded by entrenched interests in the farm, agro-industry, banking and public sectors that are politically very difficult to challenge at the present time. It will take the same kind of vision to surmount these problems and to rejuvenate the agricultural sector as it did to launch the Green Revolution some 35 years ago, that is, strong political leadership drawing on and supported by the best available current scientific knowledge.

REFERENCES

Bhalla, G.S., P. Hazell and J. Kerr. 1999. *Prospects for India's Cereal Supply and Demand to 2020.* 2020 Vision Discussion Paper 29. International Food Policy Research Institute. Washington D.C.

Candler, W. and N. Kumar. 1998. India: The Dairy Revolution. Operations Evaluation Department. World Bank. Washington, D.C.

Doornbos, M. and K.C. Nair, Eds. 1990. *Resources, Institutions and Strategies: Operation Flood and Indian Dairying.* Sage Publications. New Delhi, India.

Fan, S., P. Hazell and S. Thorat. 1999. Linkages between Government Spending, Growth and Poverty in Rural India. Research Report 110. International Food Policy Research Institute. Washington, D.C.

Fan, S., P. Hazell and T. Haque. 2000. Targeting public investments by agro-ecological zone to achieve growth and poverty alleviation goals. *Food Policy,* 25 (4): 411-428.

Gulati, A. and S. Narayanan. 2000. Demystifying fertilizer and power subsidies in India. *Economic and Political Weekly,* March 4, 784-794.

Gulati, A. and T. Kelley. 2000. *Trade Liberalization and Indian Agriculture.* Oxford University Press. New Delhi, India.

Kurien, V. 1997. The AMUL Dairy Cooperatives: Putting The Means Of Development Into The Hands Of Small Producers in India. In *Reasons for Hope: Instructive Experiences in Rural Development,* A. Krishna, N. Uphoff and M.J. Esman, Eds. 105-119. Kumarian Press. West Hartford, CT.

Pingali, P. and M.W. Rosegrant. 2000. Intensive Food Systems in Asia: Can the Degradation be Reversed? In *Tradeoffs or Synergies? Agricultural Intensification, Economic Development and the Environment,* D.R. Lee and C.B. Barrett, Eds. CABI Publishing. Wallingford, UK.

Rosegrant, M.W. and P. Hazell. 2000. *Transforming the Rural Asian Economy: The Unfinished Revolution.* Oxford University Press for the Asian Development Bank. Hong Kong.

Shiva, V. 1999. Betting on Biotechnology: Why Genetic Engineering Will Not Feed the Hungry or Save the Planet. Research Foundation for Science, Technology and Ecology. New Delhi, India.

World Bank. 1999. Toward Rural Development and Poverty Reduction. Paper presented at NCAER-IEG-World Bank Conference on Reforms in the Agricultural Sector for Growth, Efficiency, Equity and Sustainability, India Habitat Centre, New Delhi, April 15-16.

27 The Role of the Public Sector in Achieving Food Security

G. Edward Schuh

CONTENTS

INTRODUCTION

The World Food Summit organized by the FAO some 5 years ago set a daunting target of reducing the number of food-insecure persons in the world by half by the year 2015. A recent review of how well the world is doing in attaining this objective showed the results to be disheartening. The progress that most specialists thought would be made has not been achieved. Thus, a discussion of how we might do better is both timely and appropriate.

Two propositions provide the necessary context for this discussion. First, despite the importance of the agricultural sector to both the global food security problem and to economic growth more generally, agriculture and the rural population continue to suffer discrimination by national policy makers and by international development agencies. Second, the importance of public goods for dealing with food security and economic development issues is being sorely neglected, again by both national and international policy makers. It is especially worrisome that the U.S. Agency for International Development and the World Bank have shifted most of their declining support for agriculture away from helping to supply necessary public goods, both in the form of investments and better institutional arrangements.

Public goods are typically, though not exclusively, provided by governments. When we consider the responsibilities of the public sector for development efforts, we are concerned mostly with the roles of national governments. However, when the international scene is considered, it is not clear just what organizational structure should and can act when public-sector policies need to be improved. Because there is no international government, there is no entity truly responsible for providing public goods for the international economy. Instead, a number of organizations such as the World Trade Organization, the World Bank, the United Nations Environmental Programme and the United Nations Food and Agriculture Organization operate on a voluntary basis, without binding authority, or often financial means, to enforce any decisions. Yet there is a clear need for public goods to be provided by these or other international organizations.

Amartya Sen (1981) argued that food security is basically a poverty issue, not a food production problem. This does not mean that agriculture can be neglected in efforts to achieve food security objectives. On the contrary, agriculture and rural development are important means of alleviating poverty in most countries. Without adequate and growing supply, it becomes more difficult to redress existing food deficits. But it is necessary to understand that food security does not depend first and foremost on increasing the supply of food. People's ability to purchase food is most crucial and it will help to elicit the production needed. Without purchasing power, people's food needs will not be met or will be met inadequately.

When the World Bank's policy paper on food security was drafted over a decade ago, it was useful to make a distinction between short-term and long-term food security problems. Taking poverty alleviation as the guiding principle for addressing the food security problem, the solution to the longer-term problem is to raise the per capita incomes of the poor. The solution to short-term problems is to devise safety nets that will help carry people through short-term crises. Both dimensions of the food security problem will be addressed in this chapter, which has six parts. The first addresses the issue of providing the new knowledge needed for furthering agricultural development. A second part discusses the role of the public sector in providing education and good health — two important aspects of human capital. The third part looks at the issue of fertilizer policy. A fourth part then addresses international trade issues, while the following part offers some thoughts on rural development strategy. The last part will address the issue of safety nets. Throughout the discussion, the focus is on the role of the public sector.

PRODUCING NEW KNOWLEDGE
FOR AGRICULTURAL MODERNIZATION

New knowledge is a critical source of economic growth and development. The use of new knowledge for the modernization of agriculture is critical to the alleviation of food security problems, in large part because it is key to the alleviation of poverty — among both the rural and urban populations. Poverty in most developing countries, and this includes India, is a result of very low productivity in the agricultural sector. This not only affects food availability but keeps the price of food higher than it needs to be. The production and distribution of new production technology for agriculture is critical to eliminating widespread low productivity. While most of the world's poor continue to be in agriculture and the rural sectors, the poverty in urban centers is growing rapidly as urbanization proceeds at a fast pace all around the world. It turns out that the modernization of agriculture can contribute in important ways to alleviating urban poverty as well.

This important contribution that agricultural modernization can make to alleviating urban poverty is sadly neglected by policy makers and by those responsible for agricultural research. There is a failure to recognize that the modernization of agriculture has effects that go far beyond its direct effects in that sector. For example, given the conditions of demand for food, the introduction of new technology into the production of food staples tends to lead to declines in the real price of those commodities. Any decline in food prices is equivalent to an increase in the real income of consumers. The more widespread the consumption of a particular commodity produced more efficiently, the more widespread are the benefits of the modernization process. Moreover, the benefits will tend to favor the poor, because they tend to spend a larger share of their income on food.

The introduction of new technology into the production of tradable commodities — export crops and crops that compete with imports — can make similar contributions to alleviating poverty. In these cases, increases in productivity improve the competitiveness of the sector and enable the country to earn or save foreign exchange. Those exchange earnings can be used to finance greater investment for a higher rate of economic growth. This economic growth will generate more employment and thus more opportunities for gainful employment on the part of the poor and disadvantaged. Moreover, many export crops tend to be labor intensive. The increased competitiveness of such sectors will generate increased employment, some of which will be among the poor and disadvantaged.

The key issue in the modernization of agriculture is the production of the public goods needed to bring it about. One of those critical public goods is new knowledge, typically embedded in new production technology. A critical role for the public sector is to generate new biological innovations in the form of improved varieties needed to raise productivity. Mechanical innovations in the form of capital inputs will, for the most part, be provided by the private sector.

Biological technology (commonly referred to as biotechnology) and intellectual property rights have significantly changed the conditions under which biological innovations are produced. Biotechnology is equivalent to technological progress in doing biological research. When combined with the creation of intellectual property rights, it has significantly expanded the role of the private sector in producing new technology for agriculture. In effect, these two developments have at least partially transformed what was once inherently a public good into a private good that the private sector can provide.

This does not mean, however, that there is no role for the public sector in producing new knowledge or production technology for agriculture, despite the growing tendency to draw that conclusion. Substantial areas remain in which privately funded biotechnology research is not likely to produce enough or appropriate biological innovations to reduce poverty and enhance food security. The private sector does not have incentives to make many of these investments. Moreover, it is not likely to undertake the social science research needed to facilitate modernization, nor is it likely to fulfill the need for basic and strategic research. Its priorities will be, understandably, to meet the needs and demands of the better-endowed, because this will be more profitable.

The failure of governments to invest at socially optimal levels in the production of new technology for agricultural modernization constitutes a major failure to address the food security issue in a farsighted and efficient manner. It also means passing over what could be a powerful source of economic growth. Sadly, the U.S. Agency for International Development, other bilateral development agencies and the World Bank have all turned away from their past commitments to agricultural research. This failure to make the investments that could be such a powerful source of the economic growth needed to alleviate poverty, especially among the poor and disadvantaged, is laying the ground for a serious food security problem in the future.

Those who are complacent on this issue argue that the private sector will pick up the responsibility for producing the needed new technology. Obviously, the private sector can provide a significant part of it, especially if intellectual property rights can be assured. But the kinds of research the private sector can and will undertake may not be the most socially optimal or relevant to addressing the food security or general economic development issues.

Similarly, some have proposed that producers organize their own associations to provide the financial support needed for the research of interest to them. Two problems make this an unrealistic alternative. The first is the classic free-rider problem. Once created, knowledge tends to be widely available and users can benefit from the new technology without paying for the cost of creating it. But the fact that producers tend to receive only part of the benefits from most new technologies is even more important. If they should somewhat fortuitously support the research in proportion to the benefits they receive, they would still underinvest relative to the total social benefits generated if the many gains to consumers and those from alleviation of poverty are taken into account.

Producers are not the only beneficiaries of an agricultural technology, or even necessarily the primary beneficiaries. If, by raising productivity, new technology leads to lower prices, consumers may benefit more from that technology than do producers. Indeed, productivity increases can raise supply enough so that market prices fall below producers' costs of production. Accordingly, consumers should share in the cost of creating and using technology. These considerations lead to the conclusion that the public sector should make significant investments in agricultural research if the production of new knowledge for the continuing modernization of agriculture is to be sustained at a socially optimal level. To the extent that consumers are the beneficiaries of new technology, one can justify public expenditure for its development and dissemination.

EDUCATION, VOCATIONAL TRAINING AND HEALTH

Investments in education, vocational training and health are highly complementary to investments in the production of new knowledge. Cognitive skills and literacy are needed if the producer is to decode, understand and adopt the new technology as it is released to the agricultural sector. An important means of developing cognitive skills is through formal schooling. As noted in my introductory comments, governments everywhere underinvest in the education of their rural populations.

Vocational skills are also important, not only in terms of the skills needed for agricultural production, but to seek gainful employment in activities outside of agriculture. For example, it is now well known that even in poor regions of the world a significant, even major, share of the incomes of farm people comes from off-farm employment. Providing the vocational skills for these employment opportunities can make an important contribution to raising the per capita incomes of rural or agricultural families.

Perhaps the more important reason for providing both formal schooling and vocational education is that it is inherent in economic development that members of the labor force have to leave the sector to seek gainful employment elsewhere if their incomes are to keep up with those in the urban sector. This need for sectoral reallocation of productive resources is rooted in Engel's law, which states that, as per capita incomes rise, consumers will spend a smaller and smaller share of their income on food. Thus, as average incomes rise, the demand for agricultural products increases at a slower pace than the demand for nonfarm goods and services and more and more labor is needed in the expanding nonfarm sectors. In addition, as the process of modernization continues, the quantity of food demanded can be supplied by an ever-smaller labor force, especially in the case of the food staples. Thus, a double squeeze is placed on the agricultural labor force to shift to alternative employments.

It is well recognized that general education produces benefits to the society that are far larger than can be captured by the private individual. This is true for all levels of education and explains why education tends to be provided by the public sector at all levels. Even when the educational system is privatized, there is still a role for public investments.

The provision of vocational training is more of a mixed bag. Private companies that hire workers have some incentive to provide certain kinds of vocational skills to their employees. However, they will tend to provide only those skills that are specific to the firm. Helpful as that may be, it is still inadequate compared with the need to provide employment alternatives for those exiting the agricultural sector. Hence, even in the case of vocational training, there is an important role for the public sector.

Sound nutrition and health are also critical to raising per capita incomes among the poor. In the first place, agriculture tends to be a physically demanding activity. the workers' productivity is greater if they are adequately nourished and healthy. But good health and nutrition are also important for their impact on the ability of individuals to absorb and develop both cognitive and vocational skills. In fact, there is an interesting and important complementarity in the modernization of agriculture; sound nutrition and health are needed if the workers are to absorb cognitive and vocational skills and, in turn, cognitive and vocational skills are needed if the workers are to adopt and use the new knowledge produced by agricultural research. Providing all forms of human capital assures a higher rate of return on all of the investments.

Significant externalities result from sound health. Lack of widespread good health makes a population more vulnerable to contagious diseases. The effects are far-reaching in a society, which means that the benefits from good health are far larger than individuals can capture from their own investment in good health. Thus, health services become a public good and the role of the public sector in providing them at socially optimal levels needs to be recognized.

The importance of sound health as a component of agricultural modernization has been sorely neglected in past agricultural development programs. A short book edited by my colleague Vernon W. Ruttan called our attention to this deficiency a few years ago (1974). However, this issue has still not received the attention it deserves.

FERTILIZER POLICY

Fertilizer policy has become a controversial issue in many parts of the world and especially in the low-income developing countries. The controversy grew out of the attempts by international development agencies such as the World Bank to help developing countries deal with their adjustment and stabilization problems that arose as a consequence of the economic crisis of the 1980s. The need for sound fiscal policies led to policy recommendations that opposed the use of any subsidies. Subsidies for the use of fertilizers were high on the list of those expenditures recommended for elimination.

While sympathetic to the need to reduce the drain on the public coffers in many parts of the world, I am also cognizant of the desirability of subsidies being phased out in an orderly fashion. I believe the positions taken by these agencies, unequivocally pressing for subsidy elimination, have been too stringent and single-minded. There are two reasons for reconsidering the case. First, a subsidy can be socially optimal in cases where there is an information asymmetry between scientists and farmers in the knowledge about what fertilizer can contribute to production. Under these conditions, the use of the subsidy can accelerate the adoption of the new technology and thus accelerate the process of modernization. In such cases, the use of a subsidy should be phased out as soon as the knowledge of a technology's value becomes more widespread among farmers.

The other reason arises where there is no adequate credit system to enable the producer to purchase and use fertilizer, for example. The use of fertilizer can be a privately profitable and a socially valuable investment for farmers. However, they may need to borrow money to be able to make this investment. If there is no appropriate credit system in operation, not only will individual farmers be worse off because of the income they sacrifice, but the use of this modern input will be less than optimal for the society.

The first-best solution to this problem is to develop a sound agricultural credit system. With such a system, farmers will be able to borrow the resources they need to purchase the fertilizer at a privately optimal rate. In the absence of such a system, and credit systems for agriculture are notably deficient all around the world, the use of subsidies for fertilizer may be justified. The importance of this issue can be seen by noting that the use of modern fertilizers is critical to the adoption of improved varieties and, without the use of fertilizers, improved varieties will not be adopted. So the loss to society can be substantial.

This issue has come to my attention in a special way in my work in sub-Saharan Africa over the past decade. African soils have been leached of many of their nutrients by torrential rains over many, many years. Making this situation worse, farmers have mined these soils of nutrients for a long period of time. Even with an adequate credit system designed to provide short-term credit, the ability of individual farmers to restore the level of soil nutrients to their optimal level would be limited. Yet, reaching that level would be essential to allow agricultural modernization to proceed in an optimal way. Subsidies for building up soil fertility through some combination of inorganic and organic nutrients on applications could have very beneficial effects on the agricultural sector and on society.

There are undoubtedly other parts of the world in which these same conditions prevail. The role of the public sector in assuring that fertilizer is used to the optimal level is critical to both economic development and to addressing the food security issue. This is not to argue that fertilizer use should be generally subsidized. The point is that public-sector policies should be based on local conditions and should be viewed from the perspective of what is socially optimal. At certain times and in certain places, subsidies are very justifiable, such as when they can compensate for poorly performing or nonperforming credit institutions.

INTERNATIONAL TRADE

International trade has two important roles to play in dealing with the food security issue. The first is important for addressing short-term food security issues that arise as a result of crop failures. The second is important in generating economic growth in the longer term and thus addressing the poverty issue. In both cases, public policy and thus the public sector are important.

To consider the short-term problem first, policy makers in developing countries are prone to follow two kinds of policies to address their short-term food security problems. The first is to pursue food self-sufficiency policies by erecting barriers to trade and encouraging domestic production. India is an example of a country that has pursued such policies. The problems with such an approach are now evident for everyone to see. India has technically been self-sufficient in food production for some years now, but many millions of people in this country continue to be food insecure. Self-sufficiency policies do not address the basic issue of poverty that lies at the root of food insecurity.

A second policy followed by many countries is to construct storage facilities and carry food reserves to offset any shortfalls in production when they occur. Such a policy has two problems associated with it. The first is that building storage and carrying reserves is very costly and immobilizes large amounts of capital that might be used more productively in other ways. Moreover, the management of the reserves to provide true security is a very complex issue in an uncertain world. The inability to predict what the weather will be even months, if not years, in advance makes it difficult to know when and how much of the reserves to accumulate or release at any given time.

A more efficient solution to the problem would be to maintain an open economy and purchase food in the international market. In years in which output is greater than domestic demand, the surplus can be exported to earn foreign exchange. In years of shortfall, imports can be purchased from abroad to fill the gap. The foreign exchange earnings from the years of large production can be held in reserve to acquire the imports in the years they are needed. The foreign exchange can be invested in government bonds of other countries such as the United States and be earning an income. The large amounts of capital invested in granaries and reserves yield no return in most years, especially if the stocks are managed poorly.

Trade policy can thus be used to address the short-term food security problem. This policy inherently involves public-sector decision making, not only in pursuing an optimal exchange rate policy, but also in lowering protectionist barriers to facilitate an open trading system. An appropriate information policy is also important if the private sector is to play its role in importing and exporting the supplies of food.

International trade policy is also critical in addressing the longer-term food security problem, for international trade can be an important engine of economic growth. Recent developments in endogenous growth models have focused attention on this issue as a component of development policy. In fact, an important contribution of endogenous growth models is that they have shown the link between international trade policy and economic development policy.

The role of international trade in promoting economic development is rooted in the traditional concepts of the division of labor and specialization and in capitalizing on comparative advantage. Adam Smith, whose famous book *The Wealth of Nations* was published in 1776, addressed the issue of the division of labor and specialization as a source of what he referred to as economic progress. Recall that, at the time he wrote, the pace of technological change as we know it today was hardly noticeable, yet nations still experienced economic progress. Smith attributed this to specialization and the division of labor among members of the labor force. He argued that there was a limit to the benefits of this source of economic growth — the extent of the market. Given the importance of transportation costs in that era, this meant that small countries or those isolated from the international economy had a limit to the extent to which they could experience economic growth and development, especially if they were constrained by a relatively autonomous development policy.

Allyn Young (1929), rejuvenated Smith's division of labor and specialization by casting it in a somewhat larger context. Young wrote about the *sectoral* division of labor and specialization and referred to the increasing returns from such specialization. He was concerned with the fact that, as an industry expands, subsectors of it spin off as separate sectors of the economy. For example, automobiles are no longer entirely produced in one company or plant. The basic components, such as wheels, tires, generators, batteries and so on get spun off as separate industries. Young cogently argued that this sectoral specialization led to efficiencies in production and eventually to reductions in the cost of the various components. To the extent that the outputs of the subsectors were used in other sectors of the economy, the benefits of specialization and division of labor could be pervasive. We thus have a very powerful source of economic growth.

The endogenous-growth modelers have linked these benefits to international trade. If a country is willing to specialize and trade according to its comparative advantage, a limit to its economic growth is no longer imposed by the extent of the market. On the contrary, if a country is willing to specialize and integrate itself into the international economy, despite the problems that might be associated with such specialization, the potential to raise the incomes of its population is indeed great.

Opening one's economy to international trade provides other benefits, and it was to these issues that the endogenous growth modelers gave a lot of attention. Opening an economy to international competition creates pressures to search for more efficient means of production. In addition to making more efficient economic use of the nation's resources, the private sector goes in search of new production technology. That, of course, is also a powerful source of economic growth and development. Moreover, the more efficient policies that tend to be associated with having a more open economy, together with an inflow of new production technology, give greater incentive for capital to flow into the economy. The country is no longer limited in its economic growth by its domestic savings; it can draw on savings from the international economy. Thus, trade liberalization can be a powerful source of economic growth and provide the means for addressing the food security problem.

Two additional comments are pertinent here. The first concerns the role of the public sector and public policy in promoting increased competitiveness. The investments referred to in earlier sections — in the production of new technology and in education, vocational training and health — are all critical to making a country more competitive in the international economy. In fact, in today's world, a nation no longer needs to take its present comparative advantage as a given, or as determined by some "original" endowment of resources. The above investments can help it change its competitive advantage in very important ways.

Further, a nation's international trade policy is not the only component shaping its ability to address its food security problems via trade policy, although it is usually a useful place to start. The international environment in which trade takes place is also important. If we could generally succeed in lowering the barriers to trade, we could do much to alleviate food security problems all around the world. The protection that the developed countries of Europe and the United States provide to their agricultural sectors has especially pernicious consequences for low-income developing countries.

Globalization has many contemporary critics, many of them seemingly wanting either to stop the process or return to an earlier period. Such arguments fail to recognize that globalization is being driven by three technological revolutions — in the transportation, communication and information-technology sectors. These revolutions have greatly enhanced the scope of markets as the means of organizing economic resources and are driving the growth in international trade and financial flows. Society is not likely to give up the benefits of these technological breakthroughs. On the contrary, the fact that these technological revolutions have only begun to reach the developing countries, where 80% of the world's population live, and the previously centrally planned economies, suggests that the process of globalization is likely to expand and become more complex.

RURAL DEVELOPMENT

Rural development is finally returning to the agendas of international development agencies as a relatively high priority. This is long past due, because, as noted above, a major share of a nation's poor continue to be concentrated in rural areas. When the process of agricultural modernization, which is so critical to the alleviation of food security problems, is successful, the need to facilitate the exodus of labor from agriculture into other sectors is critical both as an equity and as an efficiency issue.

The rapid pace at which the process of urbanization is taking place around the world is of special concern from a poverty perspective. In the first place, this geographic migration has some very deleterious effects associated with it, despite the wide belief that migration is the only way to reduce wage differentials among regional labor markets.

Geographic migration is a highly selective process, with the migrants tending to be the young, the better-educated, the healthier and the more entrepreneurial. Hence, it involves a drain of human capital from the supplying region or sector. Rather than narrow the wage differential between two sectors or regions, this exodus of labor from rural areas decimates the supplying area and leaves it in a weakened position to develop local alternative employment opportunities.

The difficulties do not stop there, however. The accumulation of this labor in large urban centers imposes large negative externalities in those centers as well, in the form of congestion, pollution, rising costs for public services and the need for expensive transportation systems. It can also contribute to social conflict and crime. It is difficult to imagine a more counterproductive way to deal with the adjustment problem that an economy experiences as economic development proceeds, because both parts of a national economy lose rather than gain when population movement is driven more by desperation than by positive opportunities for greater productivity.

Contrary to what many believe, this process is not due just to the natural functioning of a market economy or some invariable pattern of economic development. Misguided economic policies contribute to the problem in important ways. On the agricultural or rural side, policy makers typically underinvest in agricultural technology, in the education of the rural population and in the physical infrastructure in rural areas. Each of these failures in policy weakens the performance of the rural and agricultural sectors and makes them unattractive places for private investment. On the urban side, policy makers tend to subsidize the location of private activities in such centers by a number of means. For example, transportation costs are reduced to private individuals by allowing them to pay only part of the costs they impose on society. Additionally, tax or fiscal incentives for the location of economic activities in such areas are provided and public services such as water and sewage are subsidized for urban residents but not for rural ones.

The solution to this problem is to reverse these policies. The elimination of subsidies for the accumulation of populations in urban centers is an important first step. Additional benefits would come from strengthening the educational and health care services for rural populations and by strengthening the physical infrastructure in rural areas. An industrial policy is not needed. The key issue is to eliminate present distortions in public policy. By so doing, rural areas will become more attractive to the private sector as a place for investment and thus create more employment for the rural workforce. In addition to alleviating the problem of rural poverty and thus contributing to an alleviation of food security problems, such policies will also make for a more efficient use of the nation's resources. In most parts of the world, agriculture is inherently a part-time activity. Taking the jobs closer to the worker will enable the country to make fuller use of its supply of labor.

SAFETY NETS

This analysis makes clear that promoting economic growth is essential to alleviating the poverty that is the cause of food insecurity. Whether one focuses on the short- or long-term food security issue, raising the incomes of poor people is the key to addressing food security problems. However, neither in the short term nor in the long term will this be sufficient to solve these problems. There will always be some who need assistance in a time of crisis, or who are not able to participate effectively in the market economy. For these people, safety nets are needed if our obligation to fellow human beings is to be met.

Safety nets come in various forms, including targeted feeding programs. There is a wide variety of programs from which to choose. The United States for a long time used the food surpluses that were generated by its misguided commodity programs to support a program that issued "food stamps" to the poor that could be used to buy food. That program has, for the most part, been replaced by more general poverty alleviation programs that provide cash payments. Other countries such as India use low-price food stores as a means to get food to the poor on convenient terms.

If it is not to become excessively costly, almost any such program requires a means test, screening out persons who could afford to buy food for themselves. The administrative capability for such a system, particularly for screening, is often not available, or is available only for the urban populations. However, if a nation is serious about addressing its food security problems, some form of "safety net" is essential to ensure that all those in serious need of basic nutrition have adequate and real access.

In addressing these problems, it is important to avoid inappropriate interventions in the working of the market economy. When the World Bank's policy paper on food security was being prepared, a fairly common means for addressing the food security problem was to distort all food prices downward so that the poor would be better able to acquire their food with their limited incomes. The disadvantage of that approach is that it subsidizes the rich and poor alike, at the same time leading to inefficient use of the nation's resources.

CONCLUSIONS

The widespread shift to privatization and dependence on markets in recent years has brought many gains by increasing the efficiency with which a nation's resources are used. That increased efficiency is important as a source of economic growth that increases per capita incomes, which is essential to reducing food insecurity. However, neoclassical orthodoxy often takes over in promoting dependence on markets beyond what would be socially optimal; government that intervenes the least in the economy is viewed as intrinsically the best government.

This perspective misses a very important point. Certain economic activities will not be undertaken by private enterprises, or will be undertaken at less than socially optimal rates. This can be seen from both theory and practice. Thus, the challenge is to identify those areas in which the private sector will have the incentive to produce needed goods or services and those areas where the public sector has an important contribution to make. Finding a good balance between the two sectors is the key issue. Economics as a discipline strives to promote optimization in resource use, which is unlikely to be attained by any one approach stressing a single sector rather than by some mix or blend of sectors.

To promote food security, we have argued that public investments in agricultural research, in education, in vocational training and in health are all essential activities. In addition, certain institutional policies and policy changes are crucial, such as subsidies for accelerating the use of fertilizers, facilitating freer trade, promoting rural development and providing safety nets. Domestic public sectors are the front line for most of these policies, but the international community has a role to play in agricultural research, in health care and in providing an environment in which international trade can take place to the benefit of rich countries and poor countries alike.

REFERENCES

Ruttan, Vernon W., Ed. (1974) *Health and Sustainable Development: Perspectives on Growth and Development.* Westview Press. Boulder, CO.

Sen, Amartya (1981). *Poverty and Famines: An Essay on Entitlement and Deprivation.* Clarendon Press. Oxford, UK.

Young, Allyn (1929). Increasing returns and economic progress. *The Economic Journal,* 18:152, 527-542.

28 Global Food Supply and Demand Projections and Implications for Indian Agricultural Policy

Luther Tweeten

CONTENTS

Aligning food supply and demand at acceptable prices to farmers and consumers is a challenge in any country, but, though essential, it is an especially daunting task in India. Certainly India has made great progress in increasing food production and no longer depends on food aid as in the 1960s. The Green Revolution's high-yielding rice and wheat varieties coupled with investments in irrigation, roads and other rural infrastructure have enabled India to meet its food needs as grain production has nearly tripled since 1965–66 (Kumar et al., 1995).

Yet, overall, Indian agricultural performance remains disappointing. Two thirds of its people remain employed in agriculture (compared with 43% in China) and many of those constitute the number of all Indians who are chronically food insecure (*The Economist*, 2001). Abject poverty continues to plague several hundred million Indians. With population growing at a brisk 1.7% per annum, India will surpass China as the most populous nation by year 2050. Moreover, crop yield increments continue to fall, in part because of severe soil erosion, salinization and waterlogging of irrigated lands, and because government investments in research and extension to improve crop productivity are lagging.

This chapter proposes options for India to address its food supply and demand problems, including poverty, food insecurity, environmental degradation and population growth. The analysis shows that if the Indian government could successfully resolve its food supply and demand challenges chiefly by making wise investments in education, agricultural research and infrastructure, it could let markets mostly determine whether the food needed will be produced at home or abroad. The key question for India is not whether it can produce enough food to feed itself, but whether the world will produce enough food and Indians will have the means required to purchase it.

Hence, this analysis does not project Indian domestic food supply and demand; instead, it assesses future prospects for global food supply and demand. Four studies undertaken at The Ohio State University contribute to this analysis. The first projects long-term global food supply and demand trends to provide a basis for judging whether India can draw on other countries for commercial food imports, if necessary, while it pursues a market-based policy of increasing real income by producing goods and services in which it has a comparative advantage. The second study addresses the relationship between environmental protection and income growth to determine whether the pursuit of economic growth protects or destroys the environment. The third and most important study outlines, based on global historical experience, a standard economic model of public policy to see how India can successfully address its food security, environmental, poverty and population problems. The final study lists high-payoff public investments for India — again based on global historical experience.

FUTURE GLOBAL FOOD
SUPPLY–DEMAND BALANCE

This section draws on a study by Tweeten and Zulauf (2001) that traces historic trends and projects long-term global food supply and demand. Principal conclusions of this study are:

- Rates of increase in global yields of major crop and livestock groups are falling. Cereal grain yield trends since the 1950s, as recorded by the Food and Agricultural Organization (FAO) of the United Nations, have fluctuated around a remarkably linear, straight-line trend (see Figure 28.1). Cereals account for over half of all calories, so it is not surprising that the percentage annual increases depicted along the horizontal axis are nearly the same for cereals and for an aggregate of all foods. Annual trend yield increments fell from 3.13% in 1961 to 1.92% in 1981 and further to 1.43% in 1999.
- Land area in crops, after increasing globally by approximately 1% annually for several decades, has remained quite stable now for a decade. Gains from new cropland in Brazil and elsewhere have been offset by losses of cropland to urbanization and from environmental degradation throughout the world. It is a sobering prospect that, with constant land area and

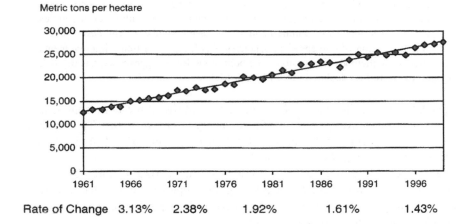

Metric tons per hectare

Rate of Change 3.13%   2.38%   1.92%   1.61%   1.43%

FIGURE 28.1 World cereal yield trend, 1961–1999.

    slowing yield growth, global food production could start to grow more
slowly than population.
- This Malthusian specter of rising real food prices (of special concern to
  countries such as India) is likely to be only narrowly averted by a profound
  shift in global demographics. All major demographic projections look to
  a continued slowing of global population growth until the total population
  begins to decline within the next 50 to 80 years.

    Since the 1950s, the total fertility rate (TFR, the expected number of children that
a woman will bear throughout her life) has fallen in every region of the world. From
an average of nearly six children per woman in 1950–55, by 1990–95 TFR had fallen
to 3.4 in India, 3.5 in the rest of Asia and 3 in Latin America and the rates continue
to fall (United Nations, 1998). TFRs in the 1990–1995 period were already below the
2.1 average number of children per woman needed to sustain population levels over
the long run in Europe (1.57), China (1.92) and North America (2.02).

    The medium population projection by the United Nations (UN) is a widely used
demographic forecast, but it assumes, apparently unrealistically, that TFRs will
converge to 2.1 in both developed and developing countries. That assumption overestimates future population, according to Lutz et al. (1996: 365):

> "The United Nations and other institutions preparing population forecasts assumed that
> fertility would increase to replacement level and that subreplacement fertility was only
> a transitory phenomenon .:.. It is difficult, however, to find many researchers who
> support this view. Too much evidence points toward low fertility. Many significant
> arguments support an assumption of further declining fertility levels. They range from
> the weakening of the family in terms of both declining marriage rates and high divorce
> rates to the increasing independence and career orientation of women, and to a value
> change toward materialism and consumerism.

"These factors, together with increasing demands and personal expectations for attention, time and also money to be given to children, are likely to result in fewer couples having more than one or two children and an increasing number of childless women. Also, the proportion of unplanned pregnancies is still high and future improvements in contraceptive methods are possible. The bulk of evidence suggests that fertility will remain low or further decline in today's industrialized societies."

The UN's low-medium scenario seems more realistic. It presumes continuation of TFR trends, but converging to 1.9 TFR for all regions by year 2025. This scenario projects a peak world population of 8.0 billion people in 2050, declining to 6.4 billion by 2150 (United Nations, 1998: 14).

Several peak world population projections are summarized in Table 28.1. Most projections indicate we will reach maximum global population in less than a century, followed by negative population growth (NPG). Today's population is not expected to double before global population peaks.

Future portents for food consumers would be ominous indeed in the absence of falling fertility rates. Figure 28.2 shows projected aggregate food supply based on a continuation of 1961–1999 yield trends and no net increase in global cropland.* Alternative aggregate food demand projections from 2000 to 2150 are for the indicated population projections coupled with a 0.3% increase in food demand per capita due to income growth.**

If population and income maintain their 1995 to 2000 trend for growth, future demand would sharply outgrow future food supply (Figure 28.2). Real commodity prices would need to rise to draw additional land and other resources into food production and to restrain demand. If the United Nations' medium population projection demand or the IIASA scenario (Lutz et al., 1996) prevails instead, food demand growth can be projected to modestly outstrip food supply growth until approximately 2075, a gap that could probably be covered by small increases in real food prices. (This would, of course, have negative effects on the poor.)

TABLE 28.1
Peak World Population Projections

	Numbers (billions)	Year
World Bank (Bos et al., 1994)[a]	11.3	2128
International Institute for Applied Systems Analysis (Lutz et al., 1996)	10.8	2080
United Nations (1998) (low-medium scenario)	8.0	2050

[a]Tweeten (1998) extended the World Bank projection to ZPG by using a quadratic equation.

---

* The medium UN projection may overestimate future TFRs (2.1) in developed countries, but the low/medium UN projection may underestimate future TFRs (1.9) in developing countries. Hence, both UN scenarios are employed in projecting food demand in Figure 2.
** The 0.3% constant rate of increase in food demand per capita results because the tendency for accelerating demand as income grows in developing countries with high income elasticities of demand is offset by falling rates of income growth worldwide as nations develop.

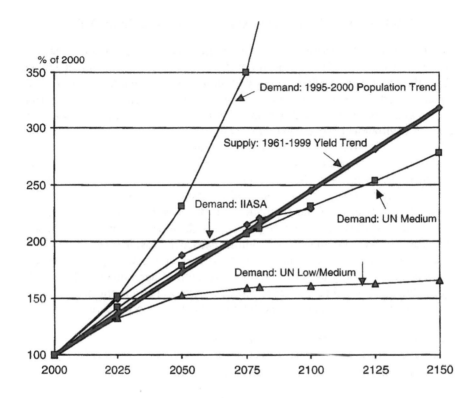

FIGURE 28.2 Projected global food supply and demand trends under alternative scenarios, 2000–2150.

If the demand scenario is based on the probably more realistic United Nations' low-medium population projection, food supply grows somewhat faster than food demand. This outcome would allow real farm commodity prices to continue to fall and would perhaps even accelerate the rate of decline (Figure 28.2). It would also help to redress the nutritional situation of the world's poor. It follows that NPG may narrowly avert rising real farm commodity and food prices in the first half of this century. However, the food balance could be so tight that the world would welcome emerging technologies such as genetic engineering.

The less likely scenario of rising real food (commodity) prices at the farm level would draw additional land and other resources into food production, probably less productive and environmentally stable than are utilized at present. Such a price increase, though not imminent, would hardly be noticed by consumers in the United States, who spend only 2% of their income on farm-supplied food ingredients. The situation would be different in India, however, where food prices heavily influence real income of consumers. The number of food consumers exceeds that of food producers in every country and poor consumers in India who purchase food in the market would be most adversely affected.

ECONOMIC PROGRESS AND THE ENVIRONMENT

The conclusion from the above analysis, that the global food supply–demand balance could be tight in future decades, need not distract India from pursuing economic progress through producing those goods and services for which it has a comparative advantage. Food should be available to countries that have purchasing power from rising incomes. At issue is whether rising national income is compatible with protecting the environment as well as food security.

A study by Hervani and Tweeten (2001) indicates that fortuitous interaction between declining population growth and high income ultimately can save the environment, assuming that effective policies are enacted to establish appropriate incentives and institutions and that ecological situations are not worsened irreversibly in the process. Lower birth and population growth rates, brought about in part by higher income, will eventually reduce pressure on the environment. Higher income allows people to save current income no longer needed to provide necessities, which, in turn, can be used to finance investment in science and technology that reduces pressure on the environment. Rapid productivity gains of agriculture made feasible by agricultural research financed out of economic growth permits cropping of fewer and environmentally safer acres while freeing cropland for grass, trees, recreation and biodiversity.

Education and research made possible by economic progress promote awareness of environmental hazards that can generate active policy protection of the environment. Higher-income consumers demand greater efforts to protect ecosystems because, once their basic needs are met, they manifest a high-income elasticity for environmental quality. Furthermore, as their incomes rise, consumers spend a larger share on services, whose production and disposal are less detrimental to the environment than are the production and disposal of goods. The result is a trajectory of increasingly less environmental damage and greater environmental preservation per unit of national income and per capita as income levels rise.

The relationship between per capita income and environmental degradation has been conceptualized as an environmental Kuznets curve (Seldon and Song, 1994). Empirical studies of the relationship between economic growth and environmental degradation, based on historical data, have supported an inverted U-shaped relationship (Grossman and Krueger, 1995; Hervani and Tweeten, 2001; Ruddle and Manshard, 1981; Seldon and Song, 1994). That is, environmental degradation first rises and then falls as per capita income rises under economic growth. Whether this relationship can be projected indefinitely into the future is, of course, beyond current empirical validation.

Results from Hervani and Tweeten are presented in Table 28.2. They regressed environmental and natural-resource variables on income per capita, population density and selected other variables for approximately 120 countries, using data from the 1990s. The numbers presented in Table 28.2 are long-run elasticities, showing the impact of a 1% increase in per capita income (from the specified income level shown at the top of the table) on the variable in the left-hand column. The geometric progression of the specified income benchmarks is well within the range of historic data except for the highest number. Only one country, Japan, had a gross domestic

TABLE 28.2
Income Elasticities at Selected Income Levels per Capita

Variables	Low Income			Middle Income		High Income	
	$500	$1,000	$2,000	$4,000	$8,000	$16,000	$32,000
Greenhouse Gases and Particulates (per capita)							
CO_2	0.27	0.52	0.98	1.76	2.66	1.93	-9.72
NO_2	0.08	0.15	0.29	0.49	0.65	-0.05	-10.24
SO_2	0.11	0.22	0.41	0.71	0.93	-0.04	-7.70
Methane	0.21	0.40	0.78	1.45	2.44	3.06	-1.16
Suspended Particulates	0.02	0.03	-0.02	-0.32	-1.75	-1.98	-33.88
Resource Depletion (per capita)							
Energy Use	0.60	1.18	2.25	4.07	6.40	5.91	-15.79
Oil Use	0.28	0.54	1.02	1.81	2.72	1.80	-10.94
Phosphate Fertilizer Use	0.78	1.50	2.79	4.70	5.94	-1.98	-59.42
Organic Water Pollution	0.09	0.17	0.33	0.60	0.93	0.81	-5.58
Population Growth							
Total Effect	-0.06	-0.08	-0.15	-0.32	-0.54	-1.21	-2.46

Source: Hervani and Tweeten (2001).

product in excess of US$32,000 per capita in 1995. Income of $32,000 per capita is included to anticipate future outcomes as per capita income grows. However, because this relationship rests on few empirical observations, it must be interpreted with care.

To illustrate the interpretation of the numbers in Table 28.2, consider the most abundant greenhouse gas (GHG), carbon dioxide (CO_2). From a base of $8,000 per capita income, a 1% increase in per capita income raises CO_2 emissions per capita by 2.66%. This elasticity declines to 1.93% at an income of $16,000 per capita and it becomes negative at $20,000 per capita. Thus, beyond some point, emissions per capita fall as income rises to high levels.

Other environmental and natural resource variables also show the anticipated pattern with respect to per capita income. Emissions per capita reach a peak (elasticity and rate of growth are zero) at approximately $15,000 per capita for NO_2 and SO_2, at approximately $25,000 per capita income for methane and at approximately $1,500 per capita for suspended particulates.*

Use of energy and oil increases to approximately $18,000 per capita, while use of phosphate fertilizer begins to decrease at approximately $14,000 per capita.

---

* SO_2 (sulfur dioxide) is another important greenhouse gas, and its elasticity is negative for higher income levels. Low SO_2 implies less acid rain, an environmental hazard. However, a negative elasticity does not necessarily imply less global warming because SO_2 reduces rather than induces global warming.

Higher levels of organic components in water indicate both water resource degradation and natural-resource depletion, the latter because resources are depleted to
produce the fertilizers and other chemicals found in water. Fertilizers, in turn, raise
levels of algae and other organisms in water. The income elasticity for organic water
pollution implies that added per capita income above approximately $20,000 reduces
organic water pollution in the long run.

The total emissions for a country are a product of per capita emissions times
population. To illustrate for CO_2, at $16,000 per capita, adding the elasticity of
population growth with respect to income (-1.21) to the elasticity of CO_2 with
respect to income (1.93) indicates that a 1% rise in income at $16,000 raises total
national CO_2 emissions by 1.93–1.21 or 0.72%. Similar results are evident for the
other variables in Table 28.2. Higher per capita income and slower population
growth can work interactively to hasten the move not only to lower per capita
emissions and use of natural resources, but also to lower total emissions and
natural-resource depletion.

Hervani and Tweeten did not examine the relations of soil erosion to income
and population growth. However, data for the United States indicate that soil
erosion falls as per capita incomes rise and finance new technology, which raises
yields and permits the "luxury" of cropping only environmentally safe lands. In
the U.S., water erosion per hectare of cropland fell from 22.2 tons of soil in 1938
to 5.2 tons of soil in 1997 (Tweeten and Amponsah, 1998: 49; USDA, 1999: 10).
Wind erosion data are not available from the 1930s, but levels fell from 3.5 tons
of soil per cropland hectare in 1982 to 2.0 tons per hectare in 1997 (USDA, 1999:
10). By improving yields and hence freeing cropland for recreational, forest and
development uses, agricultural technology has greatly benefited both the environment and society.

In summary, rising income per capita is seen as unfavorable for most of the nine
environment variables considered by Hervani and Tweeten, up to per capita income
levels of approximately $15,000 to $20,000. As income rises above these levels (and
at lower per capita income when reduced population growth is considered), all
environmental variables improve. These results are consistent with those from
numerous other studies reviewed by Hervani and Tweeten. Rising per capita income
reduces population growth, thus reinforcing the per capita environmental benefit of
rising income at higher income levels.

The conclusion that environmental protection predominates only at quite high
per capita income is troublesome, however. Countries such as India cannot easily
or quickly achieve the per capita income levels that protect the environment. The
economic growth-environmental protection dilemma is especially acute for the
world when large countries such as India and China attempt to pass through the
environmental transition to eventual success in preserving the environment. Considerable resource degradation and depletion are likely to occur in the process. It
follows that special efforts at education, technology development and protective
policies are required for India (and other countries) to make the environmental
transition with as little injury to the environment as possible. Such an effort will
require investment of public resources that, in turn, requires a substantial tax base
for public-sector revenue.

USING THE PROVEN STANDARD ECONOMIC MODEL TO ALLEVIATE POVERTY, FOOD INSECURITY AND ENVIRONMENTAL DEGRADATION

The third, and arguably most important, study provides compelling evidence that any nation, including India, can end poverty and food insecurity while protecting its environment if it follows appropriate policies (Tweeten, 1999). This study begins with the objective of food security, defined as access by all people at all times to a diet needed for a healthy and productive life. Food insecurity is seen as mainly due to poverty, which can be ended by following the standard economic model. The principal impediments to adopting this model are political and these political impediments can be traced to attitudes and institutions.

The standard model is described in detail in Tweeten (1999). Its rudiments are as follow:

Principal reliance is placed on markets that allow supply and demand to set prices and guide what, when, where and how to produce and market goods and services, including food.

A lean but effective public sector doing a few things well is essential for markets to work effectively. The government provides public goods and also a safety net for those who do not have access (called "entitlements" by Nobel laureate Amartya Sen) to essential goods and services.

Key public goods for this process to work include the rule of law, sound macroeconomic policies (balanced government operating account, money supply expanding with output only, etc.), protection of private property, open trade, and, in general, an environment where business decisions can be made and carried through based on efficiency. Key services provided by the public sector are noted in the next section.

Promoting High-Payoff Public Investments for Broad-Based Productivity Gains

The standard model emphasizes broad-based development, that is, development that raises living standards for women as well as for men, for minorities as well as for majorities and for children as well as for adults. A fourth study identifies, based again on historic experience over many countries, where economic equity as well as efficiency are served (Tweeten and McClelland, 1997). In addition to appropriate policies in general, public services with especially high payoffs for economic growth include investment in:

- Basic education. Beginning with literacy and numeracy, education is extended to higher levels as the nation's wealth builds over time.
- Agricultural research. Payoffs are unusually large for technologies that help plants and animals cope with the stresses of disease, pests, drought, cold, heat and the like.
- Investment in infrastructure. High payoff investments include roads, bridges, port facilities and irrigation structures.

IMPLEMENTING THE STANDARD MODEL

If the standard model is a proven means of alleviating poverty, food insecurity and environmental degradation, why is it not adopted by all countries? The answer is partly that many persons who make policy decisions are not aware of the model's record of success. More importantly, however, many persons benefit from inappropriate policies. If those who gain from these policies are in positions of power and authority, policy reform is strongly resisted.

Sound policies would deprive hundreds of thousands of politicians and bureaucrats of the economic rents (payoffs) afforded by government regulations, licensing, permits, trade barriers and the like (see Pinto, 1992). Where a country is living beyond its means, reforms can, unfortunately, hurt some disadvantaged people. Freer trade raises the income of a nation, but workers in protected industries will lose jobs and many will need training and other help to find new and hopefully more productive and better-paying jobs. Firms formerly protected from international competition may fail rather than restructure to compete efficiently in domestic and international markets under free trade. National fears may be heightened by fears of "invasion" by foreign firms once a country is opened to direct investment from abroad. In short, reform is not painless. But failure to follow the standard model entails far more pain in the long run.

India's current policies come at high cost. To illustrate, consider some comparisons with China. After losing 17 to 30 million people to starvation in 1958–1961 (Dreze and Sen, 1989: 210), China belatedly instituted agricultural reforms, creating market incentives for its farmers by the late 1970s. As a consequence, the number of rural Chinese below the poverty line fell from 200 million in 1979 to 70 million in 1986 (Dreze and Sen, 1989: p. 216). For males and females in 1980, life expectancy in China was 65 years, while in India it was only 52 (Dreze and Sen, 1989: pp. 218–222).

Lower life expectancy in India can be traced to poverty, inadequate medical services and lack of schooling, especially notable among females. The adult literacy rate was 56% for both females and males in China, while it was only 26% for females and 55% for males in India (Dreze and Sen, 1989: p. 222). Education of women is especially important for providing broad-based access to health care and food. The failure to follow measures even as favorable as China's mediocre economic and health policies costs India the excess mortality of 3.9 million people each year, according to Dreze and Sen (p. 265). The authors concluded that, "India seems to fill its cupboard with more skeletons in eight years than China put there in its years (1958–61) of shame."

A more proper comparison for India, however, may not be China but its own state of Kerala. That state, though poor, has aggressively supported education and health care for men and women. Kerala's adult literacy program rate is 71% for females and 86% for males — well above the numbers for China and India presented earlier (Dreze and Sen, 1989: p. 222). The result was a life expectancy for females of 68 years in Kerala — somewhat above that of China and well above the average for males and females in all of India.

India has the natural resources and industrious and enterprising people to be an economic Singapore, Hong Kong, Taiwan, or even Japan. Hence, the proper comparison for life in India is not China, but the income, wealth, life expectancy, food security and environmental protection of these other countries. Much time would be required, but India has the potential to equal the outcome for these more developed countries by following the standard economic model.

The following observations concern public policy that could help to reconcile food supply with demand.

With greater income more widely shared among its people through broad-based economic growth, India could be in a position to meet its demand for food needs even in years of short domestic supply by purchases in international markets. Food has been available in that market every year since World War II (the coefficient of variation of world food production around trend is only 1%, according to FAO). Economic growth and access to international markets under the standard model can earn the foreign exchange necessary for international food access. In addition, open trade encourages foreign and domestic market competition that is healthy for the Indian economy.

The synergy between national economic growth and agricultural productivity must be exploited if India is to successfully redress its poverty, food insecurity and environmental problems. Farms in India are small and shrinking in size each generation as high birth rates and lack of alternative employment create pressure to subdivide landholdings. For food security and economic efficiency, small farms need to increase in size as farm youths also find economic opportunities off the farm made possible by national economic growth through following the standard model. The economic reforms of 1991 were a useful beginning, but did not go far enough to sustain essential growth momentum.

Enough food stocks need to be stored in regions of India to supply food needs until imports can arrive, but most such domestic stocks can be held in private hands. The Food Corporation of India has been an assured buyer of wheat and rice, but needs only a modest reserve to provide a food safety net for the very poor. Procurement practices that have built massive reserves of grain and potatoes, only to have them rot or be consumed by weevils, rats and the like, could be terminated (*The Economist*, 2001). Laws that forbid private accumulation of food stocks need to be terminated if markets are to work.

Instead of subsidizing private commodity markets with price supports and stock accumulation and instead of subsidizing electricity, irrigation water and other inputs that would best be regulated by markets, the government of India needs to provide public goods and services such as basic agricultural research, extension, education and information to improve productivity and market efficiency.

Greater national income creates a tax base that can be used to protect the environment. Having fewer, better-paid, more able civil servants and government officials should reduce their incentives for corruption and enhance professionalism in protecting forests, biodiversity and gene banks. Governments that can afford to hire able civil servants can be more creative in establishing markets that protect the environment.

Economists can be helpful in identifying the tradeoffs between economic growth and the height and breadth of the safety net, but this height and breadth are ultimately political decisions. Targeting only the most disadvantaged individuals can increase the efficiency of the safety net.

The politics that forestall application of the standard model often spring from attitudes. Gurcharan Das (2001) describes the "old India" as being characterized by discrimination against women, caste barriers, Hindu chauvinism, official corruption, advancement based on patronage and, for businesses, profits without competition. The view that profit is a "dirty word" and that capitalists are "plunderers" is a part of free speech and thought (Das, 2001: p. 86). But those who hold, or might hold, those attitudes need to be reminded of their high cost in lost growth.

Corruption, like sin, cannot be eliminated but it can be reduced (see Agarwal, 1995). The most effective way is to minimize regulations, licenses and other impediments to markets that generate economic rents and the corrupt behavior attending such rents. As Adam Smith contended years ago, markets can turn private greed into public good. Measures to reduce corruption also include a free press, a vibrant democracy, checks and balances among branches of government and civil service merit hiring with periodic review.

CONCLUSIONS

On average, the global food supply–demand balance may become tighter in the next several decades compared with recent decades. That conclusion need not attract India to a policy of self sufficiency. The surest means to successfully address India's population, food, poverty and environmental problems is through broad-based economic growth.

The most important development in economics in the past half century has been compelling evidence from numerous countries that following a standard model for economic growth ensures economic, social and environmental progress. No country follows it exactly and it does not come without growing pains. The telling superiority of the standard model over its alternatives does not mean the end of economics. Economic policy will need to continue evolving based on what works rather than on ideology. Market principles should be a framework, not a fetish. But the model's performance does establish that no nation need remain poor, food insecure and unable to protect its environment.

India, foreseeably the world's most populous nation, has great unrealized potential. Although India is often compared to China, the standard model features of Hong Kong, Singapore and Taiwan — all of which are food secure — are more appropriate for India. The real test is not food self sufficiency, but rather whether a country follows economic policies that provide the wherewithal to operate in world markets — as a food buyer, if necessary.

All dimensions of the standard model need not be embraced, but critical elements are essential to broad-based development. The major impediment to implementation is a lack of economic education. The challenge for educators is to better inform people, including leaders, of the potential opportunities available to India by following sound economic policies.

Finally, more-developed nations such as the United States can help to ease India's food supply–demand challenges by sharing technology (particularly results of basic research), by opening markets to textiles and other exports from India, by not subsidizing their own food production in unfair competition with developing countries and, in general, by following sound economic policies that help to stabilize and grow economies around the world.

REFERENCES

Agarwal, M. 1995. Politics of crime, corruption and waste, caste and creed. *Indian Journal of Public Administration*, 41:3, 462-471.

Bos, E., M. Vu, E. Massiah and R. Bulatao. 1994. *World Population Projections, 1994-95 Edition*. Johns Hopkins University Press. Baltimore, MD.

Das, Gurcharan. 2001. *India Unbound*. Knopf. New York.

Dreze, J. and A. Sen. 1989. *Hunger and Public Action*. Oxford University Press. New York.

FAO. 2000. FAO Statistics Database: http://apps.fao.org. Food and Agriculture Organization. Rome, Italy.

Grossman, G. and A. Krueger. 1995. Economic growth and the environment. *Quarterly Journal of Economics*, 57, 353-377.

Gupta, D. 2001. *Mistaken Modernity: India between Worlds*. Harper Collins. New York.

Hervani, A. and L. Tweeten. 2001. Kuznets curves for environmental degradation and resource depletion. Working Paper from the Department of Agricultural, Environmental and Development Economics, The Ohio State University, Columbus, Ohio.

The Economist. 2001. Indian Agriculture. February 17, p. 46.

Kumar, P., M. Rosegrant and P. Hazell. 1995. Cereals Prospects in India to 2020: Implementation for Policy. Brief 23. International Food Policy Research Institute. Washington, D.C.

Lutz, W., W. Sanderson, S. Scherbov and A. Goujon. 1996. World population scenarios for the 21st century. In *The Future Population of the World*, Wolfgang Lutz et al., Eds. International Institute for Applied Systems Analysis. Laxenburg, Austria. Ch. 15.

Pinto, M. 1992. Liberalization of the Indian Bureaucracy. *Indian Journal of Administrative Science*, 3:1-2, 125-131.

Ruddle, K. and W. Manshard. 1981. *Renewable Natural Resources and the Environment: Pressing Problems in the Developing World*. Tycooly International Publishing. Dublin, Ireland.

Seldon, T.M. and D. Song. 1994. Environmental quality and development: Is there a Kuznets curve for air pollution emissions? *Journal of Environmental Economics and Management*, 27, 147-162.

Tweeten, L. 1998. Dodging a Malthusian bullet. *Agribusiness*, 14, 15-30.

Tweeten, L. 1999. The economics of global food security. *Review of Agricultural Economics*, 21:2, 473-488.

Tweeten, L. and W. Amponsah. 1998. Sustainability: The role of markets versus the government. In *Sustainability in Agricultural and Rural Development*, G. D'Souza and T. Gebremedhin. Eds. Ashgate Publishing. Brookfield, VT. Ch. 3.

Tweeten, L. and D. McClelland, Eds. 1997. *Promoting Third-World Agricultural Development and Food Security*. Praeger Publishers. Westport, CT.

Tweeten, L. and C. Zulauf. 2001. The economics of agriculture and the environment in a world of falling population. Working Paper of the Department of Agricultural, Environmental and Development Economics, The Ohio State University, Columbus, Ohio.

United Nations. 1998. World population projections to 2150 (ST/ESA/SER.A/173). United Nations. New York.

USDA. 1999. Summary Report, 1997 National resource inventory. Natural Resources Conservation Service, U. S. Department of Agriculture. Washington, D.C.

29 Context, Concepts and Policy on Poverty and Inequality

Fred J. Hitzhusen

CONTENTS

> *It would be desirable to be quite sure*
> *Just who are the deserving poor,*
> *Or else the state supported ditch*
> *May serve the undeserving rich.*
>
> **—Kenneth Boulding (1977)**

INTRODUCTION

The opening paragraph of a special issue of *Finance and Development* (December 2000), on How We Can Help the Poor, states:

> "Of the world's 6 billion people, 2.8 billion, almost half, live on less than $2 per day and 1.2 billion, a fifth, live on less than $1 a day. In the poorest countries, as many as one in every five children does not reach his or her fifth birthday and as many as half of the children under 5 are malnourished.

> "Although human conditions improved more in the twentieth century than in all of the rest of history, the distribution of global gains has been extraordinarily unequal. The average income in the richest 20 countries is 37 times the average in the poorest 20 — a gap that has doubled in the past 40 years.

> "Seventy percent of all people living on less than $1 per day are in South Asia and sub-Saharan Africa. And conditions in some parts of the world — notably the countries

of the former Soviet Union — have been getting worse: the number of people living in poverty in Central Asia and some European countries rose by more than 20 times between 1987 and 1998."

What makes this statement especially noteworthy is that *Finance and Development* is published by the International Monetary Fund, hardly a liberal or alarmist organization. Poverty and inequality are thus important issues in a global context. This concern is not simply a matter of Northern Hemisphere countries being the "haves" and Southern Hemisphere countries being the "have-nots." One must recognize that significant inequality of incomes exists *within* both high and low per capita income countries and, at least in the United States, this inequality has increased over the last 20 years.

Because India is the focus of this volume, a few Indian facts on poverty and inequality are in order. There is debate in India regarding the statistics on poverty and inequality relating in part to changes in methods of data collection, e.g., moving from 7 to 30 days for survey respondents' recall. However, there is some consensus that:

- During the 1980s, there was a 10% reduction in the number of persons living in poverty, about 50 million.
- With higher economic growth rates in the 1990s, however, most of these gains were wiped out, particularly in the rural areas. Dr. S.P. Gupta (1999) suggests that India went from a "trickle-down" strategy to "trickle-away."
- There has been some slight reduction in inequality.

Table 29.1 provides some evidence comparing India with other countries.

Some of the reasons suggested for these poverty and income inequality trends include:

- Slower growth in the poorer states of India
- Decline in investments in agriculture and infrastructure
- Less spending on rural development services
- Increased prices of food to consumers
- Decrease in the availability of rural credit

Zheng (1994) has written about the difficulty of comparing increases or decreases in poverty as a concept of absolute disadvantage, e.g., meeting basic needs, with income inequality as a relative concept of deprivation. Can Indian policy makers take comfort in some reduction in income inequality when absolute poverty has not been reduced? Alternatively, if poverty has been reduced but income inequality has increased, can one make any generalizations about what has happened to overall economic well-being? What is the relationship of poverty and inequality to food security and environmental quality? The following section delineates some of the conceptual issues involved when addressing these and other questions.

TABLE 29.1
Rank Order of Income Concentration for the 20 Less Developed Countries
with the Most Unequal and the Most Equitable Concentration Shares*

	Lowest Income Share Accruing to the Poorest 20%			Highest Income Share Accruing to the Poorest 20%	
	Country	Share		Country	Share
1.	Sierra Leone (Afr.)	1.1	1.	Slovak Rep. (EE)	11.9
2.	Guinea Bissao (Afr.)	2.1	2.	Czech Rep. (EE)	10.5
3.	Guatemala (LA)	2.1	3.	Rwanda (Afr.)	9.7
4.	Paraguay (LA)	2.3	4.	Hungary (EE)	9.7
5.	Panama (LA)	2.3	5.	Laos (Asia)	9.6
6.	Brazil (LA)	2.5	6.	Bangladesh (Asia)	9.4
7.	Niger (Afr.)	2.6	7.	Pakistan (Asia)	9.4
8.	Lesotho (Afr.)	2.8	8.	Slovenia (EE)	9.3
9.	South Africa (Afr.)	2.8	9.	Poland (EE)	9.3
10.	Colombia (LA)	3.1	10.	India (Asia)	9.2
11.	Senegal (Afr.)	3.1	11.	Romania (EE)	8.9
12.	Chile (LA)	3.5	12.	Sri Lanka (Asia)	8.9
13.	Mexico (LA)	3.6	13.	Egypt (Afr.)	8.7
14.	El Salvador (LA)	3.7	14.	Belarus (EE)	8.5
15.	Nigeria (Afr.)	4.0	15.	Bulgaria (EE)	8.3
16.	Zimbabwe (Afr.)	4.2	16.	Latvia (EE)	8.3
17.	Zambia (Afr.)	4.2	17.	Lithuania (EE)	8.1
18.	Dom. Republic (LA)	4.2	18.	Indonesia (Asia)	8.0
19.	Russia (FSU)	4.2	19.	Vietnam (Asia)	7.8
20.	Venezuela (LA)	4.3	20.	Nepal (Asia)	7.6
	10 Latin American countries			8 Asian countries	
	9 African countries			10 eastern European countries countries	
	plus Russia			2 African countries	

* Data from early to mid-1990s

Source: Derived from World Development Report 1999, World Bank, 1999. Washington D.C.

CONCEPTS OF POVERTY AND INEQUALITY

Although poverty and inequality have been longstanding concerns in economic theory and practice, they have been given minimal attention compared with economic efficiency considerations. This may partly reflect the lesser attention given to poverty and inequality in public policy, but it may also relate to some conventions in neoclassical economic theory. For example, applied welfare economics assumes constant marginal utility of money income, costless transfers and hypothetical compensation, all questionable assumptions.

In addition, there is a tendency among economists to accept the notion of trade offs rather than any complementary relationships between equity and efficiency (e.g.,

Okun, 1975) or between meeting basic needs and increasing economic growth. In fact, the earlier Harod-Domar growth model, and the growth-weighted project evaluation procedures that followed in the 1960s, assumed that income inequality was a stimulus to economic growth, i.e., the rich and government recipients of project benefits were assumed to have higher "marginal propensity to save" (MPS) and thus stimulate investment and growth (see Chapter 23).

This can lead to the conclusion that poverty and inequality concerns are more subjective and normative or are detrimental to the desire for economic growth. However, when one looks at the assumptions of neoclassical welfare economics regarding constant marginal utility, costless transfers and hypothetical compensation, these appear no less subjective or even speculative. Numerous polls suggest that most citizen-consumers believe that a marginal dollar to a poor person is worth more than a marginal dollar to a rich person. Transfers to compensate for income losses or to reallocate income are not likely to be costless, and hypothetical compensation is unlikely to be politically feasible. It is also likely that satisfying the most basic of needs in human health and nutrition (e.g., Hicks, 1980; Dasgupta and Ray, 1986, 1987 and Alesina and Radrick, 1994) is a complement or precondition, rather than a substitute, for economic growth. Finally, the MPS arguments can fail on two counts: MPS rates may not be much higher for the rich and, even if they are, they may not result in higher rates of investment in the poor country. Thus, efficiency analysis and pronouncements are also not free of normative judgments.

The World Bank, the Food and Agricultural Organization of the United Nations and the U.S. Agency for International Development all define food security in terms of access by all people at all times to sufficient food to meet dietary needs for a productive and healthy life (USAID, 1992). Food-insecure people spend 60–80% of their income for food, live mostly in rural areas of developing countries, work mostly in agriculture-related occupations and tend to be net buyers rather than sellers of food. If they are food producers, they are subsistence farmers, are disproportionately women and children and are disproportionately in sub-Saharan Africa, even though their largest absolute numbers are in South Asia. One estimate is that, although food-insecure people numbered 950 million in 1970, that number is currently around 800 million (Tweeten, 2001). Rather strong consensus supports the notion that the primary underlying cause of food insecurity is poverty.

The linkage of poverty to environmental quality is also debated among economists and environmentalists. More conservative voices extend the earlier Kuznet curves relating stages of economic growth and income inequality to environmental quality. Increases in environmental degradation may be observed in the early stages of economic growth. The debate is over whether environmental quality will begin to improve beyond some level of per capita income. (Tweeten has estimated this level to be about $15,000 per person annually, when protection of the environment is assumed to become a higher priority with higher incomes.)

Even if such a dynamic can be expected, the serious question is whether global per capita incomes can be raised to that level without experiencing irreversible environmental consequences such as global warming and loss of biodiversity. In addition, cross-section empirical support for the Kuznets curve is weaker than it was earlier and the time series support has never existed (Anand and Kanbur, (1993a,b).

This and the foregoing concerns suggest the need to delineate concepts of poverty and inequality more carefully, so as to be clearer about their implications for economic growth and environmental sustainability.

The first major distinction is that poverty generally refers to a threshold defined by basic needs or the income required to purchase these. Inequality, on the other hand, refers to the relative position of individuals in terms of either wealth or income. Maslow (1970) described a cross-cultural universal hierarchy of human needs that starts with *security needs* such as self-preservation and protection from immediate physical danger and *physiological needs* such as satisfying hunger, thirst and body warmth. These two combined categories define what the World Bank, The World Health Organization and others mean by basic needs. Maslow's hierarchy continues with *belonging* or acceptance by others, *self realization* in terms of freedom, justice, stability and independence and *self gratification* including recognition, prestige and success. Exactly where to draw the line on which needs are "basic" can be contentious, but considerable consensus exists on the validity of the concept.

Examples of poverty measures not as closely related to income include social exclusion, life chances and absolute capabilities. Bhalla and Lapeyne (2000) define the first, *social exclusion* as extending the notions of advantage and disadvantage beyond an immediate accounting of income to issues of economic security, civil rights and social integration. Uphoff argues that poverty is most significant in terms of what it does to people's *life chances*, i.e., their opportunities to get educated and have food, shelter and clothing to meet basic needs and the opportunities of their children to do the same. This views poverty with a time dimension, considering intergenerational social mobility and immobility. Sen's concept of *absolute capabilities* (1999), for example, the ability to be seen in public without shame or the ability to take part in the life of the community, regards people as living in unjustifiable deprivation if these abilities are not attained.

Although one version of equity or equality may be to assure that no one falls below a certain level of goods and services or income ensuring basic needs, it is more likely to refer to one's relative position. Quantifying relative well-being can be both difficult and contentious. The most common practice of economists has been to assess inequality in quantitative terms by plotting the distribution of income in terms of the cumulative percentage of income (on the vertical axis) against the cumulative percentage of families or households (on the horizontal axis) having that much income, as shown in Figure 29.1. A line of 45° from the origin defines the line of perfect equality and a line tracing the actual distribution of income over households is known as the Lovenz curve. The Gini coefficient as a measure of the degree of inequality, calculated as a ratio of the area between the 45° line and the actual Lovenz curve (A + B compared with the entire triangular area under the 45° line, A + B + C). Paglin (1975) challenged the lack of age adjustment with the Lorenz and Gini measures and proposed an age-adjusted measurement, shown also in Figure 29.1. This measure takes into account that individuals' income usually first increases with age and then decreases. Including children, adults and retirees all in the same distribution distorts the statistic.

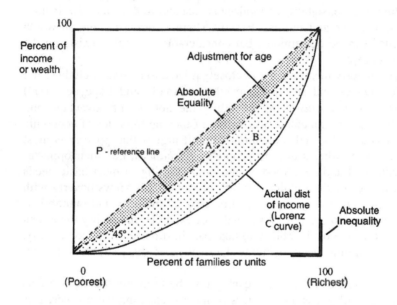

Gini Ratio = ratio of shaded area $\dfrac{(A + B)}{(A + B + C)}$
Triangle or area
under 45° line

Paglin Gini Ratio = ratio of shaded area $\dfrac{(B)}{(B + C)}$
Area under
P-reference line

Age Gini Ratio = $\dfrac{A}{A+B+C}$

° The U.S. until tax law changes in the 1980s and 1990s showed unchanged size distribution of family income since the 1940s, i.e., the bottom 20% got 5% of income, and the top 20% got 40% of income

FIGURE 29.1 Lorenz vs. Paglin measures of inequality.

There is a long-standing concern that neither the Lorenz or Paglin calculations consider in-kind transfers because of their focus on cash income. Kristol (1975) found that this approach resulted in a significant overstatement of inequality. Utilizing U.S. data from the mid-1970s, an adjustment for in-kind transfers increased the "income" share of the poorest 20% from 5 to 12% and reduced the "income" share of the richest 20% from 40 to 33%.

Economists have proposed several other alternative methods for evaluating income inequality and distribution impacts including:

- Shadow pricing under- or unemployed labor
- The constrained maximum or minimum targets approach, which refers to setting an economic efficiency minimum and maximizing redistribution or setting a redistribution minimum and maximizing economic efficiency

- Unweighted distribution of net benefits by income, region, ethnic group, etc.
- Provision of alternative weighting functions and assessment of their distribution consequences for decision-makers
- Explicit weighting of net benefits by income class based on past tax or expenditure decisions including switching or equity weights (Ahmed and Hitzhusen, 1988)
- Survey techniques to directly elicit income distribution weights.

(For a more detailed discussion of each approach, see McCullough et al., 1986.) Unfortunately, there has been rather limited empirical application of these methods.

IMPLICATIONS FOR FOOD SECURITY AND ENVIRONMENTAL QUALITY

Pinstrup-Andersen and his colleagues at the International Food Policy Research Institute (IFPRI) argue that, in dealing with food insecurity, the poverty issue must be directly confronted by expanding investments in less favored geographical areas with agricultural potential but also irregular rainfall patterns, fragile soils and many poor people. The IFPRI researchers also call for investments in primary education and health care, clean water and sanitation, empowerment of women, improved access to productive resources and expanded employment. If one adds population programs, local participation and open trade and investment policies, these recommendations are very similar to those in the World Bank World Development Report (1992) on development and the environment.

In the World Bank's World Development Report 2000–01, the focus is on poverty as seen through the eyes of the poor. The report concludes that "poverty is the result of economic, political and social processes that interact with each other and frequently reinforce reach other in ways that exacerbate the deprivation in which poor people live. Meager assets, inaccessible markets and scarce job opportunities lock people in material poverty." The multiple deprivations of poor people include living without fundamental freedoms of action and choice that the better-off take for granted. The report recommends stimulating economic growth, making markets work better for poor people, building up their assets, facilitating the empowerment of poor people and enhancing their security.

It is also important to recognize that, while poverty reduction generally leads to an increase in food security, it also may lead to a reduction in a number of local water and air pollution problems from illicit burning and cutting of forests, improper disposal of human waste, soil erosion, etc. However, as incomes increase, so does the per capita consumption of fossil fuels, global warming, air pollution, etc. Thus, to suggest bringing the developing countries up to the consumption standards of the developed world as the solution to environmental degradation is a stretch if not a distortion of the evidence and realities of ecological sustainability.

Certainly more attention needs to be given to the contributions that better-defined and enforced property rights and markets can make to the protection of the ecosystem service flows of global environments and natural-resource endowments. Water markets can allocate this critical resource beyond the entitlements that societies may choose to set aside for meeting basic needs. Safe minimum standards (SMS) coupled with the sale of rights of assimilation of air and watersheds show promise for control of greenhouse gases and some forms of water pollution such as sediments and nitrate-nitrogen. Ecotourism, if done sustainably, can include significant benefits to the poor who live in or near the ecosystems that are being showcased. The U.S. Agency for International Development has developed some success stories in this area.

The foregoing environmental-quality policy recommendations reflect the fact that considerable progress has been made in the economic conceptualization and measurement of the environment. The field of environmental economics and the related fields of natural-resource and ecological economics involve large numbers of applied economists working throughout the world. Figure 29.2 summarizes the categories of economic value, the common measures or standards and the policy instruments and options central to this fast growing field. Considerable empirical evidence on the economic value of environmental service flows, as well as the

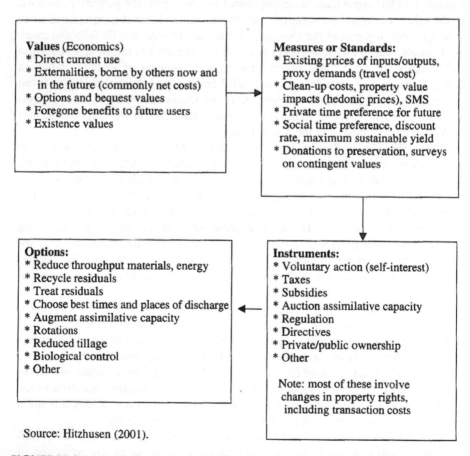

Values (Economics)
* Direct current use
* Externalities, borne by others now and
  in the future (commonly net costs)
* Options and bequest values
* Foregone benefits to future users
* Existence values

Measures or Standards:
* Existing prices of inputs/outputs,
  proxy demands (travel cost)
* Clean-up costs, property value
  impacts (hedonic prices), SMS
* Private time preference for future
* Social time preference, discount
  rate, maximum sustainable yield
* Donations to preservation, surveys
  on contingent values

Options:
* Reduce throughput materials, energy
* Recycle residuals
* Treat residuals
* Choose best times and places of discharge
* Augment assimilative capacity
* Rotations
* Reduced tillage
* Biological control
* Other

Instruments:
* Voluntary action (self-interest)
* Taxes
* Subsidies
* Auction assimilative capacity
* Regulation
* Directives
* Private/public ownership
* Other

Note: most of these involve
changes in property rights,
including transaction costs

Source: Hitzhusen (2001).

FIGURE 29.2 Monetizing environmental service flows and implementing change or reform.

importance of these flows to economic development and food production, is also being developed. For example, a study by Zhao et al. (1991) focused on identifying the factors that determine the agricultural production growth rate and on testing the effects these factors have on agricultural growth in developing countries. Specifically, this study involved statistical estimation of an aggregate agricultural growth function based on cross-country data for 23 developing countries. Special attention was devoted to environmental degradation and agricultural pricing policy and to the policy implications resulting from the effect these variables have on agricultural and food production growth.

CONCLUSIONS

The data suggest increasing evidence of inequality and poverty in the world. However, it appears that both policy makers and the general public — at least in the United States — are generally less concerned about inequality than about poverty (Samuelson, 2001). A normative stance on this question is taken in a recent editorial in *The Economist*, which states, "Helping the poor, the truly poor, is a much worthier goal than merely narrowing inequalities." Two exceptions are given for emphasizing poverty vs. inequality, where opportunities are not genuinely equal and when power is abused to raise prices or exclude competitors (*The Economist*, June 16, 2001).

Aggregating the concepts of poverty and inequality is more difficult given the absolute vs. relative aspects of these terms. It would seem useful to do some simple indexing of both concepts for a large sample of developing countries where a country's rank on each of the concepts determines the magnitude of the index. Some difficulties may be encountered in standardizing basic needs, but these problems are manageable. Much more analysis of the poverty and income distribution impacts (both weighted and unweighted) of development projects is possible if the political will and professional inclination are present. This is not a new concern. Kenneth Boulding, in his 1968 AEA presidential address, spoke at length about the limitations of welfare economics and the need to incorporate notions of malevolence and benevolence in utility theory and weights in income distribution.

The results of the Zhao et al. (1991) analysis show that price distortions in the economy and land degradation had statistically significant negative impacts, while the change in arable and permanent land was positively related to the growth of agricultural production and food production in 23 developing countries from 1971 to 1980. These results, plus the increasing evidence from conceptual developments and empirical studies on the economics of the environment, emphasize the importance of "getting prices right," particularly on environmental service flows. This, in turn, supports implementation of sustainable land and water management practices if future growth in food and agricultural output is to be sustained and reduction in food insecurity is to be realized in developing countries.

A cautionary note is to beware of large development schemes purported to be in the interest of the poor and poverty reduction. This has been particularly true of large dam projects. A recent study by the World Commission on Dams concluded that between 40 and 80 million people have been physically displaced by dams worldwide and millions of people living downstream from dams — particularly those dependent on natural flood plain function — have also suffered serious harm to their livelihoods and had the productivity of their resources placed at risk (WCD, 2000). Many of the displaced people have neither been resettled nor compensated (Venkateson, 2001). Proponents of dams counter that services of dams include 12 to 16% of world food production and 19% of world electricity supply. A case in point is the very controversial and very large Sardar Sarovar Dam in India. Analysts must look carefully at the full social benefits and costs of these projects, including environmental impacts as well as their impacts on poverty and inequality.

Finally, on a positive note, I find increased interest among applied economics graduate students in problems of poverty, inequality, food security and environmental quality. This may lead to more future involvement of professional economists in these important matters for public policy.

ACKNOWLEDGMENTS

Helpful comments were received from OSU colleagues Priyo Banerjee, Radha Ayalasomayajula and Joshua Templeton.

REFERENCES

Ahmed, H. and F. Hitzhusen. 1988. Income Distribution and Project Evaluation in LDCS: An Egyptian Case. Paper for International Meeting of Agricultural Economists, Argentina.

Alesina, Alberto and Dani Rodrik.1994. Distributive politics and economic growth, *Quarterly Journal of Economics*, 108, pp. 465-490.

Anand, Sudhir and Ravi Kanbur. 1993a. The Kuznets process and the inequality–development relationship, *Journal of Development Economics*, 40, p. 25-52.

Anand, Sudhir and Ravi Kanbur. 1993b. Inequality and development: A critique, *Journal of Development Economics*, 96, pp. 19-43.

Bhalla, Ajit S. and Frederic Lapeyne. 2000. Social exclusion: Toward an analytical and operational framework. In *The Political Economy of Inequality*, Ackerman, Goodwin, et al., Eds. Island Press. Washington, D.C.

Boulding, Kenneth E. 1977. Summary, in *Economic Development, Poverty and Income Distribution*, William Loehr and John P. Powelson, Eds. Westview Press. Boulder, CO.

Dasgupta, Partha and Debraj Ray. 1986. Inequality as a determinant of malnutrition and unemployment: Policy, *Economic Journal*, 96, pp. 1011-1034.

Dasgupta, Partha and Debraj Ray. 1987. Inequality as a determinant of malnutrition and unemployment: Policy, *Economic Journal*, 97, pp. 177-178.

How we can help the poor, *Finance and Development*. 2000, Special Edition, 37:4.

Gupta, S.P. 1999. Trickle down theory revisited: the role of employment and poverty. Lecture to Indian Society of Labor Economics, reported in *The Hindu*, Dec. 30, 1999.

Hicks, Norman. 1980. Is there a trade-off between growth and basic needs? *Finance and Development*, June, 17-20.

Hitzhusen, Fred J. 2001. Environmental concepts and evidence for sustainable development. In *Problems and Policies in World Population, Food and Environment: An Integrated Approach*, F.J. Hitzhusen, Ed. 133-151. AED Dept., The Ohio State University. Columbus, OH.

Kristol, Irving. 1976. The Poverty of Inequality. *Wall Street Journal*, July 12.

Maslow, Abraham. 1970. *Motivation and Personality*, 2nd ed. Harper and Row. New York.

McCullough, C., M. Gowen, F. Hitzhusen, C. Feinstein and S. Pintz. 1986. COMPRAN: A Computerized Project Analysis Package for Developing Countries, Department of Agricultural Economics and Rural Sociology, The Ohio State University and the Resource Systems Institute, East-West Center, Honolulu.

Okun, Arthur M. 1975. *Equity and Efficiency: The Big Trade-Off*. Brookings Institution. Washington, D.C.

Paglin, Morton. 1975. Measurement and Trends of Inequality. *American Economic Review*, 65:4, 598-609.

Samuelson, Robert J. 2001. Indifferent to Inequality? *Newsweek*, May 7, p. 45.

Sen, Amartya. 1983. Poor, Relatively Speaking. *Oxford Economic Papers*, 35, 153-169.

The Economist. 2001. Does Inequality Matter? (editorial), June 16, 9-10.

Tweeten, Luther. 2001. The Economics of Global Food Security. In *Problems and Policies in World Population, Food and Environment: An Integrated Approach*, F.J. Hitzhusen, Ed. 84-101. The Ohio State University. Columbus, OH.

USAID. 1992. *Definition of Food Security*, PD-19. U.S. Agency for International Development. Washington. D.C.

Venkateson, V. 2001. The Debate on Big Dams. *Frontline*, February 2, 73-74.

WCD. 2000. Dams and Development: A New Framework for Decision Making. Report of the World Commission on Dams. Earthscan Publications. London.

Zhao, F., F. Hitzhusen and W. Chern. 1991. Impact and implications of price policy and land degradation on agriculture growth in developing countries. *Agricultural Economics*, 5, 311-324.

Zheng, Buhong. 1994. Can a poverty index be both relative and absolute? *Econometricia*, 62, 1453-1458.

30 Sustainable Development: Some Economic Considerations

Alan Randall

CONTENTS

INTRODUCTION

Sustainable development is a term that has come into widespread use in the last decade. Many scholarly disciplines have something to say on the topic, each from its own perspective. Among society at large, environmentalists want environmental systems sustained, consumers want consumption sustained, workers want jobs sustained and local officials want all of this accomplished locally. In a memorable attempt to cut through the confusion, the Bruntland Commission defined sustainable development as "meet(ing) the needs of the present without compromising the ability of future generations to meet their own needs" (World Commission on Environment and Development, 1987). This definition does a nice job of recognizing the delicate balance between present and future interests, but the key word "needs" accommodates a variety of views as to what sustainability entails. Four distinct possibilities are identified:

1-5667-0594-0/02/$0.00+$1.50
© 2002 by CRC Press LLC

1. Sustainability is a matter of sustaining human welfare through multigenerational time. In this process, resources may provide utility in their own right (the aesthetic value of a forest) or serve as an input in the production of a good that provides utility (lumber used for building houses). What matters is not that a particular resource is used in a particular manner, but whether the broad pattern of resource use optimizes the utility of present and future generations. A high degree of substitutability among resources, commodities and locations in the satisfaction of human demands is a maintained, if implicit, assumption. Preservation of particular resources is important only if preservation provides greater utility to current and future generations than would utilization.

2. Sustainability is viewed in terms of human welfare, but more attention is paid to particular resources, some of which are viewed as essential in that other resources cannot substitute for them. For example, certain resources may be critical for life support systems and depleting these resources would threaten significant and irreversible damage to the prospects of future generations. The designation of particular resources as essential and therefore deserving of special stewardship is prudential rather than principled: the motivation is to take care of that which we might really need.

3. Sustainability may involve commitments to individual resources. A particular resource may be thought essential for production of consumption goods, for direct consumption by humans, for the fulfillment of human aesthetic or spiritual needs or for reasons independent of human valuation. This view of sustainability opens the door to nonanthropocentric views — the resource enjoys rights, or has intrinsic value, independent of human caring — as well as anthropocentric views. Fundamentally different reasons for thinking a particular resource essential invoke different kinds of commitments expected of present and future generations. For example, motivations based on welfare considerations suggest commitments to prudent conservation, whereas intrinsic value suggests a duty to preserve.

4. Sustainability may be accorded a locational dimension. While some consider sustainable development to be a global issue, others focus more on livable communities or cultural landscapes. Resources and amenities such as local ecosystems or social institutions are seen as worthy of protection so they may be enjoyed by future generations. The preservation, not only of specific resources, but specific resources in specific locations is thought necessary to meet the needs of present and future generations.

Views (3) and (4), taken together, accommodate concerns for specific resources, all manner of localism, regionalism and cultural way of life, as well as community viability concerns and a wide variety of motivations: value-based or principled and anthropocentric or nonanthropocentric. These views stand in sharp contrast with (1), which implicitly accepts a broad range of adjustments — substitutions among resources, commodities and amenities and scarcity-induced human migrations — so long as human welfare can be sustained at the global level.

Economists identify most readily with views (1) and (2), which have in common the assumption that the "needs" of the Bruntland definition are captured in the notion of assuring human welfare, but are distinguished by different assumptions about the essentiality of particular resources. When economists debate sustainability, the discussion turns quickly to reasons for believing (or not believing) that particular resources are truly critical for sustaining human welfare.

What follows will deal mostly with abstract ideas, most of which are useful in a very practical way, in helping us think constructively about sustainable development.* In particular, it will explore the economic theory of sustainability in a highly abstract economy that takes natural resources seriously; questioning some widely held beliefs about sustainability and some common policy prescriptions; recognizing reasons for optimism and for pessimism about future prospects; finding virtue in well-functioning factor, product, financial and asset markets as engines of sustainable development; but concluding that targeted precautionary policies should be held in reserve to deal with particular potential natural resource crises. Finally, it will turn to agriculture, finding grounds for optimism concerning emerging technologies, while warning against undermining these optimistic prospects with counterproductive policies.

LESSONS FROM SUSTAINABILITY MODELS

Some widely held "principles" of sustainability that deserve critical examination include:

- An economy that uses exhaustible resources is inherently unsustainable
- An economy with continually growing human population is inherently unsustainable
- The practice of discounting future prospects is biased inherently against the welfare of future generations. There are several extensions of this idea. For example, the occurrence of such a practice demonstrates that the present generation is exercising a none-too-benign dictatorship over future generations.

Highly abstract sustainability models generate insights about the nature of the sustainability challenge and the conditions that are favorable or unfavorable for sustainability — insights that challenge these widely held principles. Early and still-influential work by Solow (1974) established that welfare can be sustained indefinitely in an economy that uses exhaustible natural resources, so long as the stock of capital continues to grow as exhaustible-resource stocks decline and technological innovation keeps pace with population growth. Solow's result was not entirely comforting, however, because

---

* The conference that generated this volume had a clear focus on the role of agriculture and natural resources as a foundation for sustainable development in India. While I have spent a little time in India, and found the experience enormously stimulating, I readily concede that others at the conference and represented in this volume have vastly more on-the-ground knowledge of the Indian situation. Accordingly, my focus at the conceptual level is consistent with comparative advantage.

his model allowed a very generous form of substitutability between natural resources and capital, such that welfare would be sustainable even as the stock of natural resources became infinitesimally small so long as the capital stock grew infinitely large. In other words, it did not seem to take natural resources seriously enough. Yet, within the context of Solow's model, the response to this criticism is not obvious. If the natural resource is both exhaustible and essential (in measurable quantities), sustainability is strictly impossible — a result that is so obvious as to be uninteresting.

Farmer and Randall (1997) developed a model that takes natural resources more seriously by assuming a resource that regenerates in sigmoid fashion (so that the resource may be sustained or destroyed, depending on the rates of harvest and regeneration) and more realistic substitution possibilities between the resource and capital (substitution that is generous when both factors are fairly abundant, but less so when the factor mix becomes extremely specialized, i.e., mostly capital with only a little of the natural resource or vice versa). The Farmer-Randall model also permits consideration of the discounting question, something Solow did not consider.

In the Farmer-Randall three-generation overlapping generations model, each generation seeks to maximize undiscounted utility from consumption, C, over its three-period lifetime (young, y; middle-aged, m; and retired, r), and intergenerational asset and financial markets are fully developed.

$$\text{Max } U(C_y) + U(C_m) + U(C_r), \text{ subject to:}$$

- Production requires land (i.e., natural resources), D; capital, K; and labor, L.
- D regenerates in sigmoid (logistic) fashion.
- D and K are substitutable, but extreme factor specialization is penalized.
- The young start with a substantial endowment of L, but no D or K; to command K, they must borrow it; to command D, they must buy it.
- The middle-aged, looking forward to retirement, are willing to sell D and lend K.
- The retired live on interest and repayments of principal from loans made when middle-aged.
- All markets clear in each period.
- All budgets balance and materials balance.

RESULTS

This model generates qualitative results that cast interesting light on several sustainability issues. Consumption, $C(t)$ and hence, welfare, may be increasing or decreasing over time, depending on initial assumptions concerning resource endowments per capita, natural-resource regeneration possibilities and factor (D, K, L) substitutability (Figure 30.1a). Not surprisingly, with sigmoid natural-resource regeneration and production processes that penalize extreme factor specialization, unsustainable outcomes driven by natural-resource crises are possible. But they are by no means inevitable. Generous natural-resource regeneration, and D–K substitution and generous endowments of capital and natural resources per capita, are favorable to sustained high levels of welfare.

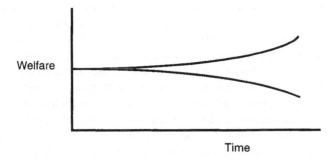

FIGURE 30.1A Welfare may inrease or decrease over time, depending on initial conditions.

While the model assumes constant population, it can be readily extended to generate Solow's result concerning population growth: population growth, *per se*, does not undermine sustainability so long as technological progress keeps pace with population. Technologies that enhance natural resource regeneration and $D–K$ substitutability are especially beneficial.

The equilibrium rate of interest is positive, despite the assumption that consumers are not impatient, because saving and investment are conducive to higher sustained welfare levels. In other words, future prospects are discounted at a positive rate in equilibrium, but this is driven, not by myopic consumers, but by the productivity of capital and the willingness of the young to borrow it. Positive interest rates contribute to sustainability of welfare by encouraging saving and investment. Consumption and welfare may be increasing or decreasing over time but, either way, a policy of interest rate suppression — as might be recommended by those who see discounting at a positive rate as inherently harmful to the future — will depress rather than elevate future prospects (Figure 30.1b).

Markets play a beneficent role. Factor and commodity prices reflect changes in relative scarcity, encouraging production, rationing consumption and promoting substitution and innovation when scarcity becomes more pressing. Fully articulated intergenerational financial and asset markets provide access to capital and resources for the young, reward saving and investment and encourage conservation of resources. In these ways, the actions of present generations take future concerns into

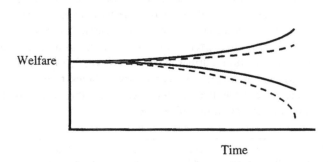

FIGURE 30.1B Interest rate suppression depresses future welfare.

account, thus providing a source of optimism about future prospects. In these findings, Farmer and Randall differ (at least in degree) from Howarth and Norgaard (1990), whose results were less optimistic in this regard, but whose model did not include fully articulated intergenerational financial and asset markets.

These results have a rather upbeat flavor. Nevertheless, there is an iron law: regardless of what else is going on, it helps to start out with plentiful capital and natural resources per capita (i.e., few people and lots of resources), and it hurts to start out with lots of people and few resources. The challenge for a country like India, then, is clear but not necessarily insurmountable. Policies should encourage saving and investment, conservation of natural resources and development of new technologies that enhance natural-resource regeneration and substitutability of capital for natural resources.

IS ECONOMIC GROWTH ENOUGH: DO WE NEED TARGETED SUSTAINABILITY POLICIES?

Imagine an opportunity that would make us enormously rich but would lead eventually and with certainty to environmental crisis. We are infinitely long-lived, so we would enjoy the riches *and* bear the disaster. Should we accept this opportunity? Two answers suggest themselves:

1. Of course we should. We'll be enormously rich when the disaster comes, so we can bribe it to go away, pay to prevent it, or pay to fix up after it. This response is in no way frivolous: being rich is an excellent defense against disasters that may befall us.
2. No, we should not accept it. Perhaps all our riches will be of no avail and the disaster just will not respond. This is exactly what we mean when we worry about possible *exhaustion* of an *essential* resource.

If we are confident in the first scenario, all that is required is a commitment to economic growth and the factor, product, financial and asset markets that encourage it. But, if we take the threat of the second scenario seriously, we need policies that directly address the problems of exhaustibility and nonsubstitutability. Some of these policies (e.g., encouraging technologies that would enhance substitutability) involve efficient investment and would be encouraged by well-articulated financial markets and realistic interest rates. Nevertheless, there is likely also to be a role for precautionary tools such as the safe minimum standard of conservation (SMS) (von Ciriacy-Wantrup, 1952, 1968) that target specific resources that are important to sustainability yet susceptible to exhaustion. Farmer and Randall (1997, 1998) define a sustainability SMS and argue that targeted sustainability policies triggered by potential resource crises make more sense than a generalized policy that would limit economic growth in the service of sustainability objectives.*

---

* Note that targeted precautionary policies such as the SMS are proposed here to deal with potential resource crises that might threaten present and future human welfare (sustainability view number 2, see introduction). The SMS could also be implemented in service of views that address particular resources for other than prudential reasons (view 3), or particular locations, communities, etc. (view 4).

A symmetry of precautions? This argument, consistent with the thrust of the economic literature on sustainability, emphasizes two types of bad outcomes to avoid — natural resource crises due to underconservation and insufficient capital accumulation due to underinvestment — and warns that actions taken to avoid one of these might leave us exposed to the other. We should be cautious to avoid natural-resource crises and we should be cautious lest actions taken to avoid such crises unduly limit our capital accumulation and hence future welfare. This is not quite a symmetry of precautions (resource crises might be abrupt and Malthusian, whereas underinvestment is more likely to result in chronic shortfalls in human welfare), but it does highlight a dilemma at the heart of the sustainability discussion.

HOW DOES ALL THIS RELATE TO THE STANDARD SUSTAINABILITY CRITERIA?

The weak and strong sustainability criteria are now well known; much of the sustainability discussion is based on these criteria and there is something approaching consensus about how these criteria are defined. How, then, do this chapter's conclusions relate to the weak and strong sustainability criteria?

WEAK SUSTAINABILITY

Weak sustainability requires sustaining human welfare and permits liberal substitution among factors of production and among commodities in consumption: an economy that ran out of some types of resources and ceased production of some commodities would be viewed as sustainable so long as substitute resources and commodities sufficient to sustain human welfare continued to be available. A forward-looking policy of weak sustainability typically involves encouraging capital accumulation while betting on future substitution possibilities; such bets are often reasonable.

STRONG SUSTAINABILITY

Strong sustainability takes particular resources seriously, often to the point of insisting on compensating resource investments: cut a tree, plant a tree; or, if a place dimension is added to the mix, cut a tree here, plant a tree here. A strong sustainability policy is cautious in what it assumes about the possibility (and perhaps the desirability) of substitution in production and consumption, but runs the risk of unduly limiting the welfare of future generations it seeks to protect.

   Where does this author's analysis come out? The business-as-usual policy should be one of weak sustainability, and well-articulated markets are potent tools for achieving that goal. But not everything is business as usual. There may be particular natural resources that should provisionally be considered essential (provisionally, because we can and should update this sort of judgment in light of emerging information, and invest in developing technologies that might hasten the day when adequate substitutes are available) and there may be natural amenities that we value very highly. In these special cases, each carefully justified, resource-specific strong sustainability policies such as the SMS should be held in reserve for use as a discrete break from business-as-usual policies, to combat impending crises.

AGRICULTURAL TECHNOLOGY: DON'T LET COUNTERPRODUCTIVE POLICIES UNDERMINE PROSPECTS FOR AN EVERGREEN REVOLUTION

The agricultural technologies of the Green Revolution have generated big increases in crop production, but there is concern that the yield increases may not be sustainable and that the technologies have tended to stress the renewable land and water resources while demanding major complements of exhaustible resources, especially fossil fuels. The hope now is for an Evergreen Revolution that would introduce technologies that are:

- Land saving, as would happen when the productivity of land increases so that land becomes relatively less scarce — this would be good, as it would tend to conserve fragile lands.
- Energy, pesticide and fertilizer saving, as would happen when the production response to these inputs becomes greater — this would be good, as it would tend to conserve fossil fuels and the assimilative capacity of the environment.
- Water saving, as would happen when the production response to water becomes greater — this would be good, as it would tend to save water and enhance its quality by reducing polluted return flows from irrigation.
- Labor saving, as would happen when the production response to labor becomes greater — this would be good, as it would tend to free labor for nonfarm uses.

Such technologies might free the later-to-modernize countries from retracing the whole path of their predecessors. To proceed directly from pre- to post-industrial technologies would enable these countries — and the world, because we are all part of a global environmental system — to avoid much of the environmental downside of the industrial age.

None of this is certain. Frequently, we have experienced land-, energy-, pesticide-, fertilizer- and water-using technologies (which demand increasing complements of the resources listed) and the Green Revolution technologies had many of these attributes. Yet, biotechnology gives us some hope that the next round of agricultural innovations will substitute for, more than complement, resources that are under stress, especially in resource-poor countries.

Imagine that we succeed in creating Evergreen technologies that have these desirable characteristics. More often than not, it seems, agricultural and related policies introduce pricing and other incentives that undermine these very benefits by encouraging production on fragile lands, high-input agricultural technologies, high use of irrigation water and the retention of labor in agriculture. Let us be very careful, this time around, to make sure that policies affecting agriculture and natural resources do not undermine the benefits that evergreen technologies promise.

REFERENCES

Farmer, M. and A. Randall, 1998, The rationality of a safe minimum standard of conservation, *Land Economics.* 74:287-302.

Farmer, M. and A. Randall, 1997, Policies for sustainability: lessons from an overlapping generations model, *Land Economics.* 73:608-622.

Howarth, R. and R. Norgaard, 1990, Intergenerational resource rights, *Land Economics.* 66:1-11.

Solow, R., 1974, Intergenerational Equity and Exhaustible Resources, Review of Economic Studies: Symposium on the Economics of Exhaustible Resources 41:29-45.

von Ciriacy-Wantrup, S., 1952 (3rd. ed., 1968). *Resource Conservation: Economics and Policies.* University of California, Division of Agricultural Science. Berkeley, CA.

World Commission on Environment and Development, 1987. *Our Common Future*, Oxford University Press. New York.

Part Six

Issues and Priorities

31 Reconciling Food Security with Environmental Quality in the 21st Century

Norman Uphoff

Food security and environmental quality go to the heart of the human condition. What could be more real than the hunger that gnaws at and heightens the misery of people who lack minimum daily caloric and nutrient intakes? What could be more visibly distressing than landscapes that have been robbed of their potential productivity by deforestation and soil erosion? What could be more wretched than foul water and polluted air that make drinking and breathing deadly?

The principal subjects of this volume are of immense importance to humankind. They are urgent matters in India, as contributions to this volume have shown, but they are also matters for the whole world to take seriously as we embark upon a new century. Already, millions of people suffer as a result of uncertain food supplies and degraded environmental quality, and there is no guarantee that the situation will get better during the coming decades.

Both food security and environmental quality are creations of the mind, abstractions that have been produced to enable analysts and policy makers to get a grasp on these immense challenges. We must take care not to conflate the real terms of these subjects with their analytical constructions — the latter are important, but they are only tools. They are a means to the end of assuring that all people will have access to healthy and sufficient food and will live in an environment that is healthy and robust.

After two decades of debate and refinement, food security is an idea whose time has come. It is on the lips and in the writings of agricultural scientists, social scientists and nutritionists as well as administrators, policy makers and policy critics. The concept was initially formulated by economists as a challenge to the proponents of "food self-sufficiency," who argued that countries needed to be able to produce enough food to feed their respective populations and that even communities and households should strive for self sufficiency.

The principles of comparative advantage argue strongly against such a view, as a less efficient way to satisfy people's food needs. The self-sufficiency approach accepts, to varying degrees, an autarkic world in which there is no trade, whether among households, communities, regions or nations. Because it is distrustful of the mechanisms and terms of trade, it is willing to forgo the production gains posited from a free-trade solution, considering that these are unlikely to benefit those most in need of food anyway.

The debate over free trade and the amount and distribution of its benefits is heating up, with many controversies now raging over the World Trade Organization and the open-markets regime it supports. Most professional and policy opinion, certainly among economists, favors a free-trade solution based on efficiency considerations. Others are less persuaded, and some are flatly opposed because of equity concerns and possible environmental effects. But what is remarkable is that all agree on the desirability and need to assure that food security, regardless of the policy and trade regime endorsed.

Food security, which started as an analytical concept and policy initiative, has become widely accepted as a human right, although many would argue that it is often honored in the breach, as are so many other human rights. Around the world, thousands of person-years are devoted to finding ways to combine supply and demand factors, augmented by logistical and storage facilities, income supplements, nutrition education, food fortification and so forth. Their goal is to assure that as many people as possible consume sufficient calories and nutrients each day to attain at least the basic level of nutrition needed for living, and hopefully, for good health and successful livelihoods.

This *ex ante* perspective on food security has informed most of the contributions to this volume. How can we understand food demand and supply dynamics sufficiently to make appropriate interventions that utilize the natural-resource base, the human resources for work and management and the stock of capital, knowledge and technology to best advantage? This is not just a matter of supply. It can also be a matter of assuring that needs are met, whether they can be expressed as effective demand with monetary backing or not.

The test of success is *ex post* accomplishment of food security, the extent to which all persons obtain the food needed for a healthy and productive life. This is not simply a result of production. It also requires appropriate distribution and utilization of food, including considerations of equitable access and bioavailability. To ensure full and sustainable food security, it is necessary to understand the operation of the entire food system.* This includes everything from the genetics of plant and animal food sources to the biochemistry of nutrient absorption in the body, soil fertility, cropping systems, harvesting and storage, marketing, food preparation and cultural influences on diet. The purpose of agricultural processes is to produce healthy people, not just food.

In this volume, contributors have reviewed the overall likely levels and interactions of food supply and demand, and considered how population growth will affect the latter. How population growth affects supply will be, at least in part, mediated

---

* Refer to the food systems analysis literature.

by environmental factors, such as how population pressures affect soil degradation, water availability and the conversion of forested area into arable land. Clearly, food security and environmental quality are interdependent, which is why this volume was framed in this manner.

Environmental quality appears to most to be more concrete, perhaps, than food security, although it refers to tangible things — soil, water, air, chemicals and vegetation — the concept and measurement of "quality" is as contingent on definitions and criteria as is the concept of "security." There are no absolute standards or time frames to be employed. Indeed, the units of analysis for "environment" are generically open systems. One can think of them as nested, in the way that river basins, watersheds and microwatersheds represent a hierarchy of subsumed units of biophysical interaction, but there are no natural boundaries for a landscape or an ecosystem. This means that environmental quality is as much a construct as is food security. It depends on many definitions and arbitrary measurements that are applied within systems of incredible complexity, which we understand very incompletely.

In this volume, we look at how efforts to raise agricultural production through intensified cropping systems affect environmental dynamics and potentials. With intensification, not only is there more pressure on land and water resources, affecting their quality, but pest and disease problems become more likely, requiring countermeasures that themselves have reduced environmental quality in the past. As consumers become more knowledgeable and concerned about health — their own and that of the environment — there is more need to devise methods of crop protection that serve both food security and environmental quality objectives.

One area of research that is not well represented in this volume, because it is an area where the agricultural and biological sciences have not made great investments in the past, concerns environmental quality and biodiversity with the soil. Relatively little is known about soil biota and their role in food production and environmental services. With new analytical techniques, including DNA analysis, researchers are starting to probe the hitherto murky processes of microbial ecology and the plant root–soil interface. In a country like India, where soils have been used for hundreds and often thousands of years, this biological dimension of food production and environmental productivity is one that we think warrants greatly accelerated investment in research and experimentation.

At the same time as we seek to advance knowledge at various microlevels, we need to pay more attention to whole systems, not just to their components. The latter are difficult enough to understand without having to consider whole sets and complexes of species and processes. Yet, we are learning that often we cannot really understand these species of processes very well when they are taken out of their ecosystem context. The concern with the structure and dynamics of larger systems that include both biophysical and socioeconomic factors can be justified simply on scientific grounds — to gain a more adequate understanding of how parts and wholes really function.

However, there is an even more important reason for looking at systems. We are concerned with the *sustainability* of development efforts. If these fail, or if they even falter, we have many more people dependent on them who will suffer, than a century ago. Our science and technology have permitted us to build larger and more complicated systems of production and distribution. They support populations more than three times larger than 100 years ago. We must take steps to ensure that this is not a scientific and institutional "house of cards" that we have constructed. Consequently, "building ecosystem vitality and productivity," as Richard Harwood subtitled his chapter in this volume, is crucial, seeking to ensure not only that these ecosystems sustain their levels of performance, but are immune to degradation and collapse. Contributors to the fourth section of this book have addressed these concerns, noting new opportunities that biotechnology, for example, may give us, while also recognizing the hazards that water scarcity could impose despite our ever-greater technological sophistication. Some of the solutions for sustainability may be, indeed, fairly simple.

While the first four sections of the book focus on mostly biophysical matters, we know that the purpose of agriculture and natural resource management is to achieve better and more secure lives for the human population, living in consonance with their environments rather than exploiting them. In the latter two sections, contributors address problems of poverty alleviation and equity. They look at criteria for evaluating progress in these areas, and particularly at the contributions that greater access to capital by the poor can make to poverty alleviation, and at the impacts of new agricultural technology on women's opportunities and burdens. Such distributional issues are not to be considered as separable consequences of development but as intrinsic causes of the ways in which development proceeds as well as of how its benefits are shared — or not.

These distributional dynamics are shaped, or at least greatly influenced, by the policy and institutional choices made at national and international levels, with the effects playing out at community, household and individual levels. The analyses in the last section of the book all agree that the performance of the agricultural sector — its rate and kinds of growth — will have dominant impacts on food security and environmental quality, but also on national and international economic success. Thus, the discussions in this volume are not just "sectoral," but speak to the welfare of whole national societies.

In the past 20 years, there has been a disposition in policy circles to choose a shrinking of the public sector in favor of an expanded private sector. While there have been some global benefits from this, it should be clear that the success and productivity of private-sector actors depend not just on their wisdom and hard work, but also on the framework of policies and investments that public-sector decision makers have created, by deliberation or by default. In particular, it is important that what economists call *public goods* be provided to the economy and, indeed, the whole society, to function effectively.

Public goods are things like public health and safety, clean air and water, the legal system and recreational facilities, which are accessible to everyone and whose use does not materially affect others' use. These are either costly or impossible for individuals to provide for themselves, so they are better provided on a broad basis. In the domains addressed by this volume, agricultural research is a prime example of public goods. It also, sadly, exemplifies the conclusion of social theorists that public goods tend to be underinvested in, for a variety of reasons. This leads to suboptimal satisfaction of human needs. When governments are faced with fiscal shortfalls, there is even more tendency to contract expenditures for public goods, but this has the effect of reducing well-being in the short and longer run. Governments are not the only source of public goods; non-governmental organizations of various sorts can provide them, as can businesses. But the latter have little incentive to do this, given their orientation to maximize private profits.

Policy makers should be aware of this unfortunate trend to underprovide public goods, especially when budgets are tight, because this only makes bad situations worse. This is not always easily seen, but it is a fact and it weighs particularly heavily on the agricultural sector, where producers are widely dispersed and unorganized and, for the most part. politically marginal.

The messages from this book are intended for a broad and diverse audience in terms of disciplines, roles and nationalities. They speak particularly to persons in policy-making or policy-influencing positions, because it is so urgent that those in responsible, authoritative posts act with clarity and a comprehensive understanding, not just some partial partisan or professional interests. This is especially true in a country like India, which is so large and complex. Overlooking any major factor in the agriculture and natural-resource arenas can have very dire consequences. Can we neglect water shortages? Can we neglect women? Can we neglect crop protection? Can we neglect access to credit?

In a world that was less complex and interdependent, with an India that had 500 million instead of its 1 billion people, there was more margin for error, more tolerance for disconnects and insularities. The world's agricultural producers, particularly in India, supported by large cadres of agricultural scientists and extensionists, have been very successful in improving the global food situation, thanks in large part to the new technologies of the Green Revolution. But this success has been uneven, and it has slowed dramatically in the last 5 to 10 years, as rates of increase have dropped and, in many countries, cereal yields have stagnated.

Part of the problem rests with policies — unattractive prices for producers, uncertainty of profitability and return — but we also see erosion of the natural-resource base, particularly as soil and water decline in quality and also per capita availability. Some of the natural-resource base is also undercut, as rural-to-urban migration depletes the stock of human talent in rural areas and endemic poverty persists. Against this, we see the continuing spread of health and education access, plus growing expectations of democratic governance supported by the electronic communication revolution. Thus, there is a growing tension between expectations and performance that has not been addressed specifically in this volume, but which contributors know lurks in the background of all their discussions.

As we embark on this 21st century, there are strong currents and pressures for democratization of public affairs and private life. Nowhere is this stronger than in India, the world's largest, though still quite imperfect, democracy. The most visible, formal hallmark of democracy, multiparty electoral competition, has had difficulty delivering on its claimed advantage over other systems — recruitment of the most talented leadership to top offices and generation of innovative and appropriate policy alternatives. This is true around the world, not just in South Asia, where such shortcomings have very high costs for society.

As yet, there are not good mechanisms for assuring that governments will proceed with as much knowledge and accountability as people expect. This volume has not gone into the realm of governance, but we recognize that this is fundamental both to formulating better policies and to implementing them successfully. What we have been able to do is examine the domains of and interactions among agricultural development, natural-resource management, sustainability, social relationships and policy choices. The analyses and conclusions from a broadly interdisciplinary set of professionals should give policy makers a better basis for thinking through their alternatives. We hope that this will be read also by a diverse cross-section of the public who can formulate clearer thinking about developmental ends and means, and thereby contribute to a climate of opinion that supports better informed and more forward-thinking actions, both privately and publicly.

Index